Mt Stromlo OBSERVATORY

It is my earnest desire that we should take our place among the great observatories of the world.

Geoffrey Duffield.

Mt Stromlo OBSERVATORY

From Bush Observatory to the Nobel Prize

Ragbir Bhathal, Ralph Sutherland
and Harvey Butcher

PUBLISHING

National Library of Australia Cataloguing-in-Publication entry

Bhathal, R. S., author.

Mt Stromlo Observatory: from bush observatory to the nobel prize/Ragbir Bhathal, Ralph Sutherland and Harvey Butcher.

9781486300754 (hardback)
9781486300761 (epdf)
9781486300778 (epub)

Includes bibliographical references and index.

Mount Stromlo Observatory.
Mount Stromlo Observatory – History.
Astronomical observatories – Australian Capital Territory – Mount Stromlo.
Astronomy – Australia – History.
Astronomy – Australian Capital Territory – History.

Sutherland, Ralph, author.
Butcher, Harvey R., 1947– author.

522.109947

Published by
CSIRO PUBLISHING
36 Gardiner Road, Clayton VIC 3168
Private Bag 10, Clayton South VIC 3169
Australia

Telephone: [+613] 9545 8555
Local call: 1300 788 000 (Australia only)
Fax: +61 3 9662 7555
Email: csiropublishing@csiro.au
Website: www.publishing.csiro.au

Front cover: Mount Stromlo Observatory (ANU/Brian Cooke).

Back cover: An ultra-deep picture of the distant universe taken by the Hubble Space Telescope (NASA).

Set in 9.5/12 Adobe Lucida and Optima
Edited by Adrienne de Kretser, Righting Writing
Cover design by Alicia Freile, Tango Media
Text design by Andrew Weatherill
Typeset by Desktop Concepts Pty Ltd, Melbourne
Index by Bruce Gillespie
Printed by Ingram Lightning Source

Published with the assistance of the Research School of Astronomy and Astrophysics, Australian National University.

Feb26_RP_ILS

Contents

Preface

This volume chronicles a century of astronomy at the Mount Stromlo Observatory. What began in support of the Federation process in Australia grew into an important astronomical research centre with a worldwide reputation for scientific excellence. A history of this evolution is timely as a contribution to the centenary celebrations of the founding of Canberra, the national capital city that the Australian public semi-affectionately refers to as its 'bush capital'.

During its early history, three forces combined to determine the character of the Observatory. First, in recognition of the economic and political importance of accurate, nation-wide time and positional standards, the Australian Constitution explicitly granted to the Commonwealth Parliament the power to make laws with respect to astronomical observations. A national observatory would be built in the new Federal Capital Territory, with the Mount Stromlo site selected for the purpose. The Observatory began its life as a government department in support of what we today would call 'nation-building'.

Second, the early 20th century saw dramatic developments in the physical sciences. In the context of astronomy, the advent of quantum mechanics would make possible the scientific interpretation of observations of the Sun and stars. The resulting excitement among scientists was contagious and would result in an enduring bias towards pure research at Mount Stromlo, a bias that would distinguish the Observatory from state government observatories in the country. By mid century, the Observatory could merge seamlessly with the Australian National University (ANU), its researchers playing constructive roles in establishing the reputation of that newly founded institution.

And third, improvements in telecommunications and in affordable travel were leading to increased internationalisation of science. The geographical latitude and longitude of Australia would encourage Mount Stromlo researchers to complement in useful and important ways observations made from Europe and the Americas. This international dimension, involving the flow of both ideas and people, ensured that Observatory staff could and would join colleagues overseas in exploring the important astronomical questions of the day.

To these ingredients was added a policy of encouraging instrumental development at the Observatory. In the earliest days this was hard necessity, but it also led to close cooperation with industry during the Second World War, to participation in the design and construction of the 3.9 m Anglo-Australian Telescope, to the development of the 2.3 m Advanced Technology Telescope, to spinning-off Auspace Pty Ltd to become Australia's premier space and satellite company. It permitted the hands-on training of generations of scholars skilled in the art of observing the distant universe, such that today Stromlo alumni can be found in influential positions in academia and industry around the world.

As the 20th century drew to a close, the policies aimed at excellence in research, training and technology, together with strong international collaboration, paid off handsomely. The scientific productivity of the Observatory reached what would be its pinnacle. Stromlo astronomers participated in major discoveries concerning the content and evolution of the cosmos, including the discovery that led to the 2011 Nobel Prize in Physics, namely that the expansion of the universe is actually accelerating.

Then, almost as retribution by the gods for so much scientific success, double disaster struck. In 2003, a firestorm destroyed much of the research infrastructure at Mount Stromlo, which was then found to have been inadequately insured by ANU. In the same period, massive cuts by ANU to the operating budget crippled any chance of rapid recovery and caused talented staff to depart.

As the Observatory embarks on its second century, however, new opportunities have arisen and recovery is well under way. Highly productive, early- and mid-career researchers have been recruited. Engineering capability has largely been rebuilt and is helping develop the billion-dollar, international Giant Magellan Telescope. Joint technology development with local industry is under way again. And national and international visitors are a constant feature at the Observatory.

This volume therefore is not only the history of one Australian institution and a contribution to the Canberra centenary, it is also a celebration of Australian scientific success on the world stage.

Harvey Butcher
Director (July 2007–January 2013)
Research School of Astronomy and Astrophysics (Mount Stromlo Observatory)
January 2013

About the authors

Ragbir Bhathal is an astrophysicist and an award-winning author who was awarded the prestigious Nancy Keesing Fellowship by the State Library of New South Wales and the CJ Dennis award for excellence in natural history writing. He is the Director of the Australian Optical SETI Project in the School of Engineering at the University of Western Sydney, a Visiting Fellow at the Research School of Astronomy and Astrophysics, Australian National University, and the Director of the National Oral History Project on Significant Australian Astronomers and Physicists, sponsored by the National Library of Australia. He is considered the father of SETI in Australia. He has written 15 books, eight of them on astronomy including two on Aboriginal astronomy. He obtained his PhD in physics from the University of Queensland on a Commonwealth scholarship. He has served as the Foundation Director of the Singapore Science Centre, as a member of the Board of Directors of the Association of Science and Technology Centres, Washington, as UNESCO consultant on science policy and science centres, and as Adviser to the Federal Minister for Science. He served as the President of the Royal Society of New South Wales and was awarded the 1988 Royal Society of New South Wales Medal for services to science and research. He is a member of the IAU World Heritage and Astronomy Working Group.

Harvey Butcher is an astronomer. He has a degree in astrophysics from Caltech and a PhD in astronomy from the Australian National University. In a career spanning more than four decades and working on five continents, he has carried out research in observational cosmology trying to understand how stars and galaxies evolve over cosmic time and why the universe around us today looks the way it does. His parallel interest in technology saw him involved with the first digital detectors and digital image processing systems in astronomy. He led technology development for advanced instrumentation for ground- and space-based platforms, including for the Square Kilometre Array radio telescope and the James Webb Space Telescope. He was Director of the Mount Stromlo Observatory (Research School of Astronomy and Astrophysics, Australian National University) for five years to 2013.

Ralph Sutherland is a Fellow at the Research School of Astronomy and Astrophysics, Australian National University, working primarily in theoretical modelling of the interstellar medium. His PhD from Mount Stromlo was granted in 1993 and, after working in the US, he returned to Mount Stromlo in 1996. He was co-editor of *Astrophysics of the Diffused Universe* with Michael Dopita.

Acknowledgements

We are indebted to Tom Frame and Don Faulkner for their earlier and excellent book on Mount Stromlo Observatory. We found it very useful when writing this book and acknowledge the use of their book in providing us with sign posts. This book would not have seen the light of day without the assistance and encouragement we received from a large number of people. We thank Professor Mike Bessell, Professor Don Mathewson, Professor Jeremy Mould, Professor Mike Dopita, Professor Brian Schmidt, Professor Penny Sackett and Professor Matthew Colless (the present Director of the Research School of Astronomy and Astrophysics) for taking the time to read parts of the book and giving us valuable comments, and in particular Professor Ken Freeman for reading the entire book and providing valuable comments and suggestions. We thank Professor Gary Da Costa and Professor John Norris for providing information about their research, and research done at the Observatory. We also wish to thank Dr Phillipa Butcher for her forbearance as Harvey 'won' the images, Johanna Bhathal and Jenny Bhathal for their patience and understanding in letting Ragbir write the book in prime family time on the weekends, Stefan Keller, David Davids, Belinda Davids, John Hart, Michelle Cicolini, Mike Fowler, Rose Metcalfe, Allegra Grevelius (Nobel Media Archive & Nobel Foundation), Joe Anderson (American Institute of Physics/Niels Bohr Library & Archives) for allowing us to use the transcripts of the interviews of Gerald Kron and Bart Bok, Rosanne Walker (Basser Library/Australian Academy of Science) for her assistance in finding the material in the Baser Library for our use, Kevin Bradley and Shelly Grant (National Library of Australia) for allowing us to use the transcripts of the interviews from the National Oral History Project on Significant Australian Astronomers and Physicists, Catherine Hobbs (College Archivist, St Ignatius' College, Riverview), Helen Bruce (Reference Archivist, University of Adelaide), Linda Thornely (Library, University of Western Sydney), Professor Brian Uy (former Head of School of Engineering and Industrial Design, University of Western Sydney), Tegan Dolstra (Communications Officer, Australian National University), Belinda Pratten for use of Brian Schmidt's photo, Sue Bowler (Editor, *Astronomy & Geophysics*), Leigh Dayton (*The Australian*), Susan Hall

(Publications Department, National Library of Australia), Chris Fluke (Swinburne University of Technology), Brad Gibson (University of Central Lancashire), John Sarkissian (CSIRO), Robert and Robyn Shobbrook, the editors of *Weekend Australian*, *Nature* and *Science*, ETHBIB Bildarchive Zurich, University of Melbourne Archives, JD Collection, State Library of New South Wales, University of Sydney Library, Dr Ragbir Bhathal Collection (National Library of Australia), Swinburne University of Technology, Australian Astronomical Observatory, Australia Telescope National Facility, National Library of Australia, National Archives of Australia, the Australian War Memorial, European Southern Observatory, Nobel Foundation and the Giant Magellan Telescope – GMT Corporation. We thank and appreciate Briana Melideo's championing the publication of the book by CSIRO Publishing. We thank Tracey Millen and the staff of CSIRO for the excellent cooperation they gave us in the publication of the book. We also wish to thank our Editor Adrienne de Kretser (Righting Writing) for critically reading and making valuable suggestions that have resulted in a greatly improved text.

1905–1923

1

A beginning in the bush

That this International Congress hears with great satisfaction of the proposal to establish a Solar Physics Observatory in Australia and expresses its decided opinion that an observing station in that part of the world would fill the gap which now exists in the system of observatories distributed over the Earth and yield contributions of great value to the study of the solar phenomena.

International Union for Solar Research,
Chateau de Meudon, Paris 1907.[9]

1905 was an auspicious year in the development of physics and astrophysics. It was the year that Albert Einstein, then an unknown clerk in the Swiss patent office, published his revolutionary paper on the special theory of relativity. The paper questioned Newton's theory of space and time and showed its limitations when objects travel near the speed of light. It also introduced a new vocabulary in physics and a new way of understanding the nature of the universe we live in. Einstein's theory was based on two simple ideas, namely that the laws of physics are the same in all inertial frames of reference and that the speed of light in vacuum is the same in all inertial frames.[1] These simple postulates led to his famous equation $E = mc^2$, which is known to every school boy and girl, and to other intriguing predictions, such as time dilation and the paradox of the twins. In 1915 Einstein introduced another revolutionary theory. Called the General Theory of Relativity, it revealed new insights into the nature of space and time and the motion of the universe. The surprise was that Einstein's new equations predicted that the geometry of the universe should be evolving with time, in contradiction to the conservative view that the universe must be static. He introduced a fudge factor (the cosmological constant) to bring his theory into line with the then accepted view of a static universe. Observations in the 1920s showed that the universe is expanding, leading Einstein to say that the introduction of the cosmological constant was the biggest blunder of his scientific career.

About 80 years later astronomers at the Mount Stromlo Observatory (now home to the ANU Research School of Astronomy and Astrophysics) in Canberra made headlines, along with astronomers from the US, when they showed that the universe not only is expanding but that the expansion is accelerating! This acceleration is being driven by an unknown pressure,

Figure 1.1: Albert Einstein (14 March 1879–18 April 1955). Photo: ETHBIB Bildarchive Zurich.

popularly referred to as Dark Energy, an anti-gravity force. One of the astronomers from Mount Stromlo Observatory, Brian Schmidt, subsequently won a Nobel Prize with his colleagues from the US for their work on the accelerating universe.[2] They gave us a new view of the universe and a conceptual shift comparable to that of the Copernican revolution in the 16th century.

In that same year (1905), an unknown research student, Geoffrey Duffield, from Adelaide attended the meeting of the International Union for Cooperation in Solar Research in Oxford and came up with the idea of establishing a solar observatory in Australia.[3] The meeting was attended by some of the leading physicists and astronomers of the day, such as George Ellery Hale, Sir Norman Lockyer and Sir Frank Dyson. Despite the fact that astrophysicists had suspected the existence of a solar–terrestrial relationship, there was no worldwide spread of solar observatories that could continuously monitor solar phenomena. Members of the International Union for Cooperation in Solar Research foresaw the importance of continuously monitoring solar behaviour, which before we had satellites in space, meant observatories around the world. The society assisted Duffield in achieving his aim of setting up the Commonwealth Solar Observatory in Australia.

Duffield had realised that there was a need to establish a solar observatory to fill the gap in observations of the Sun in the long stretch of land and sea between India and the US. Australia was in the right location both in

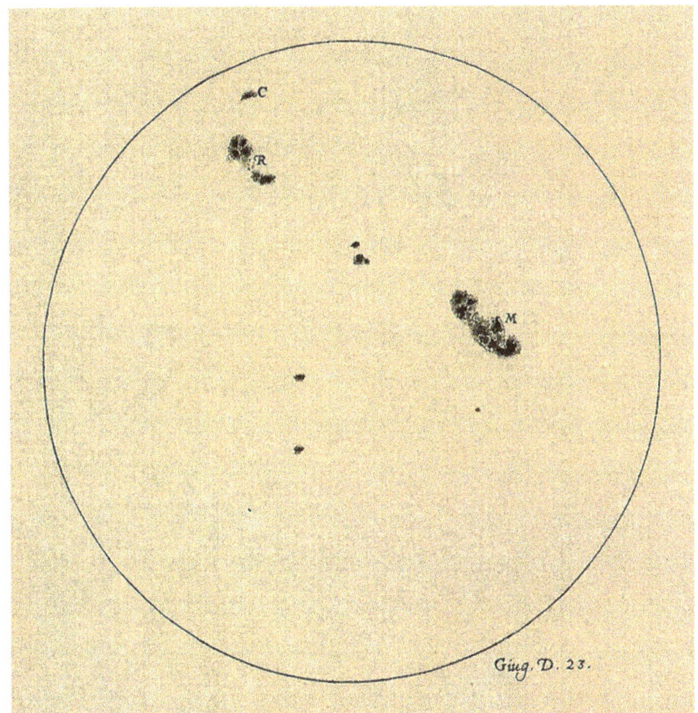

Figure 1.2: Sketch of sunspots, by Galileo, 23 June 1612. That spots occasionally appear on the face of the Sun has been known since early Chinese times. Galileo showed European civilisation how variable the Sun can be. The Sun came under intensive scrutiny at the turn of the 20th century as scientists explored its importance to life on Earth. Image: Galileo Project, Rice University.

terms of longitude and latitude to fill the gap. In the ensuing years he became a one-man international lobby group as he passionately advocated for the establishment of a solar observatory in Australia. It took almost 18 years to realise his dream. A lesser man would have abandoned this quest. Duffield skilfully used his contacts in the scientific communities in Britain and Australia and in government and business circles in Australia to achieve his ambition of setting up an astronomical institution in Australia to ride the new wave of astrophysics being pursued in the intellectual circles in Europe and the US. The new institution was to be a break away from the routine work in positional astronomy that was being undertaken by the state observatories, whose work on the long and arduous international Carte du Ciel project (the International Astrographic Catalogue) consumed available resources and eventually led to their demise. They were either closed or, in the case of Sydney Observatory, transformed into a museum.[4]

Duffield had been born with a silver spoon in his mouth and was educated at a private school, St Peter's Collegiate School, in Adelaide. He enjoyed a privileged childhood with music lessons and travel. His grandfather had built a large fortune as a pastoralist and served as a member of the South Australian Parliament. This background gave Duffield later in life the confidence to move in influential circles in government and business.

Figure 1.3: Walter Geoffrey Duffield (1879–1929). Photo: Mount Stromlo Archives.

Figure 1.4: The rich starry Australian night sky was one of the reasons why observatories were set up in Australia beginning with Dawes Observatory in Sydney in 1788. Photo: NASA.

Figure 1.5: William H. Bragg taught Duffield physics at the University of Adelaide while Duffield was an undergraduate in 1898–1900. Feeling a sense of intellectual isolation from the metropolitan centres of learning, Bragg and and his son, Lawrence, went to live in England from 1909, where their work on X-ray crystallography won them a Nobel Prize in 1915. Photo: University of Adelaide Archives.

Duffield's scientific journey began in Australia in 1898. He had been awarded a science degree in that year at Adelaide University and had been taught physics by William Henry Bragg who, with his son, was to win a Nobel Prize for their work on X-ray crystallography. Bragg was a very useful ally, who provided access to members of Britain's scientific elites in Duffield's campaign to establish a solar observatory in Australia. Duffield won the Angas Engineering Scholarship on the strong backing of Bragg. According to Bragg, 'When the question was under discussion as to whether or not he should be given the scholarship which took him to England I pleaded strongly for him against a certain reluctance on the part of the other examiners. He had not done so very well in his written papers. But I knew, or thought I did, that there was something in him which would justify an award in his favour.'[5] Having secured the scholarship, Duffield went to Cambridge University to do the undergraduate Mechanical Sciences Tripos. Graduating in 1903 with a Bachelor of Arts degree he was able to secure a Nobel Research Scholarship to work in the engineering laboratories of the Physical Laboratory in Manchester. A McKinnon Scholarship from the Royal Society enabled him to continue his studies as a research student at Owens College in Manchester, where he obtained the degree of Master of Science and later Doctor of Science in 1908 for his research in the field of spectroscopy under the

Figure 1.6: Geoffrey Duffield proposed and lobbied for the establishment of the Commonwealth Solar Observatory at Mount Stromlo. After obtaining his DSc from Manchester University he became a member of the academic staff at the University College of Reading in England. He received many awards, the Adelaide Register describing him as a 'young astronomer, with 29 letters chasing his name'. Photo: Mount Stromlo Archives.

supervision of Arthur Schuster, a leading solar physicist.[6,7,8] His research focused on the effects of increasing pressure on arc spectra, exploring the transition from emission line spectra to continuous spectra.

The Physical Laboratory also hosted the Central Bureau of the International Union for Cooperation in Solar Research. This gave Duffield the necessary background and connections to launch his idea for the establishment of a solar observatory in Australia.

Proposal for a solar observatory

The first shot was fired when Duffield wrote a letter to his mentor, Bragg, in Adelaide about his proposal to establish a solar observatory in Australia. The letter also appeared in the 3 April 1907 issue of the *Adelaide Advertiser* and attracted a lot of publicity. Duffield wrote, 'Two years ago I went to the Oxford meeting and was much disappointed to find no mention of solar work done in Australia. I was told none was done there. It seems that Australia loses a great opportunity of distinguishing herself, for with her clear skies and sunny weather she should produce results as magnificent as those of the great American observatories.'

However, the South Australian government held that the proposal should be a federal government initiative since astronomy was perceived to have

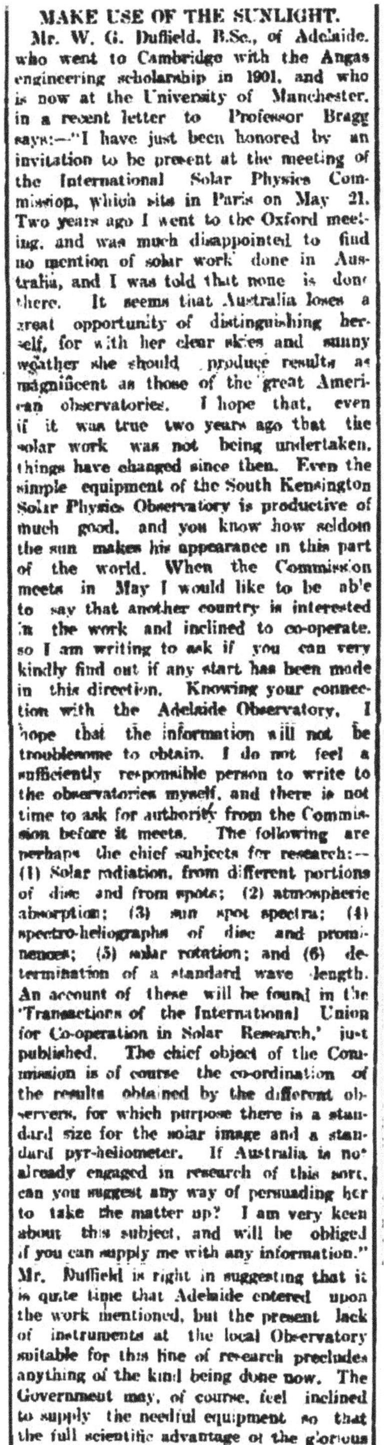

MAKE USE OF THE SUNLIGHT.

Mr. W. G. Duffield, B.Sc., of Adelaide, who went to Cambridge with the Angas engineering scholarship in 1901, and who is now at the University of Manchester, in a recent letter to Professor Bragg says:—"I have just been honored by an invitation to be present at the meeting of the International Solar Physics Commission, which sits in Paris on May 21. Two years ago I went to the Oxford meeting, and was much disappointed to find no mention of solar work done in Australia, and I was told that none is done there. It seems that Australia loses a great opportunity of distinguishing herself, for with her clear skies and sunny weather she should produce results as magnificent as those of the great American observatories. I hope that, even if it was true two years ago that the solar work was not being undertaken, things have changed since then. Even the simple equipment of the South Kensington Solar Physics Observatory is productive of much good, and you know how seldom the sun makes his appearance in this part of the world. When the Commission meets in May I would like to be able to say that another country is interested in the work and inclined to co-operate, so I am writing to ask if you can very kindly find out if any start has been made in this direction. Knowing your connection with the Adelaide Observatory, I hope that the information will not be troublesome to obtain. I do not feel a sufficiently responsible person to write to the observatories myself, and there is not time to ask for authority from the Commission before it meets. The following are perhaps the chief subjects for research:—(1) Solar radiation, from different portions of disc and from spots; (2) atmospheric absorption; (3) sun spot spectra; (4) spectro-heliographs of disc and prominences; (5) solar rotation; and (6) determination of a standard wave length. An account of these will be found in the 'Transactions of the International Union for Co-operation in Solar Research,' just published. The chief object of the Commission is of course the co-ordination of the results obtained by the different observers, for which purpose there is a standard size for the solar image and a standard pyr-heliometer. If Australia is not already engaged in research of this sort, can you suggest any way of persuading her to take the matter up? I am very keen about this subject, and will be obliged if you can supply me with any information." Mr. Duffield is right in suggesting that it is quite time that Adelaide entered upon the work mentioned, but the present lack of instruments at the local Observatory suitable for this line of research precludes anything of the kind being done now. The Government may, of course, feel inclined to supply the needful equipment so that the full scientific advantage of the glorious Australian sunlight may be reaped.

Figure 1.7: Duffield's letter to Bragg. *Adelaide Advertiser*, 3 April 1907.

become a Commonwealth responsibility. Indeed, along with powers to legislate in the areas of trade, taxes, postal service and telecommunications, defence and other practical matters, the new Australian Constitution (Chapter I, Part V, Section 51viii) gave the Commonwealth Parliament the power to make laws with respect to astronomical and meteorological observations. While not exclusive powers, this was a good excuse for the South Australian government not to invest significantly in astronomy.

Not deterred by this setback, Duffield came up with another strategy. At the meeting of the International Union for Solar Research held in May 1907 at the Chateau de Meudon near Paris, Duffield took the initiative of drafting a resolution which was put to the meeting by Sir Norman Lockyer, discoverer of helium in the Sun and editor of the influential British journal *Nature*. The resolution was very explicit: 'That this International Congress hears with great satisfaction of the proposal to establish a Solar Physics Observatory in Australia and expresses its decided opinion that an observing station in that part of the world would fill the gap which now exists in the system of observatories distributed over the Earth and yield contributions of great value to the study of the solar phenomena.'[9] A copy of the resolution was forwarded to the Colonial Office in London, which passed it on to the federal government in Australia. While not rejecting the proposal, the federal government noted that 'it is not desirable at the present juncture to incur the heavy expenditure involved'.[10]

On the instrumental front, however, the proposal met with great enthusiasm from well-heeled amateur astronomers. In 1907 Lord Farnham's estate offered a 15 cm Grubb refractor to Duffield and two years later James Oddie, a wealthy amateur astronomer from Ballarat who had built a private observatory there, offered his brand new 22 cm Howard Grubb refractor to the government of Australia on the

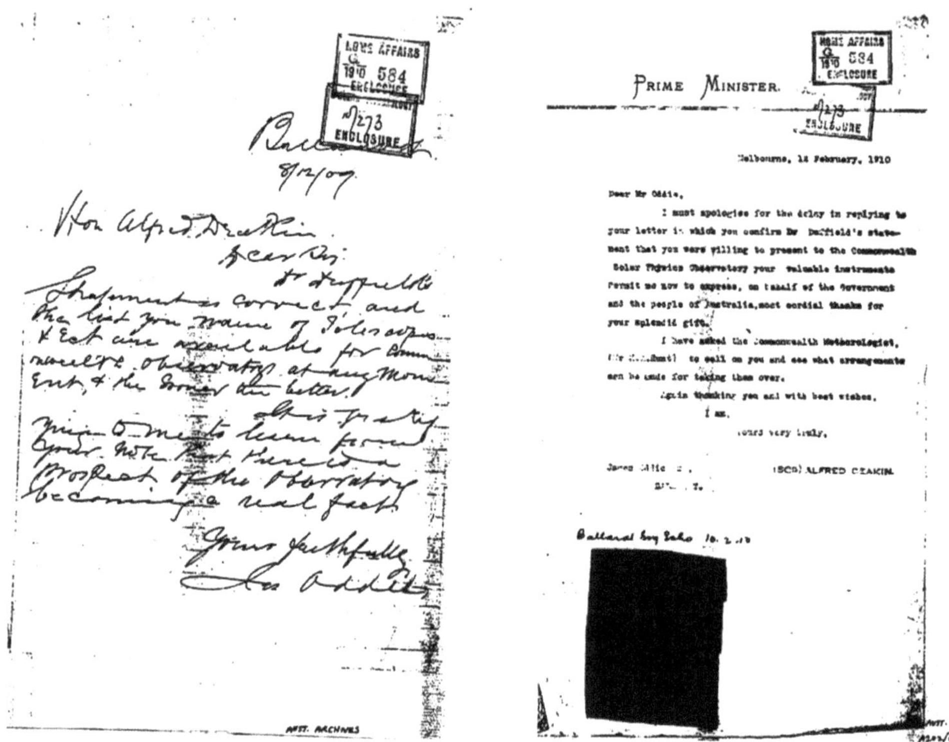

8/12/09

Hon Alfred Deakin

Dear Sir:

Dr Duffield's statement is correct and the list you name of Telescopes & ect are available for Commonwealth Observatory at any moment & the sooner the better. It is gratifying to me that there is a prospect of the Observatory becoming a real fact.

Yours faithfully

PRIME MINISTER.

Melbourne, 14 February, 1910

Dear Mr Oddie,

I must apologise for the delay in replying to your letter in which you confirm Dr Duffield's statement that you were willing to present to the Commonwealth Solar Physics Observatory your valuable instruments. Permit me now to express, on behalf of the Government and the people of Australia, most cordial thanks for your splendid gift.

I have asked the Commonwealth Meteorologist, to call on you and see what arrangements can be made for taking them over.

Again thanking you and with best wishes,

I am,

Yours very truly,

(SGD) ALFRED DEAKIN

Figure 1.8: Letters from Oddie and Deakin, supporting the proposed Observatory. In his letter to Prime Minister Deakin, Oddie wrote, 'Dr Duffield's statement is correct and the list you name of Telescopes & ect [sic] are available for Commonwealth Observatory at any moment & the sooner the better. It is gratifying to me that there is a prospect of the Observatory becoming a real fact.' Image: Australian Academy of Science.

understanding that 'it form the nucleus of a solar observatory anywhere in the Commonwealth'. In accepting the gift, the Prime Minister, Alfred Deakin, wrote, 'Permit me now to express, on behalf of the Government and people of Australia, most cordial thanks for your splendid gift'.[11] Other instruments offered to the proposed Observatory were a coelostat by Franklin Adams and a pyrheliometer by James Fowler.[3]

Meanwhile Duffield continued lobbying through his connections with Britain's scientific societies. The Royal Society and the British Association for the Advancement of Science gave their support for the establishment of the Observatory. In 1908 the Royal Society passed a resolution which had originally been written by Duffield. It noted that the Royal Society was 'strongly of the opinion that the foundation of a Solar Observatory in Australia is desirable'.[11] Duffield also spread his lobbying efforts to the scientific community in Australia. In January 1909 the Australasian Association for the Advancement of Science formed the Australian Solar Physics Committee, chaired by the influential Commonwealth Statistician, George (later Sir George) Knibbs. Its members included the Professors of Physics at the universities of Adelaide, Melbourne, Sydney and Tasmania, the Government Astronomers Pietro Baracchi (Melbourne) and William Cooke (Perth), Senator

Figure 1.9: Alfred Deakin (1856–1919) was the second Prime Minister of Australia, in 1903, and served for 217 days. He served again as Prime Minister from 1905 (for three years and 132 days) and in 1909 (for 332 days). He was a leader of the movement for an Australian federation. Deakin kept the idea of the Solar Physics Observatory alive in the House of Representatives. Photo: National Library of Australia.

Figure 1.10: James Oddie had made his fortune as an auctioneer and banker during the gold rush years in Ballarat. He was well known to Alfred Deakin, the Member of Parliament for Ballarat. Photo: Art Gallery of Ballarat.

Figure 1.11: The Oddie telescope, when first assembled at Melbourne Observatory. Pietro Baracchi, the Government Astronomer of Victoria, inspected and tested it before it was sent to Mount Stromlo. The 22 cm telescope was the first research telescope to be set up at the site. It was initially used for characterising the site's potential for astronomy, but later became an important research telescope. Photo: Mount Stromlo Archives.

Figure 1.12: The 15 cm Grubb refractor telescope that was donated by Lord Farnham's estate in 1907. It was the first telescope donated, although the Oddie telescope was first to go into service at Mount Stromlo. Photo: Mount Stromlo Archives.

Figure 1.13: George (later Sir George) Knibbs (1858–1929) was the Chairman of the Australian Solar Physics Committee that supported Duffield's plans for the establishment of a solar observatory in Australia. Knibbs was an influential public servant and was responsible for the production of the first Commonwealth Year Book, in 1908. Photo: Mount Stromlo Archives.

Joseph Keating and James Oddie. It was a well-planned Committee consisting of academics, Government Astronomers, a politician and a businessman. It represented the Who's Who of Australian society. Duffield managed to secure the position of Committee Secretary, which gave him an excellent means for pushing his idea along the channels of influence in government.

To bring his proposal to public attention, Duffield used his connections with Oddie to get the Governor-General Lord Dudley to host a public meeting

Figure 1.14: Governor-General Lord Dudley (1867–1932). Photo: National Library of Australia.

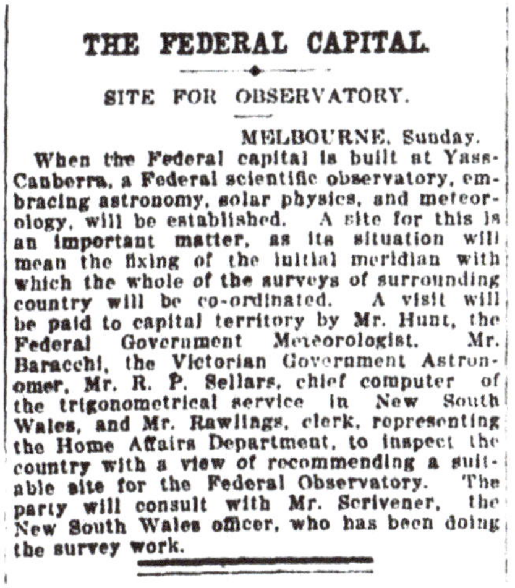

THE FEDERAL CAPITAL.

SITE FOR OBSERVATORY.

MELBOURNE, Sunday.

When the Federal capital is built at Yass-Canberra, a Federal scientific observatory, embracing astronomy, solar physics, and meteorology, will be established. A site for this is an important matter, as its situation will mean the fixing of the initial meridian with which the whole of the surveys of surrounding country will be co-ordinated. A visit will be paid to capital territory by Mr. Hunt, the Federal Government Meteorologist. Mr. Baracchi, the Victorian Government Astronomer, Mr. R. P. Sellars, chief computer of the trigonometrical service in New South Wales, and Mr. Rawlings, clerk, representing the Home Affairs Department, to inspect the country with a view of recommending a suitable site for the Federal Observatory. The party will consult with Mr. Scrivener, the New South Wales officer, who has been doing the survey work.

Figure 1.15: Announcement of the site for the observatory. *Sydney Morning Herald*, 21 February 1910.

in Melbourne Town Hall on 19 October 1909. The meeting was a great success, with Lord Dudley noting that 'it would be little short of a national misfortune if, for the sake of a few thousand pounds, Australia failed to take her place among the nations of the world in scientific research for which her geographical position had marked her out'.[3] The sentiment was well put and caught the ear of Prime Minister Deakin. It was not long before the federal government gave Duffield approval in principle for the establishment of a solar observatory and funding to cover the annual maintenance costs and modest enhancements. Very skilfully, Duffield managed to get the observatory formally incorporated into the plans for the new Federal Capital Territory, thereby ensuring its future and forestalling potential claims by state observatories for support to undertake the tasks.

Advice on possible observatory sites

In January 1910 a five-member board, including Robert McDonald (New South Wales Undersecretary of Lands), R.P. Sellors (New South Wales Geodetic Survey), Henry Hunt (Commonwealth Meteorologist), Charles Robert Scrivener (Commonwealth Director of Surveys) and Pietro Baracchi (Government Astronomer for Victoria) was established to advise on possible observatory sites within the Federal Capital Territory. The choice of Hunt and Scrivener was crucial to the success of the project. Hunt had previously vetoed the establishment of the Observatory in the Federal Capital Territory because he saw it as competition for the scarce financial resources he was trying to obtain to form a federal Meteorological Department. Duffield's first impression of Hunt was that he 'is a rough diamond, certainly rough and I suspect the diamond only. Quite pleasant but apparently uneducated in deep scientific matters'.[11] However, by getting Hunt involved in the Committee, Duffield had very cleverly neutralised Hunt and got him to support the astronomy project.

Observatory site, the First People and English settlers

The Observatory site and surrounding country had been used by the Aboriginal people, such as the Ngunnawal, Walgalu and Ngarigo, for over 20 000 years. In the summer months they gathered there for initiation ceremonies, marriage arrangements, corroborees and trade. It was also a time when the Bogong moths (*Agrotis infusa*) were in great supply. The moths were roasted and eaten whole.[12,13] They were considered a delicacy by the Aboriginal people.

After the evening meal they would sit around the camp fire and listen to the stories told by the Elders about the celestial bodies in the starry vault overhead. In the dark patches of the brilliant Milky Way they saw the shape of an emu, which has spiritual and cultural significance in their society.

The arrival of the English settlers in the 19th century pushed the Aboriginal people from their land. It was not long before large parcels of land in the Canberra region were held by prominent English families. One large parcel, which included the Mount Stromlo area, was owned by Canberra grazier Frederick Campbell. Early in 1911 the Mount Stromlo site was leased by the Commonwealth government from Campbell for an annual rent of £26 15s for a period of two years. It became Commonwealth property after that.

Figure 1.16: The Milky Way Galaxy as seen from Siding Spring Observatory. Photo: Sutherland/ Bessell. Mount Stromlo Archives.

Figure 1.17: The shape of an emu in the dark spaces of the Milky Way Galaxy, as seen by Aborigines. The Milky Way became an active area of research for astronomers at Mount Stromlo Observatory in the 20th and 21st centuries. Photo: JD Collection.

Scrivener was important for establishing the prime meridian for Australia at the site where the Observatory was to be constructed. The chosen site was a ridge of land ~1.6 km long and 200 m above the surrounding country. It was extremely important to the government and the business community

Figure 1.18: Fredrick Campbell, who rented the top of Mount Stromlo to the federal government as the Observatory site. Photo: National Library of Australia.

that the accuracy of Australian surveying be improved, as there had been a dispute about the correctness of the boundary between Victoria and South Australia – the border between the states was wrong by ~3 km. Surveying also provided the Observatory with a utilitarian purpose in the eyes of the politicians and the public.

It was on this ridge that Baracchi and J.M. Baldwin, Baracchi's assistant at Melbourne Observatory, set up the Observatory to test the site for astronomical use. The Observatory building was of simple design in keeping with the dictum of Cassini (Director of the Paris Observatory) that astronomical observatories should be of simple design and functional. The Observatory consisted of a 6 m dome resting on concrete walls. It had four small wings, which housed a kitchen, accommodation for the two astronomers (Baracchi and Baldwin) and the caretaker, and a photographic room.

Construction of the Observatory to house the Oddie telescope began in March 1911 and was completed by September 1911. By this time Duffield was back in England, as a Professor at University College, Reading. He held that post until 1923, continuing his lobbying from afar with only an occasional visit to Australia.

For one week each month over the next 12 months Baracchi and Baldwin made observations with the Oddie telescope of celestial objects, such as the Sun, the Moon, the planets, the Magellanic Clouds, the globular clusters Omega Centauri and 47 Tucanae and double stars.[14]

Baracchi was a natural choice for the site testing work as he not only had expertise in astronomy but was also an expert in geodesy and surveying.[15] Thus, he was well suited for selecting the site for the new Observatory. He was assisted by Charles Scrivener, the Commonwealth Surveyor, and Hunt,

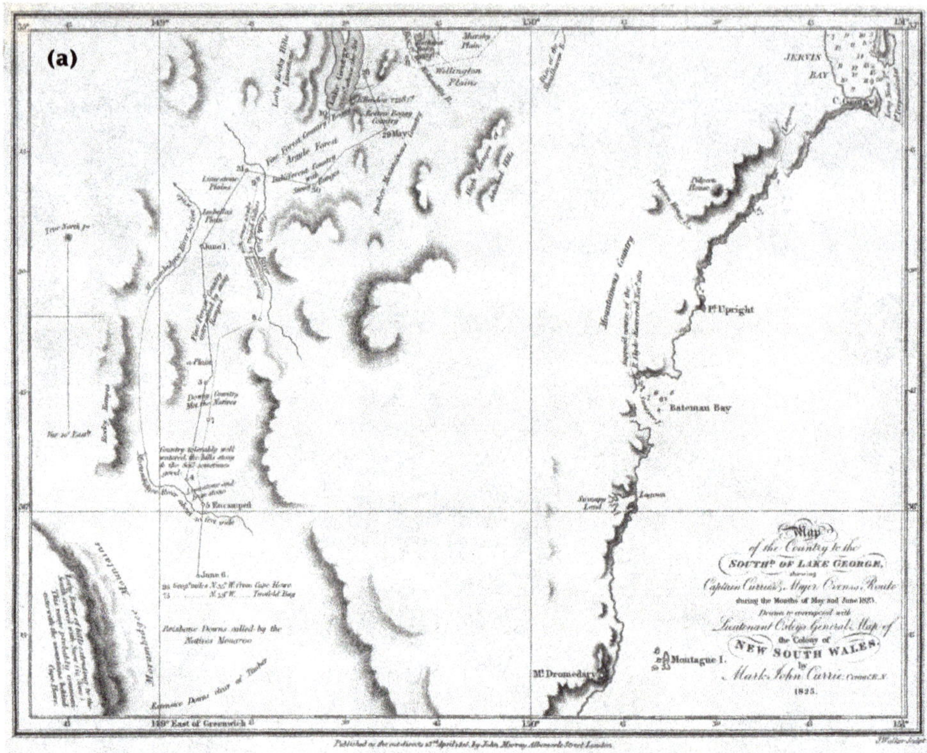

Figure 1.19: (a) A sketch map of the ACT in 1825. Mount Stromlo is not specifically marked; it is located in the blank area in the upper left corner of the map. Map: National Library of Australia. (b) Location of Mount Stromlo as shown on a current road map.

Figure 1.20: Mount Stromlo from Mount Ainslie, with Lake Burley Griffin in the foreground, July 2011. The Stromlo Observatory domes can be seen on the low hill in the middle of the picture. The hill has a volcanic origin and was formed in a single volcanic event about 420 million years ago. The hardness of the rock provides a very stable foundation for scientific equipment; however, it also presents difficulties in construction projects on the mountain. Photo: Ralph Sutherland. Mount Stromlo Archives.

Figure 1.21: Pietro Baracchi (1851–1926) was the Government Astronomer of Victoria from 1900 to 1915. He was trained as a civil engineer in Italy before emigrating to Australia. He worked as an assistant astronomer at Melbourne Observatory in 1876. He left the Observatory to become a draftsman in the Victorian Land and Survey Department, then returned to Melbourne Observatory in 1883 and rose to the position of Government Astronomer. He served as the President of the Royal Society of Victoria in 1908–1909. Photo: National Library of Australia.

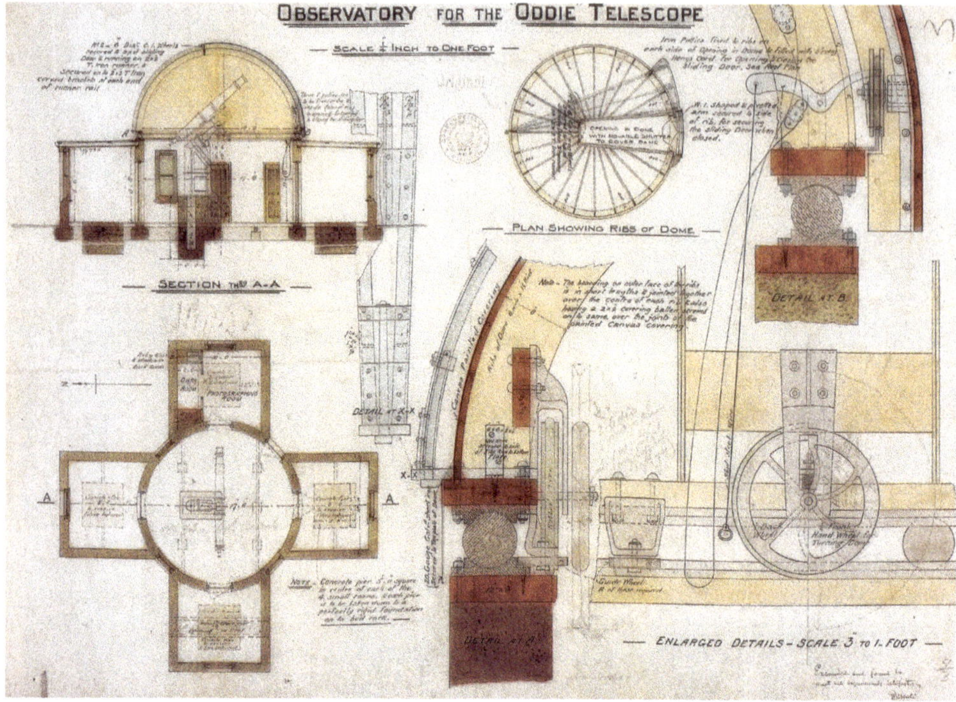

Figure 1.22: The Oddie telescope was mounted in the circular portion at the centre of the building. Photo: Mount Stromlo Archives.

the Commonwealth Meteorologist. Hunt built a meteorological station, including a kite- and balloon-launching hut, by 1914 in preparation for balloon and kite measurements of the upper atmosphere. However, after the First World War the Kite House and the hut were put to other uses. Baracchi noted that 'the atmospheric conditions were very good and that they never had less than four fine days and nights out of seven'.[3] The results of the observations were presented at the British Association for the Advancement of Science meeting held in Australia in 1914.

Figure 1.23: The earliest known photograph of the Oddie Building, 1911. It was here that the first astronomical observations were made to test the suitability of the site for the establishment of the Solar Observatory. Photo: Mount Stromlo Archives.

Figure 1.24: Alpha Centauri, Kappa Crucis (shown here) and other celestial objects were observed by Baracchi to test the suitability of Mount Stromlo as a site for the Commonwealth Solar Observatory. Photo: Bessell, Sutherland and Buxton.

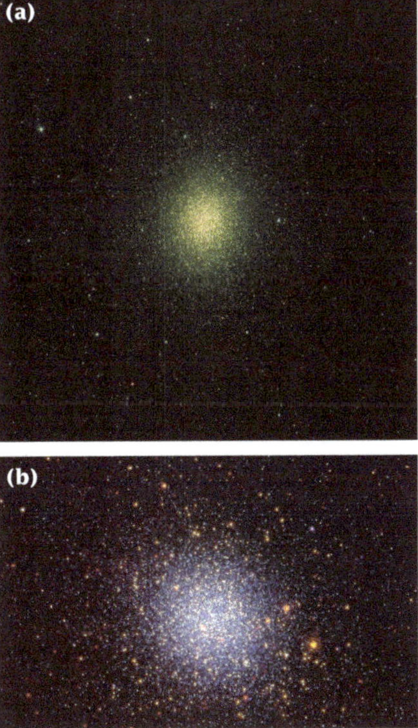

Figure 1.25: (a) Omega Centauri was one of the spectacular globular clusters observed by Baracchi and Baldwin in their site testing program at Mount Stromlo. Photo: Mount Stromlo Archives. (b) This Hubble Space Telescope image of Omega Centauri shows the dense swarm of stars at the centre of the cluster. Photo: NASA/NSF.

Figure 1.26: Scrivener (seated on the right) with his survey team. Charles Scrivener (1855–1923) was the Commonwealth Surveyor and established the prime meridian for Australia at Mount Stromlo. All Australian longitudes were to be referred to this meridian. The trig station was named the Strom Trig Station by the surveyors, to avoid a name conflict with an existing trig station on another Mount Stromlo in New South Wales, near Oberon. Photo: National Library of Australia.

Figure 1.27: Henry A. Hunt (1866–1946) was the Commonwealth Meteorologist. Hunt, E.T. Quayle and T. Griffith wrote the first textbook on Australian meteorology, *Climate and Weather in Australia.* Photo: Mount Stromlo Archives.

British Association for the Advancement of Science

In the early 20th century improvements in telecommunications and in the cost and ease of travel led to the rapidly increasing internationalisation of science. The British Association for the Advancement of Science was committed to reaching out to the farthest edges of the British Empire to promote scientific enterprise. Its report on the 1914 meeting in Australia reiterated its goals:

> *The objects of the British Association for the Advancement of Science are: To give a stronger impulse and a more systematic direction to scientific inquiry; to promote the intercourse of those who cultivate Science in different parts of the British Empire with one another and with foreign philosophers; to obtain more general attention for the objects of Science and the removal of any disadvantages of a public kind which impede its progress.*

Overshadowing the meeting in Australia in 1914, of course, was the outbreak of the Great War, which also caused considerable disruption to the planned return voyages to the UK. The report recorded (p. 703):

> *It had become known by this time [mid-August] that the three ships by which the majority of the Members had travelled out (the Orvieto, Ascanius, and Euripides), and by which many intended to return, had been requisitioned, among others, for Government purposes (conveyance of troops, &c.) in connection with the war ... The official arrangements for the visit of a party of Members to New Zealand had been cancelled (as narrated elsewhere). It therefore became necessary to the organisation that a further inquiry should be made of Members as to any change in their plans regarding their stay in and departure from Australia. This inquiry was made by means of a printed form distributed in the train between Melbourne and Albury.*

Lobbying for the Observatory

Duffield intended to use the 1914 meeting of the British Association for the Advancement of Science in Australia to pressure the federal government to confirm the establishment of the Solar Observatory at Mount Stromlo, but just before the Committee departed Britain for Australia the government made an in-principle commitment to its establishment. A few weeks later the Administrator of the Capital Territory stated that the government proposed to establish a permanent Commonwealth Observatory which would be 'adequately equipped, staffed and administered for the highest and most important aims of modern astronomical research'.[3] Interviewed by a reporter from the *Adelaide Advertiser* on 4 August 1914, on the subject of solar physics, Duffield said, 'It was important that Australia should participate in this work, because there is no observatory of the kind in the same latitude between America and India, and such an institution would fill the gap in the chain already existing around the earth, and enable the Sun to be kept under continuous observation.' It made the front page of the newspaper. More than 300 British, US, European and other foreign scientists attended the meeting in Australia. Britain's scientific elite, consisting of Ernest Rutherford, Oliver Lodge, Arthur Eddington and Henry Tizard, were among the luminaries who gave public lectures on the latest developments in science.

The Association came to Australia to spread the doctrine of imperial science which had been initiated by Sir Joseph Banks[16] and to emphasise the usefulness of Australia as a producer nation for the mother country. The congress made several recommendations, including the establishment of a solar observatory. The then Prime Minister, Joseph Cook, met a delegation from the Association to discuss the establishment of the Solar Observatory.

Figure 1.28: Sir Arthur S. Eddington (1882–1944) was Plumian Professor of Astronomy and Experimental Philosophy at Cambridge University and was elected a Fellow of the Royal Society in 1914. He conducted the expedition to observe the solar eclipse on 29 May 1919 to test Einstein's Theory of General Relativity, which predicted that light rays from stars would be bent due to the Sun's gravitational field. It was one of the first confirmations of Einstein's theory. Photo: Mount Stromlo Archives.

Figure 1.29: Andrew Fisher served as the fifth Prime Minister of Australia. He actually was Prime Minister on three occasions: 1908 (202 days), 1910 (three years and 57 days) and 1914 (one year and 41 days). Photo: National Library of Australia.

Figure 1.30: Duffield in the Royal Flying Corps. Photo: Mount Stromlo Archives.

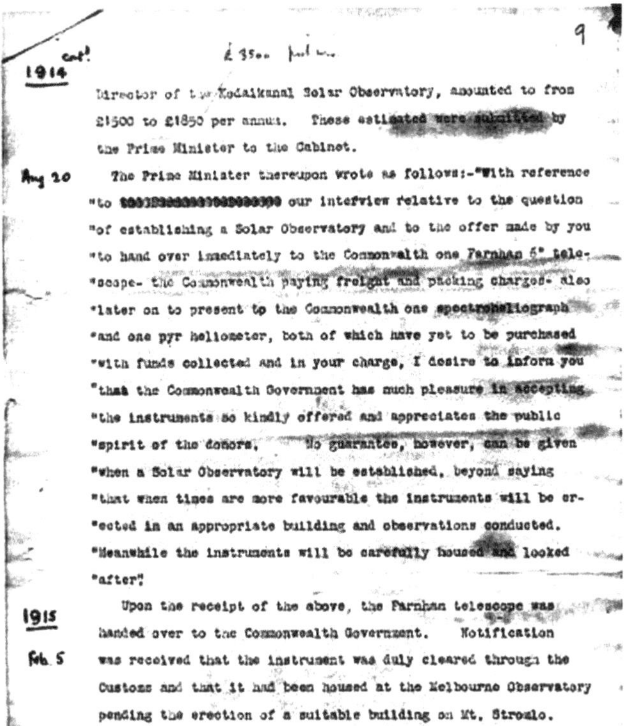

9

1914

Director of the Kodaikanal Solar Observatory, amounted to from £1500 to £1850 per annum. These estimated were submitted by the Prime Minister to the Cabinet.

Aug 20 The Prime Minister thereupon wrote as follows:-"With reference "to our interview relative to the question "of establishing a Solar Observatory and to the offer made by you "to hand over immediately to the Commonwealth one Farnham 6" tele- "scope- the Commonwealth paying freight and packing charges- also "later on to present to the Commonwealth one spectroheliograph "and one pyr heliometer, both of which have yet to be purchased "with funds collected and in your charge, I desire to inform you "that the Commonwealth Government has much pleasure in accepting "the instruments so kindly offered and appreciates the public "spirit of the donors. No guarantee, however, can be given "when a Solar Observatory will be established, beyond saying "that when times are more favourable the instruments will be er- "ected in an appropriate building and observations conducted. "Meanwhile the instruments will be carefully housed and looked "after".

1915 Upon the receipt of the above, the Farnham telescope was handed over to the Commonwealth Government. Notification

Feb. 5 was received that the instrument was duly cleared through the Customs and that it had been housed at the Melbourne Observatory pending the erection of a suitable building on Mt. Stromlo.

Figure 1.31: Duffield's account of the establishment of the Commonwealth Solar Observatory. It was written in the usual matter-of-fact voice of scientific discourse, which gave the impression of objectivity and concealed Duffield's motives and actions. Image: Australian Academy of Science.

The delegation included Duffield, Alfred Deakin (a former Prime Minister and Member for Ballarat), Sir Arthur Eddington and Sir Frank Dyson.[11] A later Prime Minister, Andrew Fisher, subsequently wrote to Duffield to inform him that 'No guarantee ... can be given when a solar observatory will be established, beyond saying that when times are more favourable the instruments will be erected in an appropriate building and observations conducted.'[17]

Work on the establishment of a Solar Observatory slowed during the First World War. Duffield joined the Royal Flying Corps, where he taught physics to the members of the armed forces.

In October 1922, Duffield returned to Australia to meet Prime Minister William Morris 'Billy' Hughes and other government ministers. However, it was not until April 1923 that the government gave a firm undertaking to establish the Commonwealth Solar Observatory 'to link up with existing institutions in England, India, America and certain European countries'.[11] As was the case in the 19th century, the Australian government sought the advice of the Astronomer Royal on the appointment of the Commonwealth Astronomer. A committee of leading British astronomers, including Sir Frank Dyson and Professor H.H. Turner, recommended Duffield be appointed the first Director of the Observatory. As to why Duffield was chosen, Turner wrote, 'Why, the man who told you there was need of it.'[18]

References

1. Young HD, Freedman RA, Bhathal RS 2011. *University Physics*. Pearson Australia, Melbourne.
2. Bhathal R 2012. Profile: Brian Schmidt. *Astronomy & Geophysics* **53**, 1.13–1.15.
3. Frame T, Faulkner D 2003. *Stromlo: An Australian Observatory*. Allen and Unwin, Sydney.
4. Bhathal R 2009. 150 years of Sydney Observatory. *Astronomy & Geophysics* **50**, 3.26–3.30.
5. Bragg WH 1929. Letter to Mrs Duffield dated 3 October 1929. Duffield papers, file no. 1–10, Box 1. Basser Library. Australian Aacademy of Science.
6. Sutherland RS, Butcher HR 2011. *An Observatory at Mount Stromlo*. Unpublished.
7. Allen CW 1981. Duffield, Walter Geoffrey. In *Australian Dictionary of Biography*. (Eds B Nairn and G Serle) p. 351. Melbourne University Press, Melbourne.
8. Love R 1984. Science and government in Australia 1905: Geoffrey Duffield and the foundation of the Commonwealth Solar Observatory. *Historical Records of Australian Science* **6**(2), 171–188. 10.1071/HR9850620171
9. Resolution of the International Union for Solar Research, Jansen Solar Observatory. Chateau de Meudon. 1907. Duffield papers. Basser Library. Australian Academy of Science.
10. Letter to A. Schuster re Solar Physics Observatory. Duffield papers. Basser Library. Australian Academy of Science.
11. Duffield papers. Basser Library. Australian Academy of Science.
12. Flood JM, David B, Magee J, English B 1987. Birrigai: a Pleistocene site in the southeastern highlands. *Archaeology in Oceania* **22**, 9–22.
13. Gillespie L 1984. *Aborigines of the Canberra Region*. Wizard (Lyall Gillespie), Canberra.
14. Nelmes M 1993. Mr Oddie's legacy. *Southern Sky* **2**, 22–27.
15. Perdix JL 1979. Baracchi, Pietro Paolo Giovanni Ernesto. In *Australian Dictionary of Biography*. (Eds B Nairn and G Serle), pp. 166–167. Melbourne University Press, Melbourne.
16. Gascoigne J 1998. *Science in the Service of Empire*. Cambridge University Press, Cambridge.
17. Duffield WG (no date). An account of the movement to establish a solar observatory in Australia. Duffield papers. Basser Library. Australian Academy of Science.
18. Turner HH 1929. Letter to the Editor: Dr W.G. Duffield's work. *The Times*, 6 August. Duffield papers. Basser Library. Australian Academy of Science.

1924–1929

2

A bush observatory

It is my earnest desire that we should take our place among the great observatories of the world.

Geoffrey Duffield (Duffield papers. Basser Library. Australian Academy of Science)

Those early days were full of excitement and it was interesting pioneering work, and my father with his charismatic personality, sense of humour and infinite patience always won through in the end [Joan Duffield].[3]

On 1 January 1924 the Commonwealth Solar Observatory was established at Mount Stromlo in Canberra. On the same day Geoffrey Duffield was officially appointed the Director of the Observatory. After almost 18 years Duffield had realised his dream of an observatory at a site 'high among the rocks and lizards and miles from anywhere'.[1] He expected to return triumphantly to Canberra from Britain after his release from the University College of Reading but was disappointed with the progress when he arrived, in November 1924. Although plans for construction of the Observatory had been approved 12 months earlier, progress had been slow. There was much to be done before it could be used for scientific and astronomical purposes.

Building Mount Stromlo Observatory

The Department of Home Affairs and Territories had arranged for Duffield's accommodation and for temporary observatory quarters at the new Hotel Canberra, which itself was undergoing construction. In fact, Duffield and his family were the first to sign the guest register. These were trying times for Duffield as he not only had to supervise the building of the Observatory according to his specifications but also to liaise with the staff of the Department of Home Affairs and Territories, which was based in Melbourne. This necessitated him travelling to Melbourne for meetings about the progress of the Observatory. Unlike Duffield, who was a man in a hurry, the bureaucrats had a rather relaxed attitude towards construction. Duffield vented his frustrations in his diary: '23 July 1925. Contrary to my previous

Figure 2.1: Canberra in 1923, looking south-west from the foot of Mount Pleasant across the Molonglo River. Parliament House is on the right. Mount Stromlo, 'high among the rocks and lizards', is in the distance with the Brindabella mountain range on the horizon. Painting by H. M. Rolland, Architect to the Federal Capital Commission. Photo: National Library of Australia.

[Reprinted from "The Official Year Book of the Commonwealth" (pp. 1033 to 1035), No. 18—1925.]

§ 6. The Commonwealth Solar Observatory.

1. **Reasons for Foundation.**—The Commonwealth Solar Observatory was established for the study of solar phenomena, for allied stellar and spectroscopic research, and for the investigation of associated terrestrial phenomena. It was founded to fill a gap in the chain of astrophysical observatories round the earth, and, with its completion, there will be stations separated by 90 degrees of longitude round the globe. Its situation in lat. 35° south places it in the unique position of being the only observatory making a feature of solar work south of the equator. It was founded chiefly for the purpose of advancing the knowledge of the Universe and the mode of its development, but also in the hope that the eventual discovery of the true relation between solar and terrestrial phenomena may lead to results which will prove of direct value to the country.

Figure 2.2: The 1925 Year Book of the Commonwealth gave the official reasoning behind the establishment of the new Solar Observatory.

Figure 2.3: The Duffield family on its first visit to Mount Stromlo. Duffield was disappointed when he first arrived at Mount Stromlo with his family. All he saw was barren ground. Photos: Mount Stromlo Archives.

Figure 2.4: In keeping with his status as Director of a national observatory, Duffield purchased a huge Crossley car. It had previously been one of the fleet used during the visit of Edward, Prince of Wales, to Australia. On Duffield's first trip with his family from Sydney to Canberra the car struck a straying cow near Bungendore during the night. They arrived in Canberra at ~2 a.m. Photo: Crossley Motor Company.

Figure 2.5: Hotel Canberra in 1924. This was Duffield's home for two years. Photo: National Archives.

Figure 2.6: Duffield visited Australia several times during his tenure at the University of Reading. An important activity on these visits was to further plans for the Observatory. Here he is shown in 1922 on Mount Stromlo examining the plans and the site with government officials (Col. Owen and H.M. Rolland). Photos: Mount Stromlo Archives.

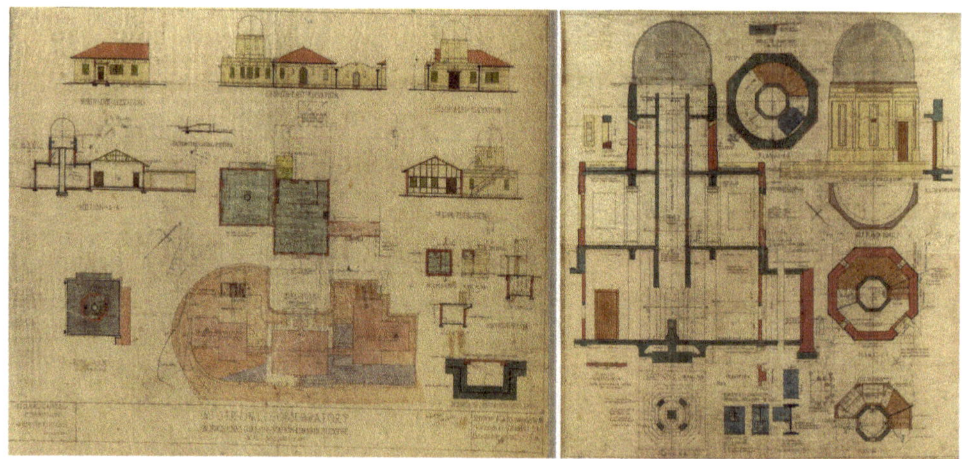

Figure 2.7: Plans for the Observatory. They were simple and functional. Photo: Mount Stromlo Archives.

Figure 2.8: Tractor bringing construction materials (1924). The route from the Hotel Canberra to the Observatory site was 'along a badly rutted and pot-holed narrow road to the foot of Mount Stromlo, and then up a very winding narrow and equally bad road to the Observatory site.'[3] Photo: Mount Stromlo Archives.

Figure 2.9: Observatory construction started in 1924. It was completed in stages, with the first assistant's cottage (shown here) the first to be ready for use. The Observatory was surrounded by a plantation of pine trees, which posed a fire danger. Photo: Mount Stromlo Archives.

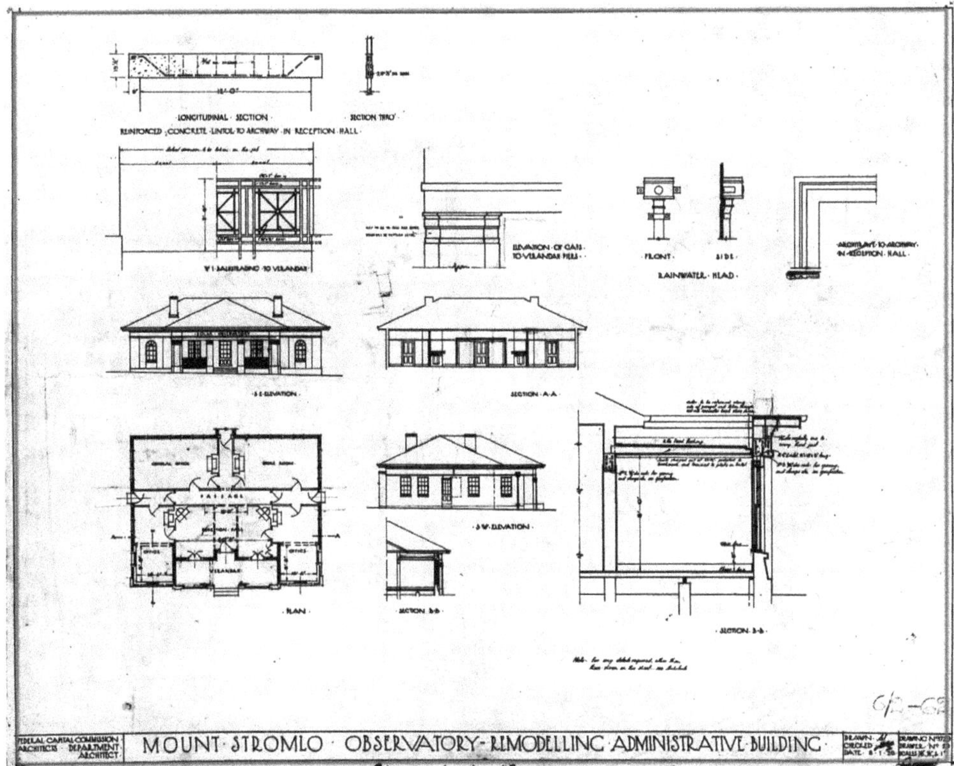

Figure 2.10: Plans for the administration building of the Observatory prepared by the Architects Department of the Federation Capital Commission, dated 8 January 1926. The plans had several iterations and modifications under Duffield's direction. Drawing: Mount Stromlo Archives.

representations to the Chief Architect there was no water in the rooms to be used as workshop and laboratory. There was no alteration to the number of windows in the Library and Director's rooms as had been requested.'[2]

Figure 2.11: A view of the administration building just after completion of the dome and annex for the Farnham telescope. Photo: Mount Stromlo Archives.

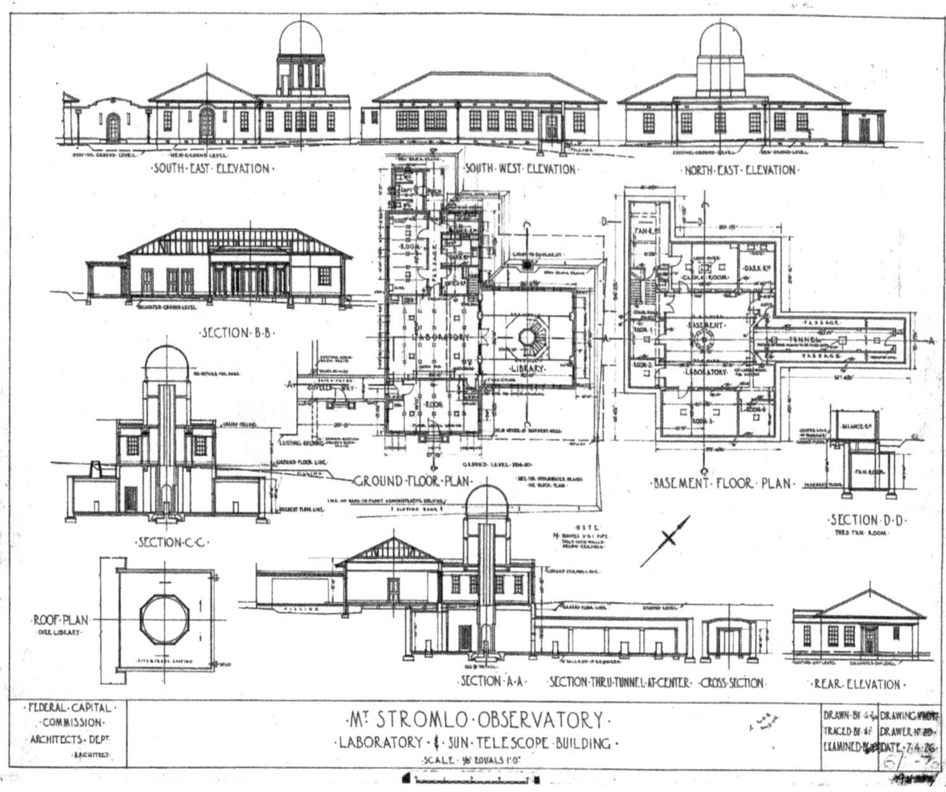

Figure 2.12: Plans for the laboratory and Sun telescope building, dated 7 April 1926. The vertical column which contained the Sun telescope beam and the small dome for the coelostat are shown in the plans. Drawing: Mount Stromlo Archives.

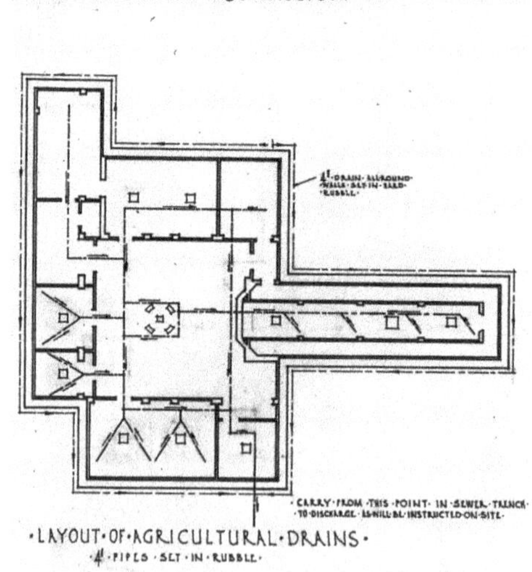

Figure 2.13: The Sun telescope included a basement that presented drainage problems. The plans for the telescope and its subterranean laboratory required an extensive, carefully constructed set of drains. Drawing: Mount Stromlo Archives.

Figure 2.14: View of the subterranean Sun telescope laboratory under construction, 1927. Photo: Mount Stromlo Archives.

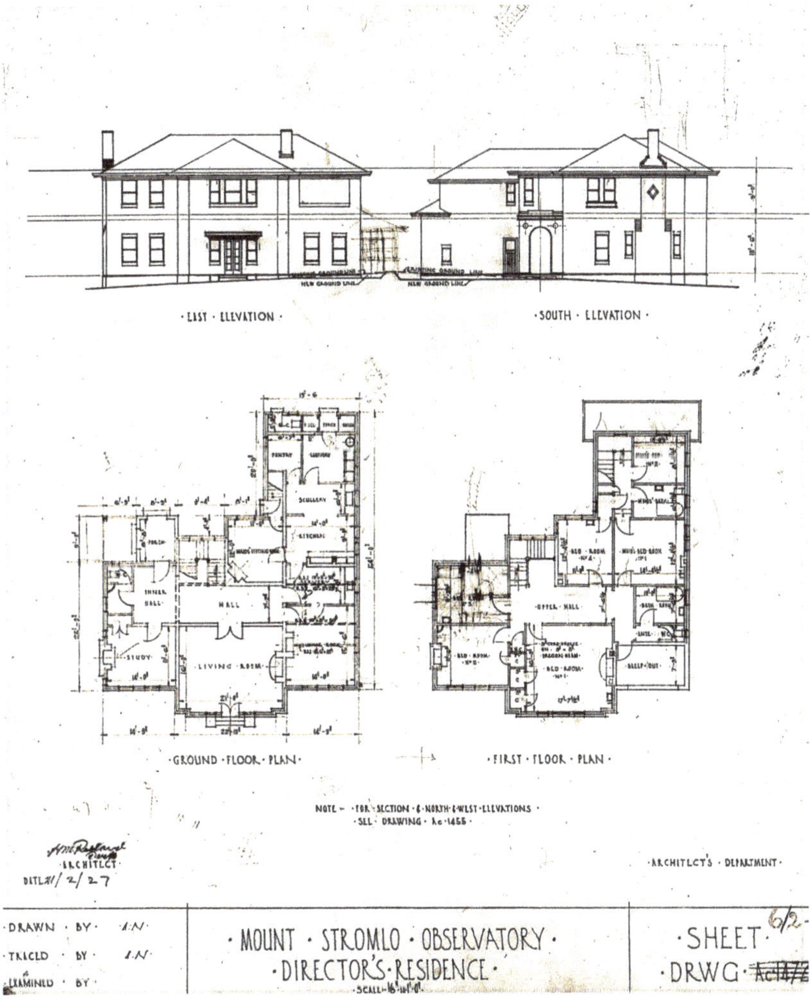

Figure 2.15: Nearly final plans for Director's Residence, 1927. Drawing: Mount Stromlo Archives.

Figure 2.16: The Director's Residence, initially called Observatory House. Duffield's children, including Joan (shown here), attended boarding school in Melbourne. The family lived first in the Hotel Canberra, then in the first assistant's house and then later moved into their completed residence. Miss Duffield has remained associated with the Observatory throughout her life, celebrating her 100th birthday on Mount Stromlo in April 2010. Photos: Mount Stromlo Archives.

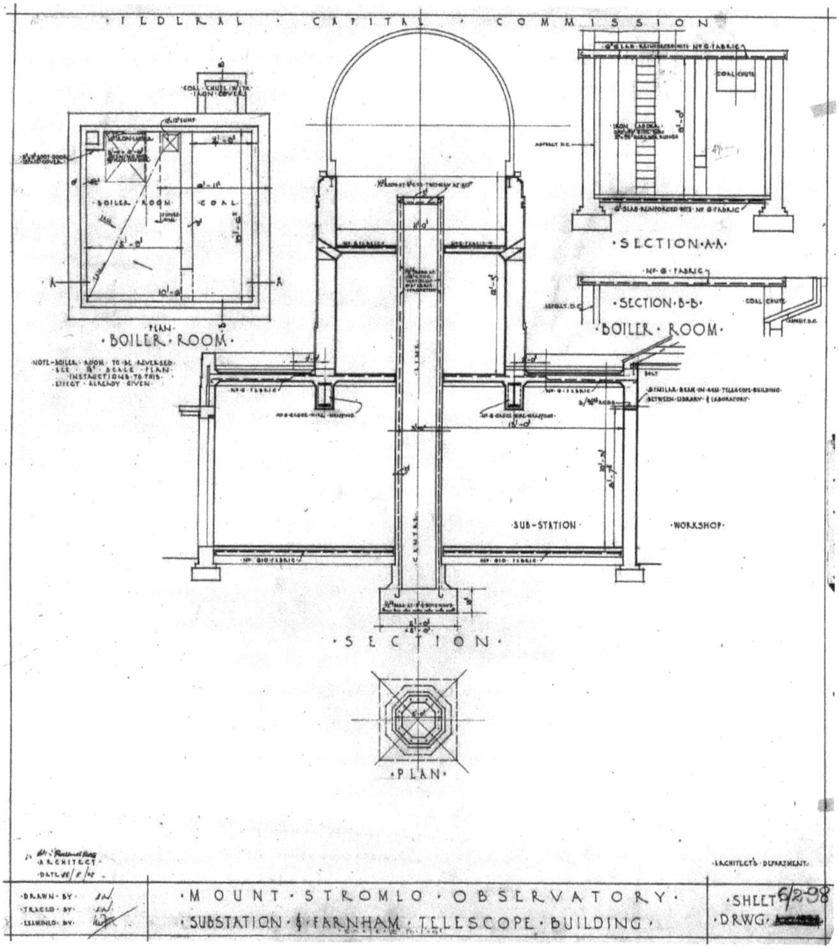

Figure 2.17: Plans for the substation and the mounting of the Farnham telescope. Drawing: Mount Stromlo Archives.

Figure 2.18: Views of the Observatory construction in 1925. Photos: Mount Stromlo Archives.

Figure 2.19: By 1928 the domes and the building of the Commonwealth Solar Observatory were fairly advanced and preparations were being made to install telescopes and research equipment. Photo: Mount Stromlo Archives.

Figure 2.20: The main building of the Commonwealth Solar Observatory was ready for occupation at the end of 1928. Photo: Mount Stromlo Archives.

Figure 2.21: At the start of 1929, landscaping around the main Observatory building was completed and the research facilities could be installed. Photo: Mount Stromlo Archives.

It was not until December 1926 that the main building and the staff houses, which were located a short distance away, were almost ready for occupation. The Director's House, called Observatory House, was completed two years later. The main works on the Observatory (the solar laboratory and the workshop buildings) continued through 1927–28. In December 1926 the astronomers began the move to Mount Stromlo. The equipment which had been used in temporary labs at Hotel Canberra was loaded on to a horse-drawn cart and taken up the rather narrow rough road to the mountain top. The east wing of the main Observatory building contained the solar laboratory, which was located in the basement. It also contained a covered tunnel for a large spectrograph. A lecture room and the library were located in the ground-level rooms. According to Cla Allen, 'A vertical concrete column through the library contained the Sun-telescope beam and was surmounted by the coelostat within a small dome. The basement laboratory was ready for all spectroscopic uses by 1928, but the Sun-telescope was not available until November 1931'.[4] The workshops and an electrical substation with transformers, a DC generator and a storage battery were located in the west wing. The dome for the Farnham 15 cm telescope surmounted the electrical substation. In addition to the Oddie telescope, the Oddie dome housed the recording pyrometer. The Kite House became the centre for meteorological and potential gradient observations.

Hiring staff

In addition to getting the buildings ready, Duffield had to hire scientific and technical staff to carry out the astronomical program he had outlined for the Observatory. The program was to include studies of the Sun, luminosity of the night sky, spectroscopy, southern sky astronomy and geophysics. It is not clear why geophysics was included in the Observatory program.

One of the first persons recruited, to work on stellar problems, was William Bolton Rimmer, a Research Assistant at the Norman Lockyer Observatory in England. He was recommended by the Astronomer Royal and the Director of the Greenwich Observatory. He had an MSc and a diploma in astrophysics from Imperial College, University of London and had published papers (with Norman Lockyer) in the journals of the Royal Astronomical

Observatory staff

Love did not stay long at the Observatory. He was reprimanded by Duffield for having a 'filthy collar'.[2] Feeling terribly insulted, he handed in his resignation.

In his place, Clabon Walter Allen (well known as 'Cla') was appointed as Research Fellow in January 1926 and was promoted to the position of Assistant in 1928. Allen had a science degree from the University of Western Australia with a Distinction in physics, which had brought him to the attention of Duffield.

Kennedy had worked at the Adelaide Observatory and with Douglas Mawson in the Antarctic. He had experience in surveys in Australia. He was a valuable member of Duffield's team in establishing the Observatory but never felt at home there and left after about three years.

Higgs had a background in physics and was involved in ionospheric studies. He had postgraduate qualifications in chemistry and had worked in industry in a research capacity.

Hogg was recruited in 1929 to work on atmospheric ion problems. He had graduated with an Honours degree in chemistry from the University of Melbourne in 1923 and obtained an MSc two years later on a Kernot Scholarship. Like Higgs, he had an industrial research background, having served as the Assistant Superintendent of Research with the Broken Hill Associated Smelters at Port Pirie in South Australia.[2]

The technical staff included B.P. Clark and H.J. 'Jim' Banham, who was later promoted to the position of workshop foreman. In later years his son, Norman, also became foreman of the Observatory workshop.

Society and the Royal Society. He was appointed as the Assistant Solar Observer in 1924 and also served as Deputy to Duffield. He was the Observatory's only trained astronomer.

The other professional staff were physicists and or geophysicists. They included A.L. Kennedy, Arthur John Higgs, William Eric Duncanson, Arthur Hogg and W.H. Love.

Having appointed the necessary scientific and technical staff, Duffield was in a position to carry out his astrophysical program and to bring the Observatory to the attention of astronomers in Europe and the US.

The new field of astrophysics

Duffield had arrived on the astronomical scene when the major metropolitan centres of astronomical research in Europe and the US were moving into the new field of astrophysics, popularly termed the 'new astronomy'. According to Agnes Clerke, in her monumental book, *History of Astronomy during the Nineteenth Century*, 'the new field of astrophysics is full of audacities, the inconsistencies, the imperfections, the possibilities of youth ... It promises everything, it has already performed much, it will doubtless perform much more'.[5]

The new field was experimentally based with not much of a theoretical foundation. In fact, in its early years many astronomers regarded

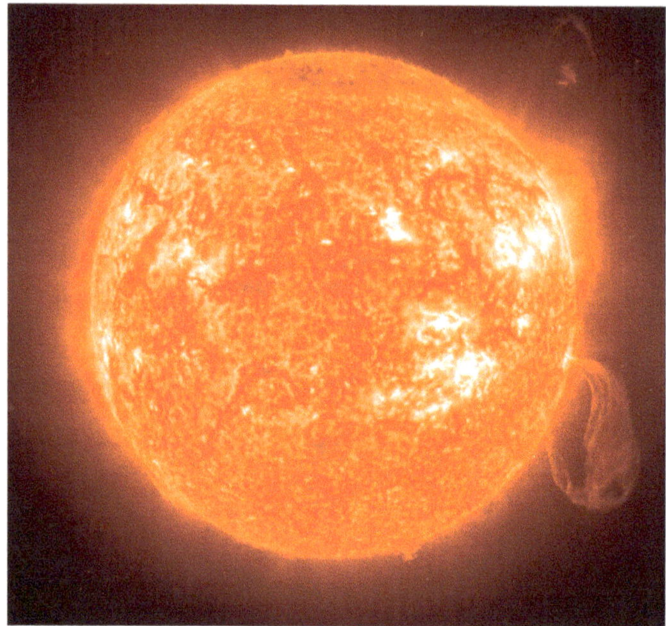

Figure 2.22: The Sun was the centre of attention in the early years of astrophysics because of its intense brightness compared to the stars. Photo: NASA.

astrophysics as inferior to classical astronomy as it did not have a solid mathematical base.[6] However, by the 1890s it had become a subject in its own right, not only with its own journal (the *Astrophysical Journal*) but with its own techniques and instruments. Interestingly, it was launched in the US, which had played only a minor role in classical astronomy.

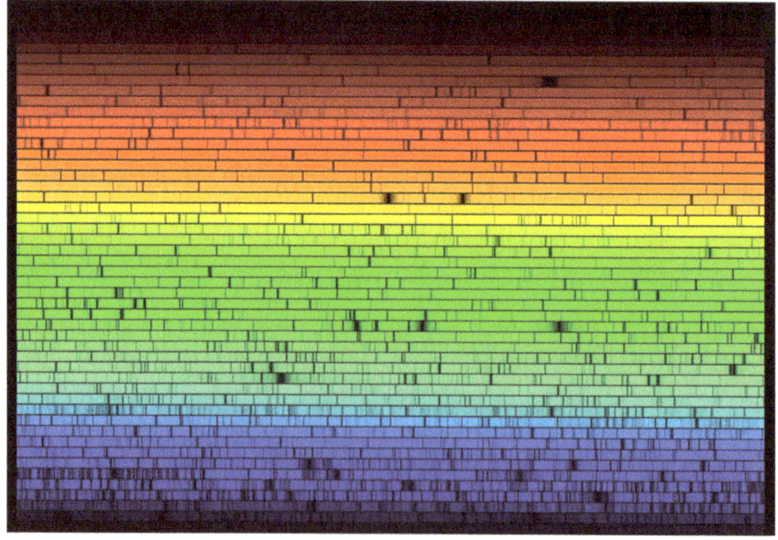

Figure 2.23: The Sun's spectrum. The solar spectrum shows the dark lines which correspond to the elements found in the Sun. Photo: AURA.

Rise of astrophysics

Astrophysics began as a hybrid science developed from the astronomy, chemistry and physics of the 19th century. The spectroscopic experiments conducted by German chemists Kirchhoff and Bunsen provided its rationale. As early as 1814 Fraunhofer had discovered nearly 600 dark lines that crossed the Sun's spectrum, later called Fraunhofer lines. Kirchhoff's analysis of the Sun's spectrum and its comparison with his laboratory spectra gave a clear correlation between the dark absorption lines in its spectrum and the chemical elements it contained. Kirchhoff also showed that when a heated source of continuous spectrum is viewed through a cooler gas, the wavelengths at which absorption occurred in the cool gas corresponded to those at which emission occurred when the gas is heated. It was soon realised that spectroscopy could provide an important tool for determining the nature and composition of stars.

Because the Sun was much brighter than the stars, it became a popular target for observation in the early years of astrophysics. In fact, the subject was initially called solar physics. Since the only way to observe the outer layers of the Sun was during a solar eclipse, expeditions to observe eclipses of the Sun became popular. A series of eclipses between 1868 and 1871 led to Lockyer's discovery of the yellow emission line of helium. It had not been seen in the laboratory, was attributed to an unknown chemical element and dubbed 'helium' (from the Greek *helios* for *Sun*) by Lockyer's collaborator, Eduard Frankland. The origin of this line was not confirmed in the laboratory until the end of the century, by William Ramsay.

Other studies of the Sun included attempts to measure the amount of heat reaching the Earth from the Sun (i.e. the solar constant), the association of sunspots with the fluctuations in the Earth's magnetic field and the relationship between activity on the Sun and terrestrial weather.

In the US, E.C. Pickering at Harvard began a large-scale investigation of stellar spectroscopy and photometry. This led to the publication, in the 1890s, of the Draper Catalogue of Stellar Spectra for the northern hemisphere. In 1901, Annie Jump Cannon and Pickering published a catalogue of bright stars of the southern hemisphere. The Draper Catalogue of Stellar Spectra introduced the alphabetical classification of stellar spectra which is still used today.

One of the problems that hindered this new field of study throughout the 19th century was the lack of a theoretical understanding of spectra. This was resolved in the 20th century by the development of quantum theory and Bohr's theory to explain the spectra of the hydrogen atom, later modified to account for atomic spectra generally.[7] In fact, in 1908 Planck wrote in his notebook, 'I'm fully convinced that the problem of spectral lines is intimately tied to the question of the nature of the quantum'[8] and in Planck's 1919 Nobel lecture he stated that Bohr's quantum atom was 'the long-sought key to the entrance-gate into the wonderland' of spectroscopy.[9] Today, spectroscopy is an indispensable tool for astrophysics.

Duffield was an experimental physicist with an engineering background, perfect preparation for participating in this new field. His early research centred on finding the effect of pressure on arc spectra. His experimental studies of spectra included iron, silver, gold, platinum and copper under pressures of ~200 atmospheres. The work was difficult and dangerous but his friends

Figure 2.24: Norman Lockyer who, with French scientist Pierre Jansen, discovered helium in the Sun. Lockyer was the founder and editor of the influential British journal *Nature*. Photo: Institute of Astronomy, University of Cambridge.

were rather irreverent about the laboratory conditions he worked in. At a theatrical party organised by his brother, he was introduced as 'the man who got up to sixty without bursting'.[10]

Duffield began his spectroscopic studies at a time of several technical developments in the study of the spectra of the Sun and stars. The introduction of rapid-dry plates had revolutionised the photographic process.[6] The earlier wet-plate process did not remain sensitive during the long exposures required for astronomical purposes. The new process enabled astronomers to efficiently record light for as long as they cared to guide the telescope to follow the stars. The sensitivity and quality of the photographic materials also increased considerably.[11]

New forms of specialised photographic instrumentation had become available. Photography became a tool of discovery. Furthermore, Rowland's invention of the concave grating, made possible by improvements in the engineering of screws, greatly improved the optical efficiency of spectrographs in critical regions of the spectrum. Duffield's engineering background allowed him to make some improvements to instrumentation as he began studying the arc spectra of metals.[12,13]

He published his findings in reputable journals such as the *Proceedings of the Royal Society of London* and the *Philosophical Transactions*. He also published in the new *Astrophysical Journal*, thus ensuring that he was known to his US colleagues. He was complimented by Bragg for his work on arc spectra under pressure. Bragg wrote, 'Your spectra are most interesting, and should lead to important deductions. The behaviour of the various series must no doubt be considered alongside Stark's results. I should think you will have a lot to say presently about the way in which light is actually emitted. I am

Stark effect

In 1913, the German physicist Johannes Stark had found that a single spectral line split into several lines when atoms are placed in an electric field.[9] The Stark effect was explained by other physicists as resulting from changes in the spacing between energy levels due to the presence of an electric field.

very glad indeed to hear such a good account of your work, and congratulate you on what you have done. It ought to attract a lot of attention'.[14] However, Duffield did not follow Bragg's suggestion.

Research activities

Duffield arrived at Mount Stromlo steeped in the latest developments in astrophysics and he was anxious to begin work, even in the temporary observatory quarters in Hotel Canberra. Initially the staff were mainly occupied with assembling, repairing and calibrating the equipment for scientific and astronomical use. Some of the equipment was found to be damaged on arrival from England. In other cases defects were found and had to be corrected. However, it was not long before the temporary observatory could be used for conducting observations. According to Cla Allen, 'In a temporary manner the Observatory proceeded to set up pyrheliometry, pyranometry, atmospheric potential gradient (electric field) recording, sunspot sketches, clocks and time signals, objective prism stellar spectroscopy, night-sky luminosity observations, microphotometry, and laboratory – all in six Hotel Canberra rooms'.[4]

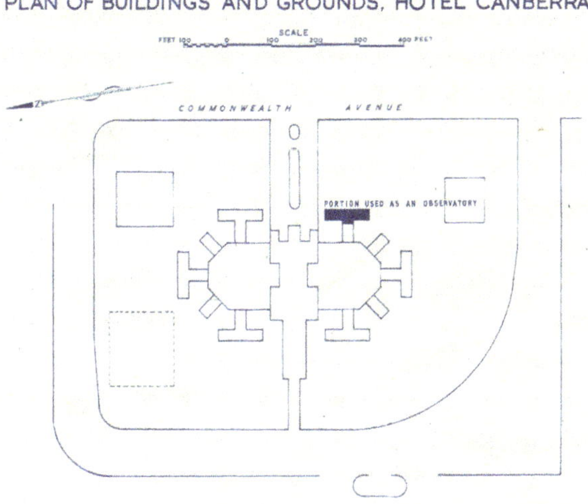

Figure 2.25: Plan showing the location of the temporary observatory in Hotel Canberra. The Hotel Canberra was established in 1924 as a government-owned hostel. In 1927, the hostel became the Hotel Canberra. Drawing: Mount Stromlo Archives.

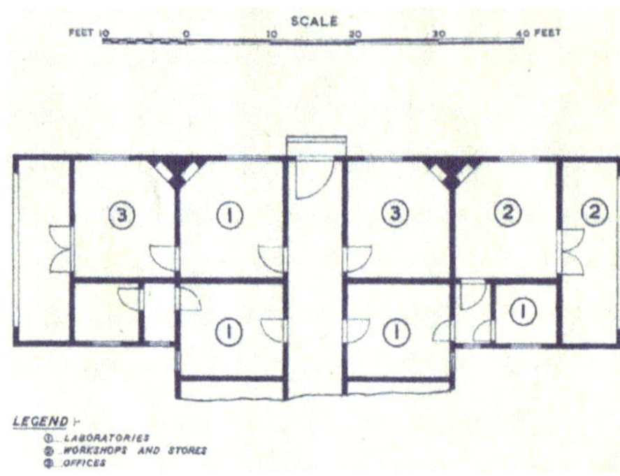

Figure 2.26: Plan of the use of temporary observatory quarters at the Hotel Canberra. Drawing: Mount Stromlo Archives.

A request from Lord Rayleigh prompted Duffield to begin a program of observations of the luminosity of the night sky between October 1925 and October 1926. The program was later continued at Mount Stromlo Observatory from October 1926 to December 1927. The observations were made with a photometer designed and supplied by Rayleigh to observatories around the world for determining the distribution of the intensity of the auroral radiation. The observations were carried out by Kennedy until October 1927, then continued by Higgs. The results were published by Duffield (1928) in the first Memoir of the Commonwealth Solar Observatory.[15] It was found, but could not be explained, that there was a general trend for the night sky to be a maximum in April–May for both Canberra and Mount Stromlo. The observations lent support to Rayleigh's division of the aurora into two types, polar and non-polar. The observations led astronomers at Mount Stromlo to realise that there was a need for a more detailed study of the fluctuations of the night sky luminosities in the southern latitudes. Further studies were carried out by Higgs and from 1934 the observations were sent to the Astronomer

Figure 2.27: Hotel Canberra in 1926 when the observatory staff used its rooms for offices and laboratories. Photo: Canberra District Historical Society.

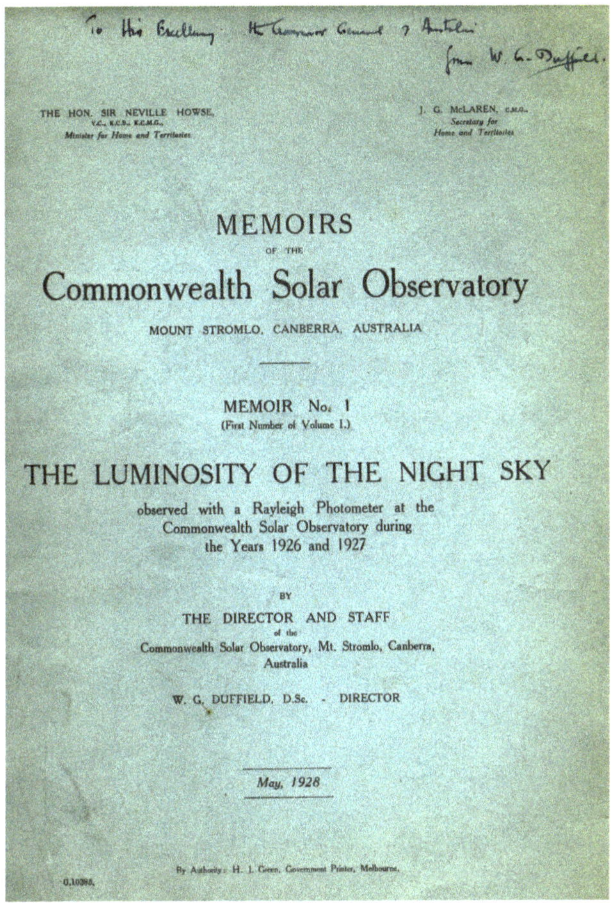

To His Excellency the Governor General of Australia
from W. G. Duffield.

THE HON. SIR NEVILLE HOWSE,
V.C., K.C.B., K.C.M.G.,
Minister for Home and Territories

J. G. McLAREN, C.M.G.,
Secretary for
Home and Territories

MEMOIRS
OF THE
Commonwealth Solar Observatory
MOUNT STROMLO, CANBERRA, AUSTRALIA

MEMOIR No. 1
(First Number of Volume I.)

THE LUMINOSITY OF THE NIGHT SKY
observed with a Rayleigh Photometer at the
Commonwealth Solar Observatory during
the Years 1926 and 1927

BY
THE DIRECTOR AND STAFF
of the
Commonwealth Solar Observatory, Mt. Stromlo, Canberra,
Australia

W. G. DUFFIELD, D.Sc. - DIRECTOR

May, 1928

By Authority: H. J. Green, Government Printer, Melbourne.
0.10285.

Figure 2.28: Memoir No. 1, written by Duffield, on the observations of the luminosity of the night sky. This is a copy he sent to His Excellency the Governor-General of Australia. Duffield ensured that his work was known by influential members of the Australian government.

Royal, Sir Harold Spencer Jones, who collated and analysed observations from several observatories around the globe.

For the spectroscopic program, Duffield had acquired the recently invented Moll microphotometer and a Hilger quartz spectrograph. Allen began spectroscopic studies in two rooms at Hotel Canberra. Grating spectra of pressurised metallic arcs, which Duffield had brought from England, were used by Allen. He measured relative line intensities with the quartz spectrograph and the microphotometer. When the three prism tunnel spectrograph became available at Mount Stromlo in 1929, Allen could also measure the line breadths. By studying the easily produced multiplets in the copper-arc spectrum, Allen was able to elucidate the abnormal behaviour of the lines in the copper-arc spectrum as opposed to the lines in the spectra of other metals.[16]

Studies of the Sun were commenced. A pyranometer from Professor Angstrom's solar physics laboratory in Sweden was erected on one of the chimneys of the Hotel Canberra. The photographic recording facility for this

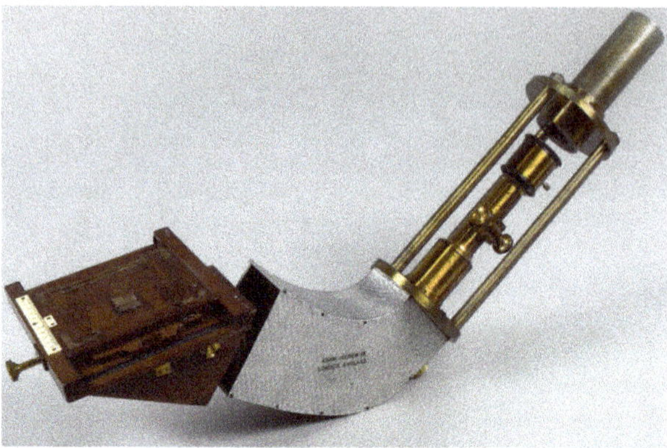

Figure 2.29: Adam Hilger prism spectrograph similar to the one used by Duffield and Allen in the late 1920s. Photo: Mount Stromlo Archives.

instrument was set up in a screened-off portion of a passage. The arrival of a pyrheliometer at a later date allowed the records to be standardised.

A program to measure the atmospheric electric potential gradient was also commenced, continuing at Mount Stromlo when the Observatory was completed.

The Oddie Observatory had been established on Mount Stromlo in 1911 for site testing and after this was completed it was mothballed. It was now found to be in need of repair and refurbishment. Unknown to Duffield, the telescope and spectrograph had not been mounted rigidly as required for the astrophysical work. This was discovered rather dramatically when the telescope toppled over in September 1925: 'The telescope had fallen towards the east wall and part of the main castings were broken and the declination axis smashed beyond the bearing.'[2] The telescope was repaired and was ready for use in February 1926.

Rimmer began an extensive program of measuring the spectroscopic parallaxes (distances) of the brightest stars in the southern hemisphere. This parallax method uses estimates of the intrinsic luminosity of stars based on

Figure 2.30: The Oddie Observatory had to be repaired and refurbished before it could be used for astronomical research. Photo: Mount Stromlo Archives.

the detailed characteristics of their spectra, then compares the apparent and intrinsic luminosities of the stars to find their distances. Rimmer photographed the spectra of 530 stars between 1926 and 1929. He initially concentrated on B type stars because they seemed particularly appropriate for the technique.

It was not until October 1928 that the dome for the Farnham telescope was ready. This telescope was planned to be used for the study of variable stars and the photometry of planets and their satellites. A stellar photometer for use with the telescope was found to be too limited in its performance and was subsequently removed.

A time service was also started with a newly available, slaved Shortt clock, which was more accurate than its predecessors and was distributed to observatories around the world. The service began operation by 1927. Several clocks were set to Eastern Standard Time and one to Sidereal Time.

Duffield led a diverse program of activities and in the first five years as Director he achieved a tremendous amount. He not only built the Observatory from ground zero but initiated an active research program in astrophysics which was to provide dividends to the Australian astronomical community in the years ahead.

Community life

Apart from his duties as Director of the Observatory, Duffield was actively involved in the community life of Canberra. He was a foundation member of Canberra's first Rotary Club and served as a member of the Canberra's University College Council. He and his wife were talented musicians. According to his daughter, Joan Duffield, 'They were both devoted to music; my mother played the violin and my father the double bass. They were instrumental in forming what they called the "Stromberra Quintet". Each week they would trundle down the mountain in the old car with the double bass slung between the front and back seats, and spend an evening with other members of the Quintet.' Musical evenings at their house were attended not only by the staff

Figure 2.31: Even after moving to the mountain, the Duffields regularly travelled down to Canberra for supplies and for music sessions. Photo: Mount Stromlo Archives.

Figure 2.32: Duffield and his wife were devoted to music. Photo: Mount Stromlo Archives.

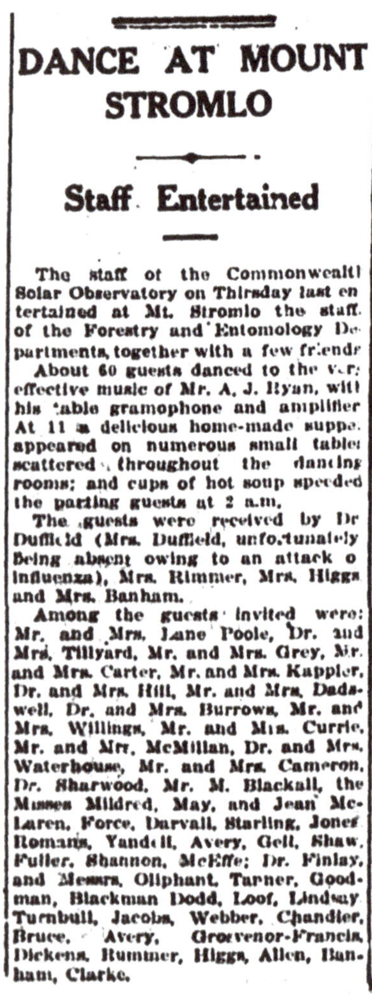

DANCE AT MOUNT STROMLO

Staff Entertained

The staff of the Commonwealth Solar Observatory on Thursday last entertained at Mt. Stromlo the staff of the Forestry and Entomology Departments, together with a few friends.

About 60 guests danced to the very effective music of Mr. A. J. Ryan, with his table gramophone and amplifier. At 11 a delicious home-made supper appeared on numerous small tables scattered throughout the dancing rooms; and cups of hot soup speeded the parting guests at 2 a.m.

The guests were received by Dr Duffield (Mrs. Duffield, unfortunately being absent owing to an attack of influenza), Mrs. Rimmer, Mrs. Higgs and Mrs. Banham.

Among the guests invited were: Mr. and Mrs. Lane Poole, Dr. and Mrs. Tillyard, Mr. and Mrs. Grey, Mr. and Mrs. Carter, Mr. and Mrs. Kappler, Dr. and Mrs. Hill, Mr. and Mrs. Dadswell, Dr. and Mrs. Burrows, Mr. and Mrs. Willings, Mr. and Mrs. Currie, Mr. and Mrs. McMillan, Dr. and Mrs. Waterhouse, Mr. and Mrs. Cameron, Dr. Sharwood, Mr. M. Blackall, the Misses Mildred, May, and Jean McLaren, Force, Darvall, Starling, Jones, Romans, Yandell, Avery, Gell, Shaw, Fuller, Shannon, McEffe; Dr. Finlay, and Messrs. Oliphant, Turner, Goodman, Blackman Dodd, Loof, Lindsay, Turnbull, Jacobs, Webber, Chandler, Bruce, Avery, Grosvenor-Francis, Dickens, Rummer, Higgs, Allen, Banham, Clarke.

Figure 2.33: The *Canberra Times* report (8 July 1929) on a dance held at Mount Stromlo.

but on some occasions by visiting dignitaries, such as Prime Minister Joseph Scullin and his wife, and the Prince of Siam and his entourage. The parties went on till the early hours and, as the *Canberra Times* of 8 July 1929 reported, 'cups of hot soup speeded the party guests at 2 am'. Duffield was a cultured man, at ease in both the scientific and musical worlds. Everything was going well for Duffield and he was riding the crest of a wave of success.

Take thou the torch

Out of the blue, tragedy struck in 1929. That year saw an extremely cold winter in the Federal Capital Territory including a very high snowfall which left Mount Stromlo isolated from the rest of Canberra. Duffield fell victim to an influenza epidemic and could not be taken for medical care to the hospital as the road was frozen and snowbound.[3] After a nine-day illness, Duffield passed away on 1 August 1929. Duffield's funeral was held on 5 August 1929. He was buried in specially consecrated ground on Mount Stromlo, 800 m along the ridge from the Observatory. Representatives from most government departments and acquaintances from all over Canberra, as well as family from Adelaide, attended the funeral. The service was conducted by the Bishop of Goulburn, and was announced in newspapers around the country.

Duffield's death was a severe blow to his family, Observatory staff and Australian astronomy. In his letter of condolence to Mrs Duffield, Bragg wrote, 'He has made a great life in the scientific world and has I believe been very happy with his work,'[17] while a note by 'H.K.E.' which appeared in the periodical, *The British Australian and New Zealander*, noted that 'Professor Duffield was one of the first to realise that it was essential in the interest of Australia and the world in general, that she have a solar observatory, and he worked as few men have worked to attain his object. Duffield was ahead of his time in science, and he hammered away at his project. Science, he realised, was to play an important part in the affairs of the Commonwealth. How right he was is now appreciated by all.'[18]

Figure 2.34: Duffield's coffin being carried to his resting-place at Mount Stromlo. Photos: National Archives of Australia.

Figure 2.35: Duffield's grave at Mount Stromlo. The inscription on the cross reads, 'Per Ardua, Ad Astra'. On the tombstone is engraved, 'In loving memory of W.G. Duffield, Aug. 12 1879 – Aug. 1 1929, Pioneer Director of This Solar Observatory 1924–1929. Take thou the torch, Carry it out of sight into the great new age I must not know, into the great new realm I must not tread.' On the small stone just before the cross is engraved, 'In loving memory of Doris T. Duffield 13.7.1881 – 12.6.1956. Beloved wife of W.G. Duffield. Loving Mother of Joan, Peter & Michael'. Photo: Mount Stromlo Archives.

References

1. Barsdell L 1947. Sweeping the skies from Stromlo. *Journal of the Royal Astronomical Society of Canada* **41**, 18–24.
2. Frame T, Faulkner D 2003. *Stromlo: An Australian Observatory*. Allen and Unwin, Sydney.
3. Duffield J (no date). Recollections of an astronomer's daughter. Basser Library. Australian Academy of Science.
4. Allen CA 1979. The beginnings of the Commonwealth Solar Observatory. *Records of the Australian Academy of Science* **4**(1), 27–49. doi:10.1071/HR9790410027

5. Clerke A 1908. *History of Astronomy during the Nineteenth Century.* Adam and Charles Black, Edinburgh.
6. Meadows AJ 1980. The origins of astrophysics. In *Astrophysics and Twentieth Century Astronomy.* (Ed. O Gingerich) pp. 3–15. Cambridge University Press, Cambridge.
7. Sutherland RS, Butcher HR 2011. An observatory at Mount Stromlo. Unpublished.
8. Evans J, Thorndike AS (Eds) 2007. *Quantum Mechanics at the Crossroads.* Springer-Verlag, Berlin.
9. Kumar M 2008. *Quantum.* Icon Books, Cambridge.
10. Love R 1984. Science and government in Australia 1905–14: Geoffrey Duffield and the foundation of the Commonwealth Observatory. *Historical Records of Australian Science* **6**(2), 171–188. doi:10.1071/HR9850620171
11. Lankford J 1984. The impact of photography on astronomy. In *Astrophysics and Twentieth Century Astronomy.* (Ed. O Gingerich) pp. 16–39. Cambridge University Press, Cambridge.
12. Duffield G 1908. The spectra near the poles of an iron arc. *Astrophysical Journal* **27**, 260–273. doi:10.1086/141553
13. Duffield G. Press cuttings book. Basser Library. Australian Academy of Science.
14. Letter from W.H. Bragg to W.G. Duffield, 2 July 1908. Duffield papers. Basser Library. Australian Academy of Science.
15. Duffield G 1928. *The Luminosity of the Night Sky. Memoir 1.* Memoirs of the Commonwealth Solar Observatory. Government Printer, Canberra.
16. Allen CW 1932. The intensity of quartet lines in the arc spectra of copper. *Physical Review* **39**, 55–63. doi:10.1103/PhysRev.39.55
17. Bragg WH 1929. Letter to Mrs Duffield dated 3 October 1929. Duffield papers, file no. 1–10. Box 1. Basser Library. Australian Academy of Science.
18. H.K.E. 1929. *The British Australian and New Zealander.* File no. 1–10, p. 15. Box 1. Duffield papers. Basser Library. Australian Academy of Science.

1930–1939

3

Caretaker

Until 1939 Rimmer was the only trained astronomer at Mount Stromlo; the other four or five staff-members were solar or geo-physicists.[1]

It was not the best of times when William Rimmer was appointed Officer-in-Charge of Mount Stromlo Observatory on the death of Duffield. About four months after he took charge, a catastrophic event occurred that had severe economic and social consequences for the whole world. The greed of the capitalist society in the US finally resulted in the Wall Street stock market crash of October 1929. It sent financial tsunamis that crashed on the shores of countries around the world. This was followed by the Great Depression, the likes of which had never been seen before. The dancing of the Charleston had come to a halt, and long queues of children and women lined up for food hand-outs. There were also long queues of men seeking jobs so they could feed their families. Australia did not escape the financial catastrophe. In early 1929 the sharp fall in the prices of wool and wheat and a 50% fall in export

Figure 3.1. A queue of men looking for work during the Great Depression. Photo: National Library of Australia.

NEW OBSERVATORY.

Dr. Duffield's Work.

SECOND ASSISTANT APPOINTED.

MELBOURNE, Wednesday.

Mr. W. B. Rimmer has been appointed second assistant to Dr. W. G. Duffield, Director of the Federal Solar Observatory which is being erected at Mt. Stromlo, in the Federal capital territory.

Mr. Rimmer, who is now on his way to Australia from England, is expected to take up his new duties at Canberra in February. Steps are being taken to make the other appointments to Dr. Duffield's staff—first assistant, a research Fellow, and a mechanic skilled in the adjustment of scientific instruments. The research Fellow will be chosen from promising graduates of Australian universities.

Mr. Rimmer is a distinguished scientist, and among the men who recommended him to the Federal Ministry were the Astronomer Royal and the Director of the Greenwich Observatory. He is a master of science, and holds the diploma of the Imperial College of Astrophysics. Before his appointment he was assisting Dr. Lockyer, director of the Lockyer astrophysics and spectroscopy, and papers written by him on those subjects have appeared in the journals of the Royal Astronomical Society and the Royal Society. Enlisting as a private in the Royal Engineers, he served during the war in Belgium with the Royal Engineers, and rose to the rank of captain.

Dr. Duffield is already at Canberra, and although the Observatory buildings at Mt. Stromlo are not yet completed, he has begun observations there.

The erection of the observatory will fill an important gap in the chain of observatories around the earth, and as the only solar observatory south of the equator, it will be of immense scientific importance. It will link Australia with India and America, enabling the sun to be kept under observation for the whole of each 24 hours.

It is believed that there is a splendid field for astrophysical research in Australia, and the Federal Ministry showed its recognition of the importance of the subject by deciding early in the history of the Federal capital to proceed with the construction of the observatory at Mt. Stromlo.

Figure 3.2: Press announcement of Rimmer's appointment. *Sydney Morning Herald*, 1 January 1925.

Figure 3.3: Rimmer (1882–1945), shown here on the right, with a colleague in front of the nearly completed main Observatory building, 1927. Photo: Mount Stromlo Archives.

prices led to a financial crisis. The national income declined. Unemployment shot up to over 20% within a year and by 1933 was standing at over 30%. Governments around the country reviewed their expenditures and capital works programs and tried to reduce deficits as best as they could. Several major capital works projects were slashed or reduced. Religious leaders told their flocks that the Depression was a sign of divine displeasure over their sins. By and large, Mount Stromlo escaped the full brunt of the Depression. However, in order to economise, Rimmer was appointed Officer-in-Charge rather than Director of the Observatory.

Rimmer had his work cut out and to his credit he not only managed to keep the Observatory functioning but achieved several significant milestones. When he first joined the Observatory his main research was to use the 22 cm Oddie refractor telescope to study the spectroscopic parallaxes (or distances) of bright southern hemisphere stars. Unfortunately, the program was terminated in 1929 after Rimmer had a bad fall at the telescope, lying unconscious on the floor of the telescope dome all night. He was found only the next morning. In early February 1930 he prepared the second *Memoir* of the Commonwealth Solar Observatory, titled *The Luminosities and Parallaxes of 350 Stars of Spectral Type B.*[2]

Solar Tower and telescope

Rimmer's most pressing problem was the construction and erection of the Solar Tower and the telescope. The Solar Tower was to be built on the eastern

Classification of stars

Astronomers classify stars into various types depending on their temperatures and luminosities. They were the astronomical accountants of the celestial objects found in the universe. In neatly organised columns and rows they wrote down the names of the celestial objects, their right ascensions, declinations and other properties. They classified the stars as O B A F G K M N, which is mostly a temperature sequence. Undergraduates remember this classification system by the rather sexist nemonic 'Oh Be A Fine Girl/Guy Kiss Me Now'. B type stars are stars with very high temperatures, while M stars are cool and have lower temperatures.

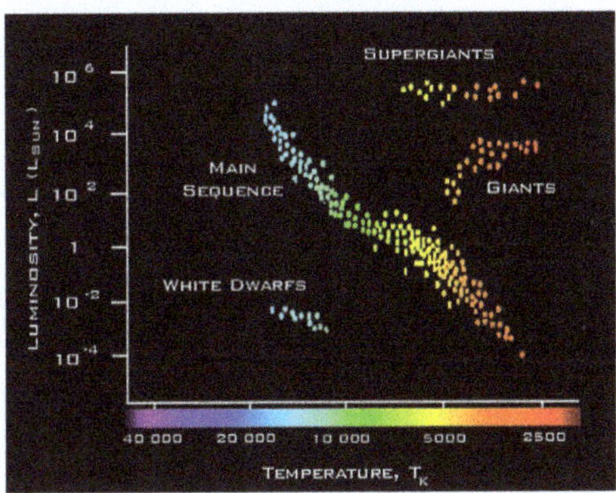

Figure 3.4: The H-R diagram is one of the most important tools used by astronomers. It shows the star's surface temperature or spectral type along the horizontal axis and its luminosity along the vertical axis. When young, most stars lie along a so-called Main Sequence, which allows their intrinsic brightness to be estimated by determining their spectral type. Rimmer studied B type stars, which are massive blue stars at the hot end of the main sequence.

Our Sun is an average G type star. According to Ben Gascoigne, 'For more than twenty years Rimmer's study of the spectra of brighter stars, especially in the southern hemisphere sky, remained the only substantial astrophysical program that had been carried out in Australia.'[1] About 25 years later Mary Lee Woods studied Rimmer's plates, which had been stored in the Observatory basement, and published her findings on 520 type O and M stars. However, by that time the Yerkes system of stellar classification by both temperature and luminosity had been published, for which Rimmer's spectra were largely inadequate.

side of the main Observatory building. The project provided many challenges that had not been foreseen by the engineers and the astronomers. The excavation of the ground for the basement was much more difficult than had been envisaged. The rock was extremely hard and the Queanbeyan firm of W.H. Mason informed the Observatory that the excavation was beyond its expertise. The problem was solved by the Federal Capital Commission,[3] which had the necessary equipment. The dome, electrical works and optical system were completed by November 1931.

The Sun's image was reflected vertically downwards to the basement by means of a system consisting of a coelostat and mirrors. A mirror in the basement directed the beam of sunlight horizontally onto a slit of a solar

THE HON. ARTHUR BLAKELEY, M.P.,
Minister for Home Affairs.

P. E. DEANE, C.M.G.,
Secretary for Home Affairs.

MEMOIRS

OF THE

Commonwealth Solar Observatory

MOUNT STROMLO, CANBERRA, AUSTRALIA.

MEMOIR No. 2

(Second Number of Volume I.)

THE LUMINOSITIES AND PARALLAXES OF 350 STARS OF SPECTRAL TYPE B

BY

W. B. RIMMER, M.Sc., D.I.C.

February, 1930.

Figure 3.5: Rimmer's publication *The Luminosities and Parallaxes of 350 Stars of Spectral Type B.*

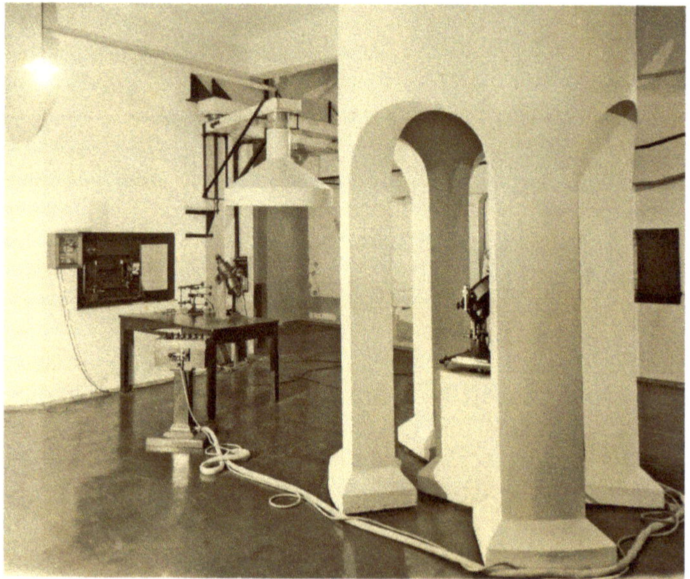

Figure 3.6: Solar telescope observing laboratory in the basement of the main Observatory building. Light from the Sun was collected by a heliostat mirror in the Sun telescope dome on the roof and passed to the spectrographs and other instruments in the observing laboratory. Photo: Mount Stromlo Archives.

spectrograph. The orientation of the mirrors was controlled electrically so that the Sun was kept in view as it moved across the sky while observations were undertaken. The spectrograph, manufactured by well-known instrument makers Hilgers of London, was mounted in a tunnel which extended away from the basement. The black Fraunhofer lines in the spectral range of 3900–6600 Angstroms were measured by a photometer. Cla Allen photographed the spectrum and published an atlas of the solar spectrum which covered the spectral lines in the range of 3600–8000 Angstroms. The atlas was published in two Observatory *Memoirs*[4,5] in 1934 and 1938. He also published his work in international journals.[6,7] It was a forerunner of the *Photometric Atlas of the Solar Spectrum* published in 1940 by Minnaert, Mulders and Houtgast. Allen's data provided a valuable source for astronomers to test theories regarding the emergence of light from the Sun's atmosphere and his atlas was a prime reference in the field of solar research for many years.

One of the problems facing solar physicists and astronomers in the 1930s was whether the Sun's total energy emission (sometimes called the solar constant) was constant. A suspicion existed that the total emission might vary with the 11-year sunspot cycle and affect the climate on Earth. Rimmer used pyrheliometers and pyranometers to study the Sun's radiation, while Allen measured the dust content of the Earth's atmosphere from the intensity of the Sun's halo[3] as seen at the Earth's surface. Allen's detailed studies of the airborne dust allowed adjustments to be made to the pyrheliometer readings and enabled estimates of the solar radiation outside the atmosphere to be calculated. Their experiments showed that that there was no evidence of any real variability (greater than ~0.1%) in the solar radiation reaching the Earth.

Sunspots had been observed for more than 100 years by astronomers and scientists but very little was known of their effect on the Earth. However, observations of the Sun seemed to suggest some correlation between

Solar constant

Today, variations in the total solar irradiance of the Earth can be measured but they were simply too small to be detected with the instruments Rimmer and Allen had at their disposal. It was 30 years later, with the advent of satellites, that the different picture emerged. Radiometers carried above the Earth's atmosphere on orbiting satellites revealed that there are small changes (~0.1–0.2%) in the Sun's irradiance.[8] These variations are due to passage of dark sunspots across the face of the Sun and of bright spots called plages that are associated with solar activity. In the end, they more or less compensate each other, leaving the net total irradiance variation very small.

The long-term weather cycles that affect agriculture in particular were of interest in Duffield and Rimmer's time. We know today these are associated with the El Niño/La Niña-Southern Oscillation effect. In today's turbo-charged climate change debate, the accuracy of the solar constant matters and it is included in all global climate models. According to Greg Kopp and Judith Lean, 'Accurate and stable irradiance measurements are critical for establishing the energy balance that determines the Earth's climate and for reliable attribution of climate change.'[9]

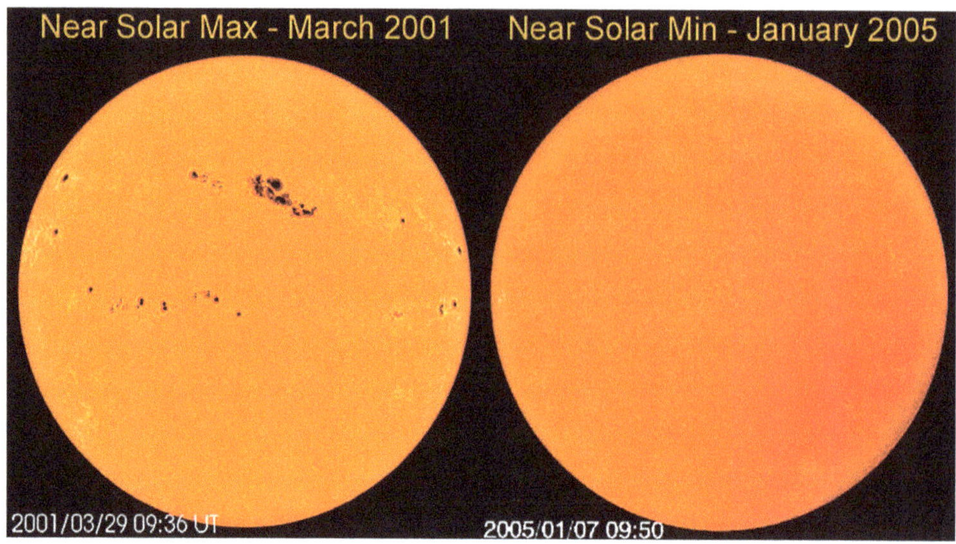

Figure 3.7: (a) Sunspots during a period of maximum solar activity. (b) Sunspots during a period of minimum activity. The discovery that sunspots are locations of intense magnetic fields was made in 1908 and added scientific urgency to the establishment of the Commonwealth Solar Observatory. Photo: NASA.

sunspot activity and various effects on the Earth, so it was not surprising that daily observations of sunspots were begun from early 1926 to find out whether there was any correlation. In order to study them, a solar image from a 7.5 cm telescope was projected onto paper. The number and sizes of the sunspots were recorded and tabulated. They supplemented the standardised sunspot data from Zurich.

In the US, George Ellery Hale had developed a spectrohelioscope to study the rapidly variable features on the Sun's surface. The Observatory acquired a spectrohelioscope and spectroheliograph to study what was taking place on the Sun. The spectroheliograph was installed in a purpose-built building and was ready for use in 1936. The program was run by Sydney Williams and Ronald Giovanelli, both physics graduates from the University of Sydney.

Giovanelli came to Mount Stromlo on a Research Fellowship in May 1937 and left in August 1939. In the short time he was at Mount Stromlo, he made some significant contributions to the study of the Sun.[10] According to him, 'My work was to study the Sun using a particular instrument, the Hale Spectrohelioscope. My first work in this area was to attempt an explanation for a very interesting and major event on the Sun, something after the nature of a storm, called in those days, an eruption, these days a flare. In the chromosphere, you could be looking through a spectrohelioscope and see these very sudden flares and in a matter of a few minutes a portion of the Sun would become extremely brilliant in light and after a while it would gradually fade back. While puzzling over the possible cause of these, I came up with the idea that they were electromagnetic in origin. I'd previously studied their association with the sunspots which were known to be regions of strong magnetic field and had found that during the growth of sunspots and also during the decay of sunspots, flares became more common. When there were

Ronald Giovanelli

Figure 3.8: Ronald Giovanelli as a young man. Giovanelli was an influential pioneer in solar research. He developed his interest in magnetic activity on the Sun while a Research Fellow at Mount Stromlo. Photo: Mount Stromlo Archives.

The Giovanelli family arrived in Australia in the 1850s. Born in Grafton in 1915, Ronald was educated at Fort Street Boys' High School and the University of Sydney. During the Second World War, with information from the National Bureau of Standards in Washington DC, he manufactured high-grade optical glass for the armed forces in Australia. With fellow scientists, he also designed and constructed a world-class filter telescope, or spectroheliograph. By the time he retired from the Applied Physics Division of CSIRO, Giovanelli had developed the first electromagnetic theory to explain the production of solar flares and was an internationally recognised expert on solar physics. He was elected a Fellow of the Australian Academy of Science in 1962 and was a founder of the Astronomical Society of Australia.

Figure 3.9: Layout of the Hale spectrohelioscope used by Giovanelli. The spectrohelioscope allowed visual observations of solar events in the light of hydrogen and other elements. Giovanelli observed 268 magnetic eruptions on the Sun with the instrument during his fellowship at Mount Stromlo. The Observatory also had a spectroheliograph, which allowed photographs to be taken of the Sun in the light of selected elements, but it was slow in operation and photographic plates not always easy to obtain. Drawing: American Astronomical Society.

no sunspots around, there were no flares. It was a natural development from this to think of the growth of magnetic fields, the escape of magnetic fields into the chromosphere and through the corona of course, as being the cause of flares. So I produced a theory of chromospheric flares, an electromagnetic theory, which certainly had quite a strong influence on people's thinking.'[11,12] He also developed theories about the motion of flares in relation to radiation pressure and suggested that the Lyman Alpha line of hydrogen provided the radiation pressure.[13]

Giovanelli and Higgs[14] also studied the association of eruptions or flares with the radio fade-outs that occur occasionally during long-distance radio communications. Giovanelli noted, 'sizeable flares produced blackouts of shortwave radio communications and this was very important. It meant, for example, that all radio communication between Europe and the United States would be blocked. They needed to know what frequencies to change to and how often were these going to occur. There was a very big effort to study the radio connection between these flares.'[11]

The ionosphere and the ozone layer

Higgs designed instrumentation that allowed continuous monitoring of the ionosphere. This provided sufficient data to carry out a quantitative evaluation of the ionising radiation emitted by the Sun and its variation with the sunspot cycle. It allowed him to predict the maximum and optimum radio frequencies which could be used for radio communications. This proved extremely useful during the Second World War, during which he was seconded to the Radiophysics Laboratory. The fade-outs were indeed found to be associated with major solar flares.[15]

At Mount Stromlo, in addition to the studies of the Sun, research programs focused on atmospheric ozone and atmospheric electricity. The studies of the ozone layer[16] were conducted by Higgs from 1929 to 1932. From his observations, he was able to measure the height of the ozone layer. By 1939 he had collected over 10 years worth of data. Unfortunately, the Second World War broke out, interrupting his research and, because the significance of the ozone layer to climate was then unknown, the data were never published.

Atmospheric electricity

Research in atmospheric electricity became the domain of Arthur Hogg from 1931 and he published several papers and *Memoir 7* on this subject.[17] He began by making regular observations of atmospheric electrical phenomena, in particular of the conductivity, ionic mobilities and ion balance in the lower atmosphere. From these measurements he was able to work out the air–earth current. He designed and built his own equipment and was not only a skilful but an ingenious experimenter. Because of his early work on the diurnal variation of conductivity and on the average lives, rates of production and mobility of small ions, he was invited to work at Kew Observatory in London by Whittle. He was at Kew from 1937 to 1938. With a fine piece of equipment he had built at Mount Stromlo Observatory, he made some of the first measurements of the intermediate atmospheric ions, which are relatively frequent in the industrial atmosphere of London: 'In this important piece of work he was able to show conclusively that these ions existed in discrete groups, composed of one or more droplets of sulphuric acid, each containing about 2,200 molecules. While at Kew he also carried out some experiments which led to a reconciliation between two apparently conflicting methods of measuring the air-current; the difficulty was traced to the local effect of radioactive matter in the soil. His final paper on the subject also dates from this time, although it was not published until 1950. It was a statistical examination of the air-current at a number of stations widely distributed over the earth. Hogg was able to derive an annual term for the variation, and to show that it lent substantial support to the thunderstorm theory, advanced by Whittle, for the maintenance of the Earth's electric charge.'[18]

Figure 3.10: Arthur Hogg was born in 1903 in Creswick, Victoria. Duffield's last duty was to sign Hogg's letter of appointment. Joining Mount Stromlo Observatory on 1 August 1929, Hogg remained there for 37 years. He played a leading part in setting up and testing the 1.9 m telescope. Photo: Australian Academy of Science.

At the 1930 meeting of the Australian and New Zealand Association for the Advancement of Science, a recommendation was made for Mount Stromlo Observatory to carry out continuous observations of cosmic rays. With his experience in the investigations of atmospheric ionisation, Hogg took on the task of conducting experiments on cosmic rays. In 1932, he began preliminary work with both Geiger counters and ionization chambers but by late 1933 he abandoned the Geiger counters as they proved to be unstable for a long-term investigation. He redesigned and built a new high-pressure ionization chamber and had it operating by late 1935. Measurements of cosmic ray intensity were made on an hourly basis for the next five years. They were stopped in 1940 because of the war. The results of the experiments and the analysis were published in July 1949, in *Memoir 10*,[19] a significant piece of work on the variation of cosmic intensity with time. 'Hogg's careful statistical analysis ... is probably the best of its kind which has been made, and led him to lifetimes and absorptions in good agreement with those determined by other methods.'[18] This work and the analysis that went with it constituted Hogg's most important work, and led to his being elected a Fellow of the Australian Academy of Science in 1954.

While Rimmer was in charge, the Reynolds telescope which was donated to Duffield in 1924 was finally given attention. It was intended to be used for photographing southern nebulae. The contract for construction of the building was awarded to W.H. Mason while the dome was constructed by James Steel Engineering Co.[3] Most of the work on the installation of the Reynolds telescope was completed by 1932. The telescope was converted to a modified Cassegrain configuration about seven years later when it was proposed to use it for photoelectric photometry and stellar spectroscopy. However, some teething problems remained in the set-up of the telescope and by the time

Figure 3.11: David Rivett was Chief Executive Officer of the CSIR and later its Chairman, an influential figure in both government and scientific circles. He was a Foundation Fellow of the Australian Academy of Science when it was established in 1954. Photo: Australian Academy of Science.

war broke out in 1939 not much productive astronomical work had been done. The Farnham telescope was limited to visual observations and showing visitors the night sky.[20] Both telescopes were used much more productively after the Second World War.

The future of the Observatory

In 1937, questions about the future of the Observatory and the need to appoint a full-time director were raised by the Secretary of the Department of the Interior. Despite its active research program, it was felt that the Observatory's research and mission lacked overall direction. The well-known and influential British astronomer, William McCrea, the doyen of British astronomy, was of the opinion that Mount Stromlo was 'a small, isolated and neglected Observatory'.[21] A two-member committee consisting of Dr A.C.D. (David) Rivett, Chief Executive Officer of the CSIR, and Professor Oscar Vonwiller, Head of the Physics Department at Sydney University, were asked to report on the scientific value of the work conducted at the Observatory, the adequacy of the equipment and whether a fully qualified Director needed to be appointed. They reported that the Observatory work was useful and the equipment was adequate. The Committee also recommended that Rimmer be promoted to First Assistant rather than Director and that a search be made for a suitable Director throughout the British Empire.

Rimmer was duly promoted to the position of First Assistant and was awarded the Order of the British Empire for services to astronomy. As was the case with colonial observatories in 19th-century Australia, the Australian government again turned to Britain for the selection of a suitable Director. A committee of prominent British astronomers – the Astronomer Royal, Sir Harold Spencer Jones, Professor Arthur Eddington from Cambridge University and Professor F.J.M. Stratton from Cambridge University – chose Richard Woolley. 'Stratton masterminded Woolley's whole professional career. That's no doubt an overstatement, but all the same Woolley once told me that Stratton suggested to him that if he were to apply for the Australian job and make a success of it, by the time the then Astronomer Royal, Spencer Jones retired, he, Woolley, would be well placed to succeed him. That was exactly how it worked out.'[11]

The 33-year-old Woolley arrived at Mount Stromlo Observatory on 4 December 1939 with the aim of changing the entire direction of the Observatory from solar astronomy to stellar astronomy. He changed the face of Australian astronomy.

References

1. Gascoigne SCB 2002. Rimmer, William Bolton (1882–1945). *Australian Dictionary of Biography.* **16**, 94–95. Melbourne University Press, Melbourne.
2. Rimmer WB 1930. *The Luminosities and Parallaxes of 350 Stars of Spectral Type B. Memoir 2.* Memoirs of the Commonwealth Solar Observatory. Government Printer, Canberra.
3. Frame T, Faulkner D 2003. *Stromlo: An Australian Observatory.* Allen and Unwin, Sydney.
4. Allen CW 1934. *Fraunhofer Intensity Tables: (a) General Intensity Table, 4277–6600 A. (b) Multiplet Intensity Table, 4036–6600 A. Memoir 5.* Memoirs of the Commonwealth Solar Observatory. Government Printer, Canberra.
5. Allen CW 1938. *Fraunhofer Intensity Table, 3924–4300 A. Memoir 6.* Memoirs of the Commonwealth Solar Observatory. Government Printer, Canberra.
6. Allen CW 1938. Fraunhofer intensities in the infrared region λ λ 8800–11830 Angstroms. *Astrophysical Journal* **88**, 125–132. doi:10.1086/143965.
7. Allen CW 1938. Fraunhofer lines in the infrared region λ λ 8800–11830 A. *Contributions from Mount Wilson Observatory/Carnegie Institution of Washington* **594**, 1–8.
8. Newkirk G, Jnr 1983. Variations in solar luminosity. *Annual Review of Astronomy and Astrophysics* **21**, 429–464. doi:10.1146/annurev.aa.21.090183.002241.
9. Kopp G, Lean JL 2001. A new, lower value of solar irradiance: evidence and climate significance. *Geophysical Research Letters* **38**, L1706.
10. Giovanelli RG 1938. Eruptive prominences and ionospheric disturbances. *Astrophysical Journal* **88**, 204–205. doi:10.1086/143975.
11. Bhathal R 1996. *Australian Astronomers: Achievements at the Frontiers of Astronomy.* National Library of Australia, Canberra.
12. Giovanelli RG 1939. The relations between eruptions and sunspots. *Astrophysical Journal* **89**, 555–567. doi:10.1086/144081.
13. Giovanelli RG 1940. Solar eruptions. *Astrophysical Journal* **91**, 334–349. doi:10.1086/144170.
14. Giovanelli RG, Higgs AJ 1939. The association of radio fade-outs with solar eruptions. *Terrestrial Magnetism and Atmospheric Electricity* **44**(2), 181–187. doi:10.1029/TE044i002p00181.
15. Martyn DF, Munro GH, Higgs AJ, Williams SE 1937. Ionospheric disturbanes, fadeouts and bright hydrogen eruptions. *Nature* **140**, 603–605.
16. Higgs AJ 1934. *Measurements of Atmospheric Ozone. Memoir 3.* Memoirs of the Commonwealth Solar Observatory. Government Printer, Canberra.
17. Hogg AR 1939. *Conduction of Electricity at the Lowest Levels of the Atmosphere. Memoir 7.* Memoirs of the Commonwealth Solar Observatory. Government Printer, Canberra.
18. Gascoigne B 1968. Arthur Robert Hogg (1903–1960). *Records of the Australian Academy of Science* **1**(3).
19. Hogg AR 1949. *Measurements of the Intensity of Cosmic Rays. Memoir 10.* Memoirs of the Commonwealth Solar Observatory. Government Printer, Canberra.
20. Commonwealth Solar Observatory *Annual Report 1930–31.* Government Printer, Canberra.
21. McCrea W 1988. Richard van der Reit Woolley 1906– 86. *Historical Records of Australian Science* **7**(3), 315–345.

1940–1945

4

Second World War: the Observatory becomes an Optical Munitions Factory

The observatory became an industrial factory effectively for making optical munitions, specifically precision optics for artillery fire control. They also made some optics for anti-aircraft predictors and for air compasses and all sorts of devices for army munitions. But our main job was sighting telescopes, artillery directors, range finders and things like that.[8]

About three months before Richard Woolley arrived at Mount Stromlo Observatory, war had broken out in Europe. There were less than 22 years between the end of the First World War and the beginning of the Second World War, which turned Europe and Asia into killing fields. The land of Mozart and Goethe had at its helm Adolf Hitler, whose philosophy of the Aryans as the master race led to the extermination of millions of Jews in the gas chambers. Not long after the British Prime Minister, Neville Chamberlain, declared war on Germany, the Prime Minister of Australia, Robert Menzies, announced on 3 September 1939, 'it is my melancholy duty to inform you officially that, in consequence of a persistence by Germany in her invasion of Poland, Great Britain has declared war upon her and that as a result Australia is also at war'.[1]

Thomas Laby, Professor of Natural Philosophy at the University of Melbourne, who was well known to every physics and chemistry undergraduate for his book *Physical and Chemical Constants* (co-authored with Kaye), wrote to the Prime Minister offering the help of physicists. He received a rather cool bureaucratic response from Menzies, but the attitude changed when Laurence Hartnett, Manager of General Motors Holden (Australia), was appointed by the government to head the Ordnance Production within the Department of Munitions. Hartnett quickly found that optical sights for the 2-pounder anti-tank guns being produced in Australia could not be sourced from Britain. He also found, rather to his dismay, that Australia did not have a large-scale optical industry that could supply the needs of the armed forces. He needed physicists and mathematicians who could design the necessary optical instruments and testing devices to kick-start an optical munitions industry. He created the Optical Munitions Panel and invited Thomas

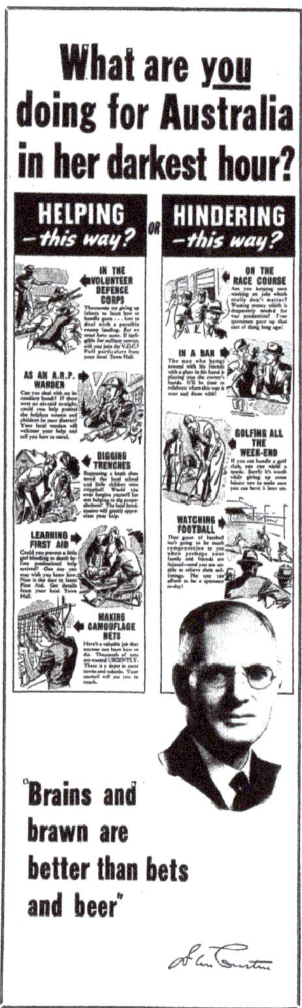

Figure 4.1: War posters such as these appeared all over Australia asking citizens to join the armed forces or assist the war effort on the home front. *West Australian*, 18 March 1942.

Laby to be the Chairman, to solve the urgent problem of developing and producing optical sights for anti-tank guns, telescopes and other optical devices. J.S. Rogers, also from the University of Melbourne, was appointed the Secretary of the Panel.

Woolley's intention to change the direction of research at the Observatory had to be put on the back-burner. He was recruited onto the Panel and into war work, in charge of all optical design and problems associated with the manufacture of optical devices.[2]

Mount Stromlo Observatory became an optical munitions factory and remained so for the duration of the war. The original workshop was extended from 55 to 110 m^2. An optical shop, assembly shop, store and heat treatment room were added. The Observatory had 138 m^2 of additional space and a range of machine tools due to the wartime program. It was not the only such factory. Rather than having a central research institution, the optical work was spread out to no fewer than 25 establishments, including several firms and university physics departments. They made and assembled the optical components. Forty-three different types of optical instruments and 26 237 individual instruments were manufactured.[3]

'The principal activity was the manufacture of precision optical instruments like gun sights, artillery directors and so on. They were built either as prototypes or in small production runs, Mount Stromlo being the only place in Australia where a complete telescope could be built under one roof, from the initial layout to final assembly and inspection.'[4]

The physicists

The team of physicists involved in war work at Mount Stromlo included Ben Gascoigne, Cla Allen, Noel Chamberlain, Walter Stibbs (Research Assistant) and Jim Dooley (Research Student). They had the varied backgrounds indicative of a shortage of skilled personnel in Australia and of the embryonic state of the educational system.

Ben Gascoigne was born in New Zealand in 1915 and obtained a PhD in astronomical optics from the University of Bristol in the UK. Back in New Zealand, in 1941 he was in charge of a practical optical workshop which made eyepieces for a mortar and optical gun sights. According to Gascoigne,

In defence of the mother country

The mother country had to be defended. If Britain was at war so was Australia.[1] By June 1940 over 100 000 young men had been recruited or had volunteered to fight the Germans on European soil. Many would never return to Australia. Even the Aborigines whose land had been stolen by the Crown took up arms to defend Australia. According to members of the Aboriginal community, the services of Aboriginal Diggers were never acknowledged after the war was over. Only recently have they been invited to join the marches on ANZAC Day.

Fearing an invasion of Australia by the Japanese, the Australian government drew the Brisbane Line. In case of an invasion by the Japanese, all European inhabitants north of the line were to be evacuated to the southern cities.[5] Aborigines would be left to fend for themselves and defend their country as best as they could.

Since the founding of the colony at Sydney Cove in 1788, Australia had relied on Britain not only culturally but economically. Britain was Australia's largest trading partner and overseas investor. The Australian government looked to Britain for leadership in foreign, defence, science and technology policies. The reliance on Britain was never questioned. As increasing numbers of young men left to fight overseas, it became clear that it was necessary to retain a sufficient number of skilled men at home, so a Schedule of Reserved Occupations was instituted.[6] Men in reserved categories were not allowed to enlist unless they were employed in their trade in the armed services. Although eager to serve, physicists were placed in the reserved category.

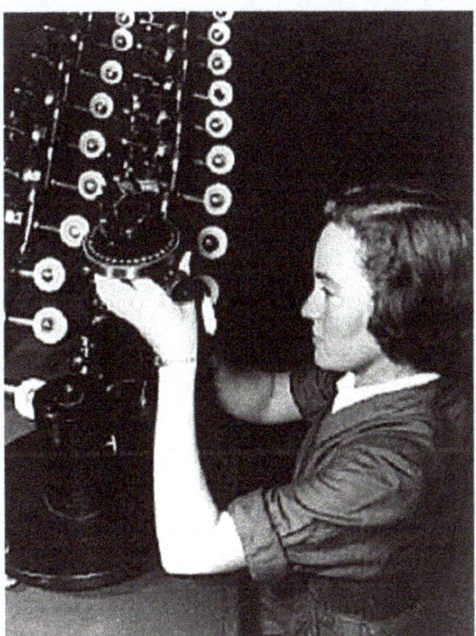

Figure 4.2: Women quickly became skilled optical workers. Photo: E.L. Cranstone. Australian War Memorial negative no. 12188.

LIST OF RESERVED OCCUPATIONS:

Laboratory staffs, to include: Chemists, Engineers, Physicists and other Scientific Workers, qualified by Degree, or Diploma, or Students in 3rd year at the University or Technical College.

Students (University): Medical and Dental. All students in 2nd and later years of Physical Sciences, Chemistry, Physics, Metallurgy and Engineering. All students in 2nd and later years of Biological Sciences, Bacteriology, Bio-Chemistry, Zoology, Botany and Agriculture. Last Year Students in Economics, Commerce, Mathematics. 3rd Year Honours Students in Languages. Post Graduate and Research Students. University Teaching Staffs.

Scientific and Research Workers: Physicists. Scientific Research Workers (full time) in a University or Technical School.

Politicians and policy-makers realised that there was a tremendous shortage of skilled labour. Australia had fewer than 20 000 engineering and other skilled tradesmen. Women were called to become industrial workers for the war effort on the home front and were quickly trained to assist in various industries.

'I had heard that the Commonwealth Solar Observatory was entering the optical business, doing optical calculations and making lenses. So I wrote to Woolley, I've had this experience at Bristol and worked with Burch. Woolley sent back a cable, *Offer immediate fellowship. As soon as I learnt you had worked with Dr Burch I had no hesitation*'.[7] Gascoigne was involved in the design department designing telescope optics.

Noel Chamberlain, who had previously worked at the Watheroo Magnetic Observatory, served as Woolley's assistant in the design, manufacture and inspection of optical instruments.

Walter Stibbs was a Deas Thomson Scholar in the School of Physics at the University of Sydney. He joined the Observatory as a Research Assistant and

Figure 4.3: Thomas Laby, Professor of Natural Philosophy at the University of Melbourne then Chairman of the Optical Munitions Panel. According to Kerr Grant, Professor of Physics at the University of Adelaide, 'The success of the Panel has been due in no small manner to the zeal and unsparing efforts of Dr Laby'. Photo: E.L. Cranstone. Australian War Memorial negative no. 6060/University of Melbourne Archives.

Optical Munitions Panel

Members of the Optical Munitions Panel included Lieutenant Colonel G.H. Adams (Assistant Director of Artillery, Army Headquarters), Frank Daley (Department of Defence, Ordnance Department, Maribyrnong), N.A. Esserman (National Standards Laboratory), Professor Sir Kerr Grant (University of Adelaide), Laurence Hartnett (Department of Munitions), Associate Professor Eric Hercus (University of Melbourne), Assistant Superintendent E.L. Sayce (Munitions Supply Laboratories), Professor Oscar Vonwiller (University of Sydney) and Dr Richard Woolley (Director, Commonwealth Solar Observatory).

Figure 4.4: Optical Munitions Panel, 1943. Front row, left to right: Lt Col G.H. Adams, G.H. Briggs, E.O. Hercus, K. Grant, T.H. Laby, J.S. Rogers, A.D. Ross, A.L. McAulay. Back row, left to right. R. Woolley, H.J. Frost, T. Harrigan. Photo: Mount Stromlo Observatory/Australian National University.

worked on optical munitions and the development of an ionospheric prediction service. Years later, in 1959, he was appointed Napier Professor of Astronomy at the University of St Andrews in Scotland.

Jim Dooley, a research student at the University of Melbourne, joined the Observatory on Laby's recommendation, to mitigate the shortage of physicists at Mount Stromlo.

The physicists were assisted by a group of well-trained and skilful technical staff: Jim Banham (Foreman of the Mechanical Workshop), Francis Lord (Optical Technician) and Syd Elwin (Assistant Optical Technician). Lord, who had studied optics in Paris under the well-known Professor Fabry, had come to Australia as a refugee from Czechoslovakia. He had found a job with the British Optical Co. in Sydney but was dismissed because he was considered an 'enemy alien'.

Desperate to find people with technical skills for his optical munitions factory at Mount Stromlo, Woolley began looking for skilled people among the European refugees who had escaped the atrocities of Nazi Germany. He came across Lord and offered him a job at Mount Stromlo. He also found Kurt Gottlieb, who had been born into a middle-class Jewish family and had escaped

Ernst Hartung and the production of optical telescopes

One of the first problems to face the Panel was the production of the Sighting Telescope 24B. An order for 3500 instruments had been placed. A specimen was requested from the War Office in London for reverse engineering, but it arrived with no specifications, such as the type of glass used, radii of curvature and thickness of the lenses. Woolley solved the problems, but when the new instrument (Sighting Telescope 124) was sent for field trials it did not perform satisfactorily. One of the problems was that the manufacturers did not have test plates with which to test their lenses, so Woolley had his workshop produce test plates and spherical test surfaces. This allowed the sighting telescopes to be produced satisfactorily by the Australian Optical Co. in Melbourne and the British Optical Co. in Sydney.

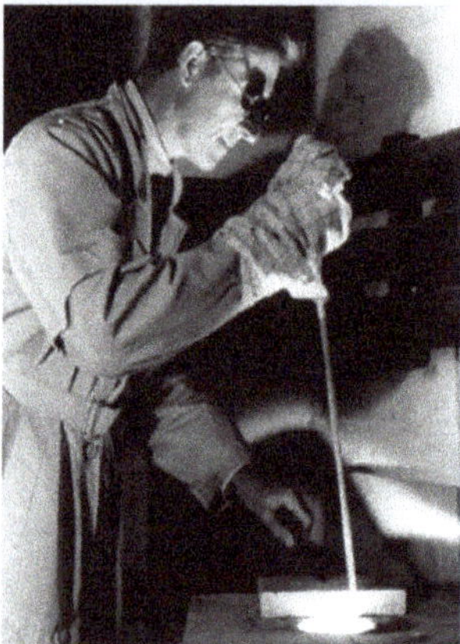

Figure 4.5: Ernst Hartung was Professor of Chemistry at the University of Melbourne and an amateur astronomer. He wrote the well-known book: *Astronomical Objects for Southern Telescopes: A Handbook for Amateur Astronomers.* He was responsible for solving the problems of manufacturing optical glass for the war effort. Photo: University of Melbourne Archives, UMA/I/1473.

The Panel also had to solve another pressing problem. The optical devices required flint glass, which is made from very special iron-free sand, and it had not been made previously in Australia. Ernst Hartung came to the rescue. He found suitable sand and began his experiments with Australian Window Glass (a subsidiary of Australian Consolidated Industries); by November 1941 the first commercial production of crown and flint glass had taken place. Hartung, writing in the British journal *Nature*, noted that there is 'an abundant supply of excellent material for munitions purposes'.[8]

Enemy aliens

Australia was not the only country that established a category of 'enemy aliens'. The US and Canada also interned persons of Japanese and German origin in 'enemy alien' camps although many of them had been born and had lived in the countries for years. In Broome, even Aboriginal women who had married Japanese local residents were considered enemy aliens.[6]

Figure 4.6: Walter Stibbs attended the Commonwealth Solar Observatory (1939–42) as an undergraduate vacation student, and later became a scientific officer there (1945–51). During the war he participated in the optical munitions work, including the design of a folded optical system gunsight and a sun-compass for desert warfare. After the war he worked with Woolley on radiative transfer problems and carried out observational work in photoelectric photometry. With Woolley he wrote *The Outer Layers of a Star* (1953). Photo: Mount Stromlo Archives.

Nazi anti-semitism in 1940. Gottlieb had an engineering degree that was not recognised in Australia, but during the war he worked as a mechanical designer and became a Research Engineer at Mount Stromlo after the war.[9] Gus Krentler, A. Nisbet, Georg Froelich and Hans Meyer were all placed in the personal custody of Woolley. Women were also employed at the Observatory.

Additional duties

As the war progressed, Woolley took on additional duties. Increasingly fearing an invasion of Australia by the Japanese, the government set up the Army Inventions Directorate to receive, develop and vet inventions related to the war. It reported directly to the Minister for the Army. Woolley was appointed the Chief Executive Officer[10] in April 1942.

Unless it was related to the war effort, most of the astronomical work at the Observatory came to a halt but in October 1940, Cla Allen and Arthur

Laurence Hartnett

Laurence Hartnett, Chairman of the Army Inventions Board, said in a press statement that 'since the directorate was established two years ago it had received no fewer than 14,500 submissions from inventors, of which 2,500 were referred to the fighting services and 64 were accepted ... this might appear to be a small percentage, but on his trips to Britain, the United States and Canada he had gathered that the percentage of acceptance in those countries was no higher'.[11]

Figure 4.7: The Optical Munitions Factory at Mount Stromlo during the war years. The Observatory was unique in the country in its ability to carry out the whole process of development at the one site, from design through manufacture to delivery. Photo: Mount Stromlo Archives.

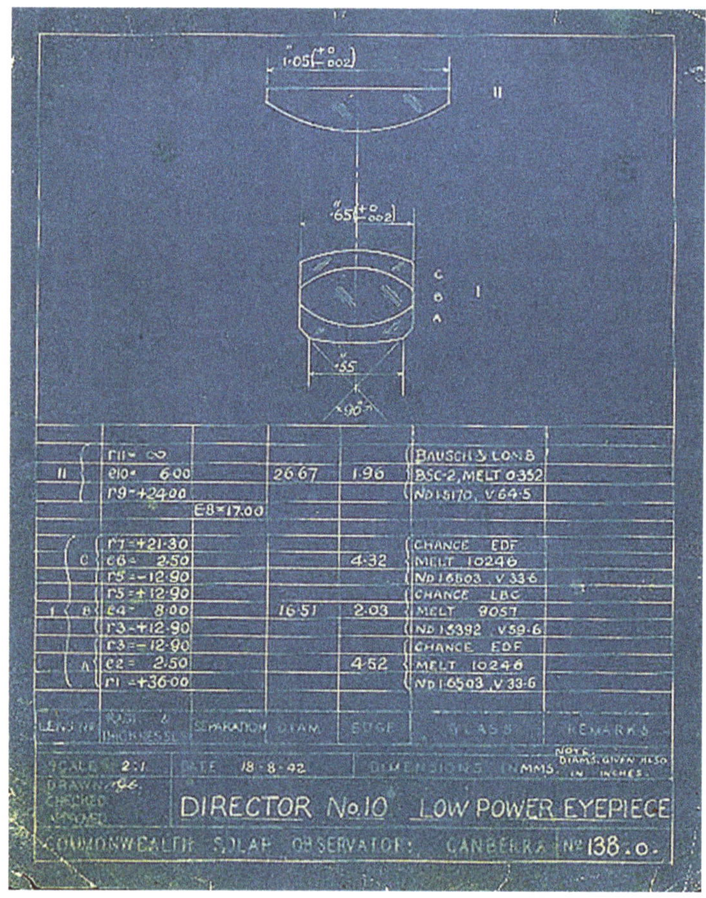

Figure 4.8: Optical eyepiece design. Optical munitions production at Mt Stromlo included precision instruments including gunsights, artillery directors and the like, which were generally made as prototypes or in small production runs. Photo: Mount Stromlo Archives.

Figure 4.9: Production of optical lenses. Following the successful R&D in Melbourne that led to production of optical glass of adequate quality, the Optical Munitions Factory at Mount Stromlo began producing lenses for the military. Photo: Mount Stromlo Archives.

Higgs were sent by Woolley to observe the solar eclipse at Victoria West, a small town ~500 km from Cape Town in South Africa. Among other things, the expedition was to study any changes that might take place within the ionosphere and their effects on long-distance radio transmissions.

The Second World War ended in 1945 when atomic bombs, which had been developed by physicists at Los Alamos, were dropped on Hiroshima and Nagasaki. The Optical Munitions Panel was renamed the Scientific Instruments and Optical Panel in the hope of developing a long-term high-tech optical industry in Australia. However, the initial enthusiasm did not last long: 'To have kept the very large wartime effort in optics going that the Panel had started and to have maintained its initiative and converted it into a modern industry with world-wide markets could have been done but it would have needed something like a National Institute of Optics, perhaps as part of CSIRO, together with a fiscal policy of protection that successive

Figure 4.10: Richard Woolley (at left) increased staffing at the Observatory from 10 to over 70 to support the war effort. The technical capabilities developed at the Optical Munitions Factory were important for the design and construction of astronomical instruments and telescopes in later decades. Photo: Mount Stromlo Archives.

The ionosphere and radio communications

Studies of the ionosphere (a region of the Earth's atmosphere extending from ~70–300 km above the Earth) became an important area of investigation because the ionosphere affected communications, which were crucial. The problem was to ascertain and provide a prediction service for usable frequencies for radio communications at different times of the day. Allen carried out the investigations.

His investigations of flares and sunspots and the correlation of these with the Earth's magnetic field allowed him to provide fairly accurate data. His work was 'greatly appreciated by all Australian services and did much to enhance the reliability of communications'.[12] His forecasts of when radio communications might be interrupted. were valuable to the military and after the war led to the establishment of the Ionospheric Prediction Service.

Figure 4.11: Investigations into the nature and function of the ionosphere were carried out at the Kite House. In 1937, Mount Stromlo had become the first ionospheric station to be equipped with an automatic recording system. Arthur Higgs was responsible for the collection, reduction and prediction of frequencies which were useful for communication purposes. Photo: Mount Stromlo Archives.

Allen's work enabled him to publish a paper on the relationship between magnetic storms and solar activity.[13] His other investigations involved the study of recurrent geomagnetic storms that were related to the period of solar rotation. The unknown regions of the Sun that produced these disturbances were known as M-regions. Allen tried to explain these regions in terms of white light coronal streamers. However, the correct explanation for the recurrent geomagnetic storms came in the late 1970s in terms of coronal holes, i.e. the dark regions of the X-ray Sun.[14]

Allen's work in solar physics kept Australia at the forefront in this area of specialisation.

Figure 4.12: Clabon Walter Allen adjusting the spectroheliograph. Allen moved from working on optical munitions to studying the effects of solar activity on shortwave radio communications. Photo: Mount Stromlo Archives.

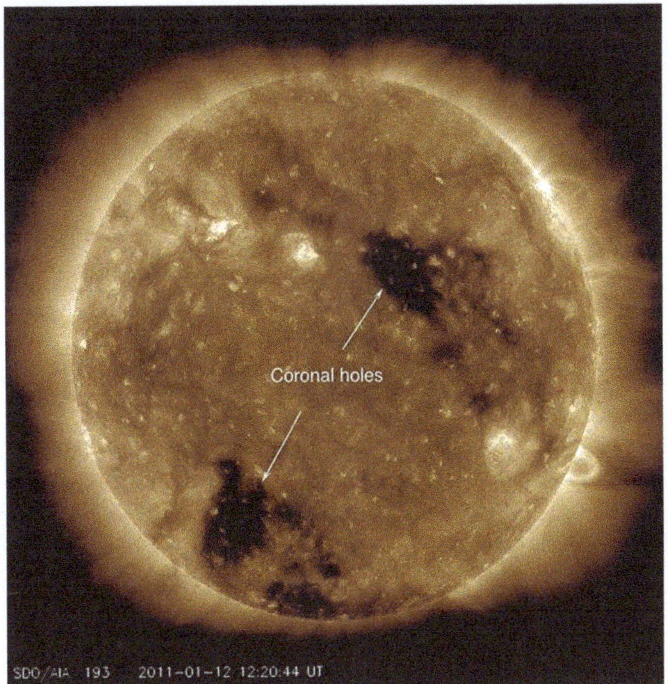

Figure 4.13: Coronal holes are areas on the Sun where the solar magnetic field has burst forth, in the process accelerating solar plasma particles to high velocities. Should these particles impact the Earth, they distort the magnetic field, cause the northern and southern aurorae and disrupt radio communications. Photo: NASA.

Figure 4.14: Ionospheric measurements from Mount Stromlo continue as part of the worldwide network of ionospheric stations. Today, the antennas have been relocated off the main Stromlo ridge to the west and the facility is operated remotely. Stromlo personnel still help maintain the facility under contract to the IPS Radio and Space Services unit of the Australian government. Photo: H. Butcher.

Figure 4.15: Aerial view of the Observatory in 1945. The numbers indicate the facilities: 1. The main Commonwealth Solar Observatory building, housing offices, workshops, the solar telescope (the right-hand dome) and the Farnham Telescope (left-hand dome). 2. The Director's House. 3. The ionospheric antennas and Kite House containing the electronics. 4. The Oddie Telescope. 5. The Reynolds Telescope. 6. Staff houses. Photo: Mount Stromlo Archives.

Australian Governments have not been willing to support. There was indeed a lost opportunity.'[3]

Valuable contacts

However, Mount Stromlo benefited from its involvement in the research, development and manufacture of optical equipment. This experience was of tremendous value in establishing the Observatory as a centre for research in optical astronomy and instrumentation in later years. Woolley had not only done an excellent job but had made valuable connections with the influential public servants, such as H.C. 'Nugget' Coombs, who had been appointed in 1943 to head the Department of Post-War Reconstruction. According to Woolley, 'The work brought me into contact with a number of powerful civil servants ... I learnt the rules of Public Service in-fighting'.[15]

These contacts were very valuable. Woolley was ready to get back to the main objective – changing the direction of the Observatory's research program – for which he had been appointed.

References

1. Gavin L 1973. *The Six Years War: A Concise History of Australia in the 1939–45 War.* Australian War Memorial/Australian Government Publishing Service, Canberra.
2. Rogers JS 1946. History of the Scientific Instruments and Optical Panel initially Optical Munitions Panel. Ministry of Munitions, Ordnance Production Directorate. Australian Archives, Melbourne. MP730/11 Box 3.

3. Bolton HC 1990. Optical instruments in Australia in the 1939–45 war: successes and lost opportunities. *Australian Physicist* **27**(3), 31–43.
4. Gascoigne SCB 1984. Astrophysics at Mount Stromlo: the Woolley era. *Proceedings of the Astronomical Society of Australia* **5**(4), 597–605.
5. Choo C 1996. The impact of war on the Aborigines of the Kimberley. In *On the Homefront: Western Australia and World War II*. (Ed. J Gregory) pp. 134–144. University of Western Australia Press, Perth.
6. Gregory J 1996. *On the Homefront: Western Australia and World War II*. University of Western Australia Press, Perth.
7. Bhathal R 1996. *Australian Astronomers: Achievements at the Frontiers of Astronomy*. National Library of Australia, Canberra.
8. Hartung E 1942. Production of optical glass in Australia. *Nature* **149**, 518–519. doi:10.1038/149518a0
9. Mervyn D, Gascoigne B 1995. Refugee engineer had precision in his sights. Obituary. *The Australian*, 4 September.
10. Frame T, Faulkner D 2003. *Stromlo: An Australian Observatory*. Allen and Unwin, Sydney.
11. Singleton Argus 1944. 8 May, p. 1.
12. Mellor DP 1958. The role of science and industry. In *Official History of Australia in the War of 1939–45*. Australian War Memorial, Canberra.
13. Allen CW 1944. Relation between magnetic storms and solar activity. *Monthly Notices of the Royal Astronomical Society* **104**, 13–21.
14. Chaisson E, McMillan S 2011. *Astronomy Today*. Addison-Wesley.
15. Woolley R 1968. Mount Stromlo Observatory. *Records of the Australian Academy of Science* **1**(3), 54–57.

1946–1955

5

The change master

This man got things done.

Melbourne *Herald*, 23 June 1955.

The choice of Woolley was critical. He was an astronomer, of the classical British school of Cambridge and Greenwich, and his avowed purpose was to change completely the direction of the observatory, by discontinuing its existing program of solar and geophysical work and concentrating on stellar astronomy. The change had momentous consequences.[28]

Two developments which took place after the end of the Second World War had a tremendous impact on the future of astronomy in Australia. They would change the entire course of astronomy in the country and propel it to the international frontiers of astrophysics.

The first came from the scientists and engineers who had been involved in the development and application of radar during the war. They were left with marvellous technical 'junk' which they turned into new machines to observe and investigate the mysterious universe and begin the new field of radio astronomy. Field stations with a variety of radio telescope designs were set up at various locations around Sydney. According to Chris Christiansen, 'Each morning people set off in open trucks to the field stations where their equipment, mainly salvaged and modified from radar installations, had been installed in ex-army and navy huts.'[1] It was at one of these field stations, in October 1945, that radio astronomy was born in Australia. Joseph Pawsey, together with Ruby Payne-Scott and Lindsay McCready, used an existing RAAF radar antenna overlooking the sea at Collaroy, a northern suburb of Sydney, to pick up solar radiation from the Sun. Later, at the Dover Heights field station south of Sydney, Pawsey used a technique known as sea interferometry to pinpoint the sunspot region from where the strong radio emission originated.[2] Pawsey's group was the first in the world to introduce the interferometer into radio astronomy. These early experiments threw up a group of brilliant young men and a woman who were propelled into international scientific fame. They included John Bolton, Joe Pawsey, 'Taffy' Bowen, Ruby Payne-Scott, Chris Christiansen, Bernard Mills, Harry Minnett and Paul

Wild.[3,4] Pawsey (1954), Mills (1963), Wild (1970), Bolton (1973) and Bowen (1975) were elected Fellows of the Royal Society of London. They bequeathed Australia a rich heritage of radio astronomy which in the 21st century enabled Australian astronomers to win a half share of the Square Kilometre Array (SKA), a project to study the early universe with radio waves and make discoveries that no one has yet imagined.

The second came from optical astronomy, which was almost single-handedly ushered in by Woolley with the assistance of the Astronomer Royal, Harold Spencer Jones. At the end of the war Woolley had not only a fully equipped optical workshop but an excellent group of technicians and a handful of enthusiastic and skilful physicists and astronomers. Over the next few years he changed the entire direction of the Observatory from solar physics to stellar astronomy although no one on the staff had any extensive knowledge of stellar astronomy. Even Woolley's expertise lay in the study of the Sun. He was not in touch with developments in mainstream astronomy in the US and Europe by astronomers such as Hans Bethe, Walter Baade, Jan Oort and A. Unsold. They were setting the pace in stellar and galactic astronomy. Woolley dropped the word Solar from the name of the Observatory. It became the Commonwealth Observatory and he became the first and only Commonwealth Astronomer. It was a slow and arduous process to wind down the old programs and initiate the new. There was also the need to upgrade the instrumentation and acquire larger telescopes. In addition, there were administrative problems to be resolved and opposition to Woolley's ideas of changing the Observatory from a government department to a university institution.

New directions of research

After the war the Observatory began a series of studies devoted to stellar optical astronomy with special reference to narrow-band measurements determined photographically and photoelectrically, spectro-photometric gradient measures, eclipsing binaries, emission nebulae and theoretical investigations of stellar atmospheres.[5,6,7] Earlier, in 1938, Hans Bethe at Cornell University had identified the source of stellar energy generation as the thermonuclear conversion of hydrogen to helium. This paved the way for understanding and estimating stellar ages and the rate of the production of heavy elements from the conversion of hydrogen to helium to carbon and the heavy elements. This led to new insights and opened a new field of study for the astronomers at Mount Stromlo. The instruments that were available to the astronomers were the Sun Telescope, the Farnham Telescope, the Oddie Telescope and the 76 cm Reynolds Telescope. The last had been donated to the Observatory by J.H. Reynolds, who had the distinction of being the only amateur astronomer to serve as President of the Royal Astronomical Society in Britain; it arrived at Mount Stromlo in 1929. The Reynolds Telescope, however, was in need of repair and refurbishment. No work had been done on it since the war broke out. After the war, the Reynolds was refurbished and used to begin fundamental research on the distance scale of the universe and on the distribution and natures of galaxies in the southern sky. It became operational in 1946 and was well used after that. It became clear that to be

Figure 5.1: Ben Gascoigne adjusting the 76 cm Reynolds Telescope, which he used extensively for his studies of stars in the Magellanic Clouds. After the war, Observatory staff quickly turned to astronomical and astrophysics research. Photo: National Archives of Australia.

competitive with its counterparts in Europe and the US, however, the Observatory needed to acquire larger, more sensitive telescopes.

Several programs were slowly brought to a conclusion as the Observatory embarked on new directions. Arthur Hogg built a photo-electric photometer and began studying eclipsing binaries.[8] When the new photo-electric 1P21 detector became available, Hogg realised its potential and became one of the early pioneers of southern hemisphere photomultiplier photometry.[9] He used the new tube on the 22 cm Oddie refractor as well as the Reynolds reflector from 1948.[10,11] His work on the eclipsing binary Zeta Phoenicis allowed him to produce some of the best masses and radii of B type stars.[12]

Walter Stibbs produced a significant piece of work by using a 1P21 photometer on the Oddie Telescope. His study of the variable magnetic star HD125248 led to his oblique rotator model, which allowed him to explain the properties of the variable star.[13] Following this, he and Woolley wrote *The Outer Layers of a Star*, published in 1953. He left the Observatory to work at Pretoria and Oxford before being appointed Director of the St Andrews Observatory in Scotland.

At the same time, Woolley and Allen were busy constructing models of the corona and chromosphere and they published several papers in this area.[14,15]

Figure 5.2: Arthur Hogg's photo-electric photometer. Photo: Mount Stromlo Archives.

Late in 1944, D.F. Martyn from CSIR (Commonwealth Scientific and Industrial Research) joined Mount Stromlo. During his stay he worked out ideas about the effects of solar tides in the ionosphere, which led to his being elected a Fellow of the Royal Society of London. He left the Observatory in 1956.

Some initial work in radio astronomy was carried out by Allen and Colin Gum. They used a 200 MHz aerial and receiver which had been acquired from the Radiophysics Laboratory to survey the emission from the southern part of the Milky Way Galaxy.[16] They produced the first map of the southern part of the Galaxy at 200 MHz. Despite this initial collaboration, the radio and optical astronomers formed separate communities, at least partly because of

Figure 5.3: D.F. Martyn in his office at Mount Stromlo. Photo: Australian Academy of Science.

Figure 5.4: The Gum Nebula extends over some 70 times the diameter of the full moon in the southern constellations Puppis and Vela. It is a cloud of diffuse glowing gas too faint to see with the naked eye. It was discovered by Colin Gum, a graduate student at Mount Stromlo in the 1950s. The Gum Nebula lies about a thousand light-years from Earth and is thought to be the remnant of an ancient exploding star. Gum's career was cut short when he died in a skiing accident in Switzerland in 1960. Photo: Mount Stromlo/Bessell and Sutherland.

Figure 5.5: The Reynolds Telescope with the spectrograph used by de Vaucouleurs for a pioneering survey of southern galaxies. Photo: Mount Stromlo Archives.

Figure 5.6: de Vaucouleurs introduced extragalactic astronomy into the Observatory's program. Photo: McDonald Observatory/UT-Austin.

some rather unfortunate remarks made by Woolley about the future of radio astronomy: at the 1947 Australian and New Zealand Association for the Advancement of Science meeting in Perth, Woolley dismissed the rise of radio astronomy by confidently noting that it would be forgotten within 10 years.

In late 1951 Allen left the Observatory to take up the newly created Perren Chair for Astronomy at the University College, London. He published the well-known book *Astrophysical Quantities* in 1973. Allen departed as the new era in stellar and galactic astronomy began to emerge at the Observatory.

Gum had joined the observatory in 1951 to undertake his PhD. With a modified spectrograph, he carried out an H-α survey of the southern Milky Way Galaxy and discovered 44 previously unknown hydrogen emission regions including the great emission nebula in the constellation of Vela,[17] which became known as the Gum Nebula. H-α regions had been recognised by Walter Baade as tracers of spiral structure. When Gum plotted his emission nebulae on the galactic plane it confirmed very nicely the pattern for the northern part of the galaxy that had just been published by Morgan, Whitford and Code.[18]

Also in 1951, Gerard de Vaucouleurs joined the staff as a Research Fellow and introduced extragalactic astronomy into the Observatory's research program. [18] While at Mount Stromlo he revised Hubble's classification scheme and introduced the concept of the supergalaxy. He was awarded the Herschel Medal for this work by the Royal Astronomical Society. de Vaucouleurs carried out the first systematic survey of galaxies in the southern sky. Not only are the nearest significant galaxies, the Large and Small Magellanic Clouds, in the southern sky, but many large systems are nearby and can be studied in detail. His main observation program was carried out with the 76 cm Reynolds Telescope. From his deep small-scale photography of the Large Magellanic Cloud, he announced that the Large Magellanic Cloud was a flat rotating system very similar to spirals. This was met with scepticism. However, with Frank Kerr he was able to use the 21 cm hydrogen line observations of the Cloud that had

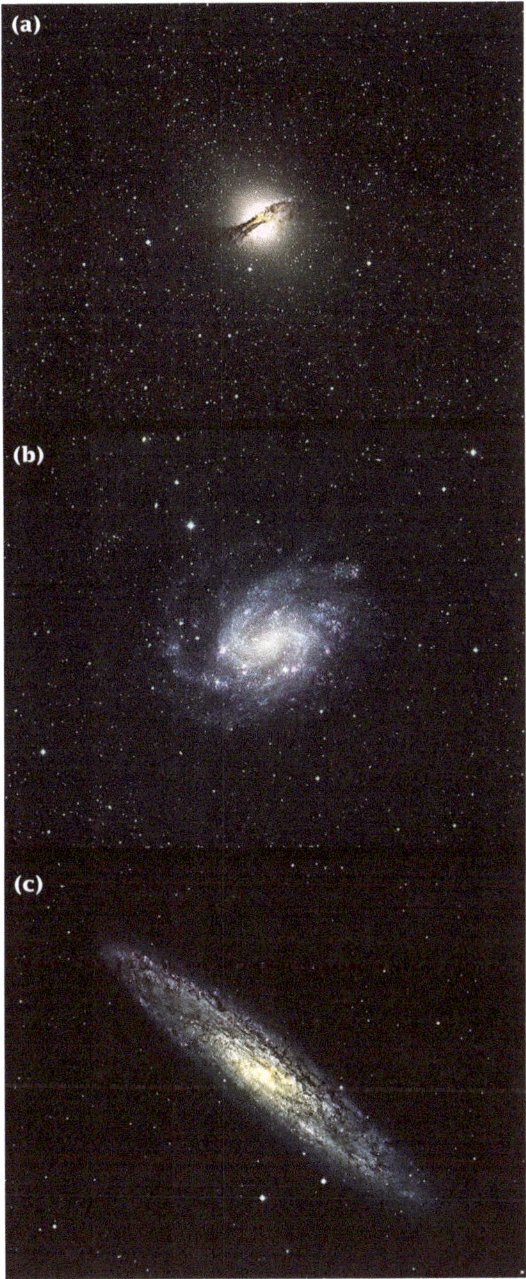

Figure 5.7: (a) NGC 5128 is a large elliptical galaxy with a massive ring of gas and dust. It was originally discovered in 1826 by James Dunlop from his home in Parramatta, New South Wales. It contains a central massive black hole and at radio wavelengths is seen to be ejecting opposing jets of relativistic plasma. (b) NGC 300 is a nearby spiral galaxy, one of the nearest systems beyond the Local Group of which the Milky Way is a member. It seems to have a binary object in its centre consisting of a massive hot star and a black hole. (c) NGC 253 is a nearly edge-on spiral galaxy in the Sculptor group, the nearest group to our own Local Group. It is remarkable in that it is forming stars at an extraordinary rate, including in several super-clusters of stars. Photos: AAO.

been made by the CSIRO radiophysics astronomers to show that his proposed model was essentially correct. In 1954 de Vaucouleurs became the Officer-in-Charge of the Yale–Columbia Telescope at Mount Stromlo.

Three years later he left Mount Stromlo for the US, where he would get involved in the 'distance scale wars' with Allan Sandage of the Carnegie Observatories. The 1954 *Fortune* magazine had called Sandage one of the 10 most promising young scientists and noted that he was 'helping to define the age and structure of the universe'. De Vaucouleurs became the most vociferous critic of Sandage's work, which gradually took on the character of personal attacks. According to Overbye, 'He [de Vaucouleurs] lectured Sandage about putting criticism in print where it could be taken apart. Sandage, he complained, once suggested that de Vaucouleurs not publish some argument because they were the only two people in the world who could understand it. Sandage reminded de Vaucouleurs of his suggestion that Sandage retire.'[19]

Figure 5.8: The 1P21 was the most famous photomultiplier from RCA, first produced in 1943. Photo: RCA.

Figure 5.9: Gerald Kron (1913–2012) made a number of sabbatical visits to Canberra, where the Observatory offered access to the southern skies. He joined the staff as a Senior Research Fellow from 1974 to 1976. He pioneered the use of photomultipliers in astronomy, installing photometers on the Oddie and later the 1.27 m telescopes. Photo: *New York Times*.

US influence at Mount Stromlo

The arrival of Gerald Kron and Olin Eggen from the Lick Observatory on extended visits in 1951 not only brought US influence into the Observatory but was a turning point in Gascoigne's career.

Gascoigne had been very impressed by Baade's work on stellar populations and its implications for the chemical and dynamical evolution of galaxies. Baade had divided the stars in the Milky Way Galaxy into Population I and Population II stars. Population I stars were young and metal-rich and belonged to the flat spinning disc of the Milky Way while Population II stars were old, metal-poor and found in the central nucleus and the spheroidal halo surrounding the Galaxy. He saw the Magellanic Clouds as a fantastic opportunity for testing Baade's ideas. The arrival of Kron and Eggen provided an excellent incentive. According to Kron,[20] they had 'gone down to Mount Stromlo at Father O'Connell's request and Woolley wanted someone down there to look at their photometric equipment and make recommendations for it'. Kron and Eggen were experts in photoelectric photometry. They had both been Joel Stebbins' postgraduate students at the University of Wisconsin, Madison, where much of the earliest work on photoelectric photometry was done.

This was the first of several extended visits by Eggen to Mount Stromlo before he became the Observatory's Director in 1966.

Kron brought his Lick photometer and, with Gascoigne, established the first red and near infrared photoelectric magnitudes in the southern hemisphere.[21,22,23] Kron was an engineer turned astronomer and was an expert in constructing instruments for astronomical purposes, especially photoelectric photometric instruments. Kron 'was a natural instrumentalist and was skilled in machine tools since his high-school days. Before entering astronomy, Kron had studied engineering, and during his work on radar at MIT [Massachusetts Institute of Technology] and on missile technology in the US Navy gave him an invaluable background in electronics'.[9]

Kron made extensive tests on the 1P21 and noted that 'the convenience offered by the multiplier will not fail to impress workers in the field of photometric photometry'. After 1946 many observatories acquired 1P21 photomultipliers on his advice. They constructed photometers from them and were able to undertake reliable photoelectric photometry for the first time. The 1P21's very low dark current and high overall sensitivity made it suitable for the measurement of extremely small quantities of light.

Figure 5.10: The Large and Small Magellanic Clouds were studied by Gascoigne and Kron. They provided new insights into the evolution of galaxies. Photo: Vince Ford Collection. Mount Stromlo Archives.

Impressions of Canberra

It is perhaps interesting to note Kron's impressions of Mount Stromlo and Canberra: 'It was a primitive place, the observatory was primitive and Canberra was completely devoid of comforts. There was no good place to eat. The houses were terrible and it was a small struggling community that did not have the sympathy of anyone in Australia. Nobody wanted it except the people who conceived the notion of having a grand capital in Australia'.[20]

Before Kron came to Australia he had begun studying red dwarf stars, but found that too few were available in the northern hemisphere. Coming to Mount Stromlo was a great way to study a larger sample of red dwarfs in the southern hemisphere. Since he wanted someone to give him a hand, he invited Gascoigne to team up with him. In return he agreed to assist Gascoigne in his observations of the Magellanic Clouds. Kron's collaboration with Gascoigne was mutually beneficial and in particular their work on the Magellanic Clouds led to significant insights. Kron was involved in the discovery that globular star clusters in the Magellanic Clouds have very different histories from similar clusters in the Milky Way, and he contributed to a revision of the scale of distances outside the Milky Way. Their observations showed that the Small Cloud cepheids were much bluer than those in the Galaxy. They also measured the colours of the globular-like clusters in the Small Cloud.

These observations had important implications, in particular that the distances of the Clouds and the extra-galactic distance scale had been underestimated by a factor of two. 'It turned out that the work in the Small Cloud was most important, that had some very important consequences. And one of them was to support an increase in distance by factor of two. Another was that Gascoigne and I found the so-called blue cepheids in the Small Cloud and that caused a real flap in the astronomical world.'[20]

Figure 5.11: Gascoigne at his desk, with a slide rule. He was given the task of overseeing the newly transferred Time Service. Photo: Mount Stromlo Archives.

Figure 5.12: Claire Wehner adjusting a clock that was part of the Time Service at Mount Stromlo Observatory. Photo: Wehner Collection/W. Pedersen photograph L.17387. Mount Stromlo Archives.

Figure 5.13: The Great Melbourne Telescope being erected in Melbourne in 1869. The telescope had a 1.2 m speculum metal primary mirror, and at the time was the second-largest telescope operating in the world, after Lord Rosse's telescope in Ireland. Photo: Mount Stromlo Archives.

They also found that there were both blue and red globular clusters in the Small Cloud while those in the Galaxy were only red. After completing his measurements on the Large Cloud and combining the results of the observations, Gascoigne concluded that the three galaxies had evolved differently. Arising from this work, Gascoigne was invited to organise a Joint Discussion on the 'Luminosity of Cepheids' for the 1958 meeting of the International Astronomical Union in Moscow. This was a great honour. The work on the clusters and cepheids established lines of work which have been followed at Mount Stromlo ever since.[18]

Larger telescopes and better equipment

In his drive to change the direction of the Observatory, Woolley embarked on the acquisition of larger telescopes. With the closure of Melbourne Observatory in 1944, Woolley acquired not only the much-maligned Great Melbourne Telescope for its scrap value from the Victorian government, but also the Time Service. Mount Stromlo was connected to HMAS Harmon, the naval wireless station at Canberra, from which highly accurate time signals were broadcast worldwide. One of the clocks kept rhythmic time with 61 'pips' per minute instead of the usual 60, thus providing a vernier function for the highly accurate requirements of navigation and land survey purposes. Slaved Shortt pendulum clocks were used to provide a national time standard

Figure 5.14: The Great Melbourne Telescope looking south-east. The instrument was housed in a building that had a roof that could be retracted on rails. Problems keeping the mirror shiny and the telescope's relative unsuitability for photography kept it from becoming an important research instrument. Photo: Mount Stromlo Archives.

for Australia. They were subsequently replaced by quartz clocks and later by atomic standards. In 1968 the Time Service was transferred to the Division of National Mapping because Eggen considered it did not contribute to the Observatory's research program. However, working with the Time Service allowed Gascoigne to acquire the necessary expertise in the algorithms which he was able to use later when the Anglo-Australian Telescope control software was installed.

The Great Melbourne Telescope had been acquired by Melbourne Observatory in 1868 to carry out observations of southern hemisphere nebulae. It was, for its time, one of the largest reflecting telescopes in the world. Its development and establishment in Melbourne involved the British scientific elite, Thomas Grubb of Dublin, a reputable and highly experienced manufacturer of telescopes, and the scientists and bureaucrats in Melbourne.

However, the telescope failed to live up to expectations. This began the 'GMT wars' in which the blame game was played out in the press in Australia and England. The President of the Royal Society, Sir E. Sabine, in his annual address of 30 November 1869, said that 'its performance since erection does not appear to have given altogether the same satisfaction at Melbourne that it did in Dublin; but the defects complained of may arise partly from

Figure 5.15: Lithograph of of Eta Carinae by Joseph Turner in 1875, made from visual observations with the Great Melbourne Telescope. Image: Museum Victoria.

Figure 5.16: Image of the Eta Carina Nebula by Bessell taken with a modern detector and filters through the 1 m telescope at Siding Spring Observatory. Photo: Bessell, Mount Stromlo Archives.

Figure 5.17: (a) Installing the polar axis of the 1.27 m at Mount Stromlo Observatory, c. 1954. (b) Installing the shortened tube. Photos: (a) Helen Beryl Collection. (b) Wehner Collection. Mount Stromlo Archives.

Figure 5.18: The Great Melbourne Telescope after its first renovation. Photo: Mount Stromlo Archives.

Figure 5.19: The telescope was equipped with modern, sensitive detectors including the photoelectric photometer shown here at the Cassegrain focus behind the primary mirror. Moonlight was found to enter the open latticework of the upper tube, causing unwanted signals in sensitive instruments, so the upper tube was enclosed as shown. The original speculum metal mirror is shown mounted on the wall behind the telescope. Photo: Mount Stromlo Archives.

Figure 5.20: Richard Woolley (left) and Sir Harold Spencer Jones, the Astronomer Royal, in the basement of the Mount Stromlo Observatory in 1947. Photo: Mount Stromlo Archives.

imperfect knowledge of the principles of instrument and inexperience in the use of so large a telescope, partly from experimental alterations made at Melbourne, and partly from atmospherical circumstances'.[24] Sabine placed the blame for the non-functioning of the telescope squarely on the inexperience of the colonials.

On arrival at Mount Stromlo the Great Melbourne Telescope was extensively modified. The original speculum metal mirror was replaced with a new 1.27 m aluminised Pyrex mirror, the polar axis was lengthened, the tube was shortened and the drive and controls motorised. Gascoigne became the first big user of the telescope for his photoelectric work on the Clouds.

But Woolley was after a bigger telescope to allow observations of more distant stars and nebulae. It was time for him to use the contacts he had made with senior bureaucrats during the war. He arranged for the Astronomer Royal, Sir Harold Spencer Jones, to visit the Observatory during the 1947 meeting of the Board of Visitors. Sir Harold recommended to the Chifley government that the Observatory acquire a 1,9 m telescope similar to that at the Radcliffe Observatory in Pretoria, so that it could complement the astrophysical research being conducted with the large telescopes in the northern hemisphere.[25] According to Gascoigne, 'Woolley then approached the Prime Minister, at that time J.B. Chifley, and as he told me afterwards, obtained the money so easily he wondered whether he ought to have gone for a 100 inch [2.54 m]'.[18] The

Figure 5.21: The 1.9 m telescope was manufactured by Grubb-Parsons in the UK. It was completed in 1951 and was displayed at the London Exhibition before being shipped. Following arrival in Sydney in 1954, it was transported by trucks to Canberra and Mount Stromlo. This image shows these vehicles on the road with Lake George in the background. Photo: Mount Stromlo Archives.

Figure 5.22: Construction of the building and the dome for the 1.9 m telescope were also delayed. They were built around the telescope. Photos: Wehner Collection. Mount Stromlo Archives.

Figure 5.23: When completed, the 1.9 m telescope was the largest optical telescope in Australia and the equal largest in the southern hemisphere. It retained that status for two decades, until the Anglo-Australian Telescope took the title. Photo: Mount Stromlo Archives.

Figure 5.24: The formal dedication ceremony for the 1.9 m telescope was held on 8 November 1955 in the dome under the telescope. Director Richard Woolley addressed the assembled dignitaries and guests. Photo: Mount Stromlo Archives.

Figure 5.25: Woolley arranged for (a) the Uppsala Schmidt telescope and (b) the Yale–Columbia telescope to be moved to Mount Stromlo. Photos: Mount Stromlo Archives.

Figure 5.26: (a) Initial attempts to stop the blaze using fire hoses were thwarted when the electricity failed and the water pumps could no longer be used. (b) Hotel Canberra sent up a barrel of beer and sandwiches for the tired staff who had bravely fought the fire. The person seated on the left is a future director, Olin Eggen, a visiting astronomer at the time. Photos: (a) Wehner Collection. (b) Eggen Collection. Mount Stromlo Archives.

Figure 5.27: Workshop foreman, Jim Banham, surveying the devastation following the 1952 bushfire. The workshop, various records and auxiliary buildings were destroyed, but luckily no loss of life or loss of research telescopes were incurred. Photo: Wehner Collection. Mount Stromlo Archives.

telescope was commissioned shortly after Woolley left the Observatory and for the next 20 years it was the largest telescope in Australia.

Woolley was also responsible for getting the Yale and Columbia observatories to relocate their 66 cm astrometric refractor from South Africa to Mount Stromlo. It was a large-format photographic telescope for measuring the parallaxes and space motions of nearby stars. It also derived orbits for the moons of Saturn for NASA. His other coup was to get the University of Uppsala in Sweden to locate its new 51 cm Schmidt on Mount Stromlo in 1957, to undertake photographic surveys of the Southern Milky Way and Magellanic Clouds. It was also the camera that took the first astronomical photograph of Sputnik. However, as Canberra grew, its lights began to interfere with observations and the Schmidt was moved to Siding Spring.

The first bushfire

On a hot February day in 1952 the Observatory suffered a setback. A ferocious fire broke out nearby and raced towards the Observatory. A spark ignited leaves in the guttering of the west wing workshop of the Observatory building, and the workshop became engulfed in flames. The fire destroyed the electricity supply to the water pumps, so there was no water in the hoses to fight the fire. The gallant staff formed a bucket brigade to save the Observatory but the workshop was destroyed, as were valuable records from the Great Melbourne Telescope, two storage buildings and machine tools.

Transfer to ANU

Woolley's lasting legacy to the Australian astronomical community was his brilliant move to have the Observatory transferred from the Department of the Interior to the Australian National University.

For this he used his connections and influence with H.C. 'Nugget' Coombs, who had been appointed the Director of Post-War Reconstruction in 1943. Coombs was a key player in the establishment of the Australian National University. According to Coombs, 'a centre for research in the physical sciences, was almost an afterthought. When the Mills Committee (this Committee was set up to prepare a report for Cabinet about the Commonwealth's role in education) was at work, I was sharing a flat with Dick (Richard) Woolley, the Commonwealth Astronomer. Woolley was critical of the idea of a research university with no involvement in the natural sciences'.[26] The original idea was to have a research university only in the areas of social sciences and social medicine.

Woolley had begun taking the necessary steps and lobbying the right people for the transfer of the Observatory well before 1953, the year that he was elected a Fellow of the Royal Society, although that gave him some leverage with the authorities. Rather surprisingly, he met resistance from his most senior staff member, Arthur Hogg. Hogg did not want to stop being a public servant. He was well versed in the public service regulations and was expert at manoeuvring his way through the Canberra bureaucracy. Being second in command of the Observatory, he was concerned that his power and influence might be eroded in a university environment. He had the ear of the Public Service Board and the Treasury and lobbied senior bureaucrats not to agree to the transfer. It was not surprising that the Minister for the Interior, Wilfred Kent Hughes, rejected the transfer of the Observatory to the Australian National University.

It took a Prime Minister to make a different decision. In November 1955 the Prime Minister, Robert Menzies, formally directed the Vice-Chancellor of the Australian National University to appoint a full Professor of Astronomy.[27] In January 1957 the Observatory was finally transferred to the Australian National University and its name changed from the Commonwealth Observatory to the Mount Stromlo Observatory.

Not yet in the first division

On 7 December 1955 Woolley resigned as Director at Mount Stromlo to become the 11th Astronomer Royal. He joined a long line of distinguished Astronomers Royal who had shaped astronomical programs not only for England but also for its former colonies. He had served as a proxy for British astronomy in Australia and was a strong advocate for the establishment of the 3.9 m Anglo-Australian Telescope at Coonabarabran as a joint venture between the British and Australian governments and their astronomers.

Woolley had been the change master of Australian astronomy. During his 16 years at Mount Stromlo he had achieved his aims. He had changed the astronomy program at Mount Stromlo from solar to stellar astronomy, increased the power of the telescopes available to the astronomers, saw the growth of some lines of innovative research which made an impact on international astronomy, began a postgraduate program and saw the transfer of the Observatory to the Australian National University. The transfer of the Observatory was his most enduring legacy to Australian astronomy.

Nevertheless, 'The Observatory was not yet in the first division'.[18]

References

1. Bhathal R 1996. *Australian Astronomers: Achievements at the Frontiers of Astronomy.* National Library of Australia, Canberra.
2. Pawsey JL, Payne-Scott R, McCready LL 1946. Radio frequency from the Sun. *Nature* **157**, 158–159. doi:10.1038/157158a0.
3. Sullivan WT III 1988. Early years in Australian radio astronomy. In *Australian Science in the Making.* (Ed. RW Home) pp. 308–344. Cambridge University Press, Cambridge.
4. Bowen EG 1984. The origins of radio astronomy in Australia. In *The Early Years of Radio Astronomy.*(Ed. WT Sullivan III) pp. 85–111. Cambridge University Press, Cambridge.
5. Woolley R 1946. The mechanism of ionisation. *Proceedings of the Royal Society of London. Series A* **187**, 403–415.
6. Woolley R 1947. Presidential Address to Section A. ANZAAS, Perth.
7. Hogg AR 1953. *The Commonwealth Observatory.* Astronomical Society of the Pacific. pp. 346–353.
8. Hogg AR 1946. Photoelectric observations of V Puppis. *Monthly Notices of the Royal Astronomical Society* **106**, 292–299.
9. Hearnshaw JB 1996. *The Measurement of Starlight: Two Centuries of Astronomical Photometry.* Cambridge University Press, Cambridge.
10. Hogg AR, Bowe PWA 1950. Photoelectric observations of S. Antliae. *Monthly Notices of the Royal Astronomical Society* **110**, 373–380.
11. Hogg AR, Hall B 1951. Narrow-band photoelectric photometry of bright southern stars. *Monthly Notices of the Royal Astronomical Society* **111**, 325–338.
12. Hogg AR 1951. The eclipsing system Zeta Phoenicis. *Monthly Notices of the Royal Astronomical Society* **111**, 315–324.
13. Stibbs DWN 1950. A study of the spectrum and magnetic variable star HD125248. *Monthly Notices of the Royal Astronomical Society* **110**, 395–404.
14. Woolley R, Allen CW 1948. The coronal emission. *Monthly Notices of the Royal Astronomical Society* **108**, 292–305.
15. Allen CW 1954. The physical conditions of the solar corona. *Reports on Progress in Physics* **17**, 135–153. doi:10.1088/0034-4885/17/1/304.
16. Allen CW, Gum CS 1950. Survey of galactic radio noise at 200 Mc/s. *Australian Journal of Scientific Research, Series A: Physical Sciences* **3**, 224–233.
17. Gum CS 1955. A survey of southern HII regions. *Memoirs of the Royal Astronomical Society* **67**, 155–177.
18. Gascoigne SCB 1984. Astrophysics at Mount Stromlo: the Woolley era. *Proceedings of the Astronomical Society of Australia* **5**(4), 597–605.
19. Overbye D 1991. *Lonely Hearts of the Cosmos.* Picador.
20. De Vorkin D 1978. Interview with Dr Gerald Kron in Flagstaff, Arizona. 20 May. American Institute of Physics.

21. Kron GE, White HS, Gascoigne SCB 1953. Red and infrared magnitudes for 138 stars observed as photometric standards. *Astrophysical Journal* **118**, 502–510. doi:10.1086/145778.

22. Kron GE, Gascoigne SCB, White HS 1957. Red and infrared magnitudes for 282 stars with known trigonometric parallaxes. *Astronomical Journal* **62**, 205–220. doi:10.1086/107521.

23. Kron GE, Gascoigne SCB 1953. Red and infrared magnitudes for 27 southern hemisphere stars of early spectral type. *Astrophysical Journal* **118**, 511–512. doi:10.1086/145779.

24. Bhathal R, Graeme W 1991. *Under the Southern Cross: A Brief History of Astronomy in Australia.* Kangaroo Press, Sydney.

25. Australian Archives 1946. CRS A431. 47/728.

26. Coombs HC 1981. *Trial Balance.* Macmillan, Melbourne.

27. Australian National University Archives 1955. Prime Minister to Vice-Chancellor, 14 November 1955. 8.8.1.0 (A) Part 1.

28. Gascoigne SCB 1988. Australian astronomy since the Second World War. In *Australian Science in the Making.* (Ed. RW Home) pp. 345–373. Cambridge University Press.

1956–1966

6

The astronomical godfather

Bart Jan Bok was a graduate of the University of Leiden's Astronomy Department which produced some of the leading Dutch American astronomers Gerard Kuiper, Dirk Brouwer, Bart Jan Bok and William Luyten. The mentors of these astronomers at the University of Leiden and the University of Groningen were such well known astronomers such as Jan Oort, Ejnar Hertzsprung, Antonie Pannekoek, Willem de Sitter and Jan Woltjer.[25]

His major achievements in Australia were the establishment of the highly successful international graduate school at Mount Stromlo Observatory and the initiation of a survey for a suitable optical observatory site which eventually led to the establishment of Siding Spring Mountain as Australia's premier observatory site.[3]

Arthur Hogg was appointed the Acting Director of Mount Stromlo Observatory when Woolley resigned in December 1955 to take up the position of the Astronomer Royal in England. Hogg served in this capacity for over a year before the Australian National University found a replacement.

Since the 19th century Australia had looked to Britain for a recommendation for directors of its observatories, but in the 1950s Australia found that it had to make the decision itself. Without being asked, Woolley had provided notes on the qualities needed for a Director of Mount Stromlo Observatory. He ended with, 'I am sure that everyone would like to [see] an Australian appointed – but not, of course, if he were not the best man available for the job.'[1] Unfortunately, there was no 'best man available for the job' in Australia at that time. Woolley favoured Olin Eggen as his successor but the radio astronomers in Sydney, Mike Bessell said, 'wanted someone sympathetic to radio astronomy and were very keen on Bok'.[2] According to Bart Bok, 'Mrs Bok and I were well known to have a great interest in the southern Milky Way. So during a General Assembly of the International Astronomical Union in Dublin in Ireland, the then Director of the Observatory, now Sir Richard Woolley, and Dr J.L. Pawsey, the second-in-command of CSIRO's Radiophysics Division in Sydney, approached us both together with the request, would we be interested in considering leaving Harvard and coming to Australia?'[3]

Figure 6.1: Bart Bok and Priscilla Bok – a husband and wife team of well-known astronomers who were deeply interested in the southern Milky Way. Under their guidance the international graduate school at Mount Stromlo Observatory became a great success. Photo: Mount Stromlo Archives.

Bart Bok, the Dutch astronomers and the Great Debate

Born in Holland in 1906, Bok had been educated at the universities of Leiden and Groningen where some of the great Dutch scientists, such as Kamerlingh Onnes, Paul Ehrenfest, J. Woltjer, P.J. van Rhijn, H.A. Kramers, Sam Goudsmit, George Uhlenbeck, J.C. Kapteyn and Jan Oort held academic posts. He began his PhD at the University of Groningen and at the age of 23 left Holland on an Agassiz Fellowship to study under Harlow Shapley, the Director of the Harvard College Observatory, who had made his name with observations of the Milky Way. There was also another reason why Bok moved to Boston. According to Mike Bessell, 'This was to get married to Priscilla Fairfield who was also an astronomer.'[2]

Shapley was well-known for his Great Debate with Lick Observatory's Heber Curtis as to whether the Milky Way was the entire universe or just one galaxy among many. Shapley had insisted that the faint, misty patches of light visible in the night sky which had been called 'nebulae' were part of the Milky Way. Heber Curtis disagreed. He favoured an 'island universe' theory, believing that the faint patches of light were galaxies in their own right. The issue was formally discussed in the Great Debate held on 26 March 1920 under the auspices of the National Academy of Science. The debate did not resolve the issue. It was left to Edwin Hubble to resolve the issue observationally in 1926, putting the island universe theory on a firm footing with his demonstration of the great distances of the nebulae.

As a young boy Bok had read Shapley's article on the Great Debate. In fact, Bok said, 'I wrote to the City of the Hague library and told them that no library in the world would be complete if they didn't have this critical volume. And the City of the Hague Library bought a copy from the National Research Council. So I became interested in it, and I wanted to go to the university and study astronomy.'[4]

Under Shapley's guidance Bok became an expert on the Milky Way. The southern Milky Way had a special attraction for him, especially the Carina region of the galaxy. In fact, based on his work at Harvard, the Eta Carinae Nebula was the subject of Bok's doctoral thesis at the University of Groningen. He and colleagues had used star counts and statistical analysis to investigate the distribution of stars in the galaxy. However, their star counting efforts were hampered by the effects of interstellar absorption. Bok became interested in this problem and called attention to the existence of small dark clouds, now called Bok globules,[5] that consist of gas and dust and are thought to be the progenitors of star formation. He made many important contributions to the understanding of the structure and physical properties of the Galaxy. However, the spiral structure of the Milky Way would be discovered

Figure 6.2: Bok globules are dense clouds of gas and dust, typically seen in silhouette against a background of stars or, as in the case of the 'caterpillar' shown here, against a background of glowing gas. They are opaque to visible light but infrared observations show that stars are being formed inside many of them. They were first observed by Bart Bok. Photo: NASA/ESA.

Figure 6.3: Barnard 335, a classic Bok globule, shown against a background of stars. Photo: Martin C. Germano, Thousand Oaks, CA. mcgermano@verizon.net.

Figure 6.4: The Milky Way Galaxy as seen over Siding Spring Observatory. Our own Galaxy is most readily studied from the southern hemisphere. Photo: © Kwon O. Chul: AstroKorea.com/kwon572.

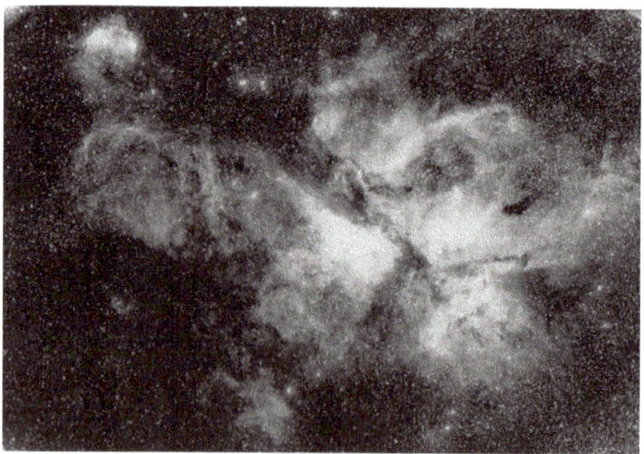

Figure 6.5: The Eta Carinae Nebula. In the late 19th century the southern star Eta Carinae became the second-brightest star in the night sky, then 20 years after it erupted it unexpectedly faded away. It was the subject of Bok's doctoral dissertation. Photo: NASA.

McCarthyism and the intellectual community in the US

McCarthyism (a political practice whereby intellectuals, scientists, artists and actors in the US were accused, without due regard for evidence, of disloyalty and subversion) began in the early 1950s as the Cold War between the US and the Soviet Union escalated after the end of the Second World War. Senator Joe McCarthy went on a vicious witch-hunt for 'un-American activities'. He accused intellectuals, artists, scientists and actors of being communists or communist sympathisers and dangerous to the country. Shapley was singled out by McCarthy. As the investigations continued, the outspoken Shapley came to be seen by top Harvard administrators as a liability. Whatever influence Shapley had was eroded by the accusations. Exhausted and intimidated by McCarthyism, Shapley retired from the directorship of the Observatory.

by other astronomers, not by Bok. With Priscilla, an astronomer in her own right, he wrote the *Milky Way* which became one of the most widely read astronomy books among both professionals and the public. It went through five editions.

Under normal circumstances Bok would have succeeded to the directorship of Harvard College Observatory when Shapley retired. But this was not to be. Things began to fall apart when Senator Joe McCarthy accused Shapley of membership or support of a subversive organisation.

Bok was a staunch supporter of Shapley and their association played a significant part in Bok being sidelined. He was told that 'life would be so much easier at Harvard if you [Bok] would stop your support of Harlow Shapley'.[6] Bok himself was hauled over the coals. His membership of the American Association of Scientific Workers, which had been infiltrated by communists, was questioned. So was his association with UNESCO. He

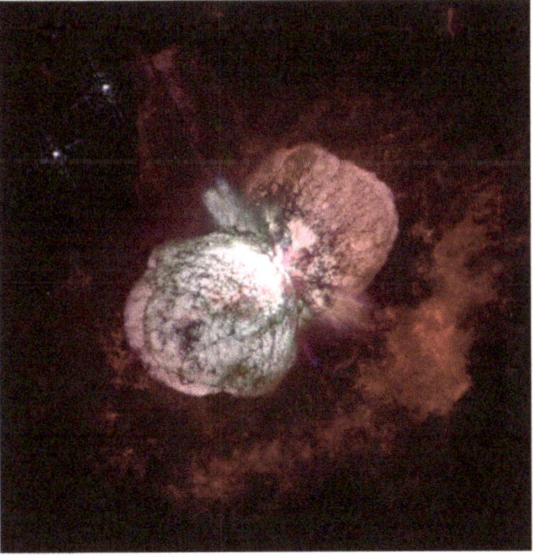

Figure 6.6: A recent composite image from the Hubble Space Telescope of the very centre of the Eta Carinae Nebula near the star itself. The double-lobed structure is believed to be material expanding from the earlier explosive eruption of the star. Photo: NASA.

challenged the investigating committee 'to bring a whole stack of Bibles: I will put my left hand on the bibles, raise my right hand, and swear that I have never been a Communist'.[6] However, the damage had been done.

The directorship of the Observatory went to Donald Menzel, a solar astronomer who not only had connections with the military establishment but who also brought much-needed funds for a university which was suffering from a financial crisis. Due to a shortage of funds the university also transferred its Boyden Station in South Africa, where Bok had spent several months studying the Milky Way, to a consortium of six institutions. Bok's work on the Milky Way was considered to be too expensive by Menzel. The next blow fell when the university invited the Smithsonian Astrophysical Observatory (SAO) to join the Harvard College Observatory under the directorship of Fred Whipple, an expert in cometary astronomy. SAO moved from Washington to Cambridge (Massachusetts) in 1955. Bok's work on the Milky Way was slowly being pushed out of the research program of the Observatory.

But as those doors closed new ones opened. The discovery of the 21 cm wavelength emission line from neutral hydrogen atoms by Harvard physicists Harold Ewen and Edward Purcell, and independently by Henk van de Hulst and Jan Oort at Leiden, saw Bok move into the new field of radio astronomy.[7] With funding from Mabel Agassiz, a member of Boston's wealthy classes, and from the National Science Foundation, Bok began to build radio telescopes for the new astronomy. With a group of young students, he designed and built the 7.3 m equatorially mounted radio telescope at Harvard's Agassiz Station. It was opened in November 1952. One of the young men was Frank Drake, who in 1960 became the first astronomer to use a radio telescope to search for intelligent life elsewhere in the universe; he later set up the SETI Institute in California ('SETI' standing for Search for Extra-Terrestrial Intelligence). Bok went on to build an 18 m radio telescope, also at Harvard, which opened in April 1956.[7] The Parkes radio telescope in New South Wales was mooted in 1956 and opened in 1961.

Figure 6.7: Bart Bok was the first Professor of Astronomy at the Australian National University. He arrived at Mount Stromlo Observatory just when it was transferred to the university from the Department of the Interior. Photo: Mount Stromlo Archives.

Bok became an expert in radio astronomy and gathered a group of enthusiastic radio astronomers. He was now conversant in both optical and radio astronomy. Furthermore, he was not only working at the frontiers of astrophysics but had the necessary connections with leaders in astrophysics in the US and Europe. Australia was looking for such a person to take up the directorship of Mount Stromlo Observatory. In particular, the radio astronomers felt he would be supportive of their community as well as of optical astronomy. Bok was offered the position in March 1956.

Figure 6.8: The Mount Stromlo Observatory staff in October 1957. Sitting (left to right): H. Gollonow, K. Gottlieb, H.J.M. Abraham, A.R. Hogg, B.J. Bok, S.C.B. Gascoigne, T. Dunham Jr, W. Buscombe. Standing: P.M. Morris, A. Przybylski, R.M. Henry, G. Hagemann, J. Crisp, D.G. Thomas, J. Banham, B. Westerlund, C. Wehner, H. Wehner, G. Oom, A.W. Rodgers. Photo: Mount Stromlo Archives.

Bok publicly announced his resignation from Harvard on 10 May 1956.[8] He was pretty much disgusted with the politics of the place and informed Harvard's President Nathan Pusey, 'I want to get the hell out.'[6] Years later Bok noted, 'Had it not been for McCarthy and his goddamned investigations, I might never have left Harvard.'[6] Bok arrived at Mount Stromlo in March 1957 as Director of the Mount Stromlo Observatory and as Professor and Head of the Department of Astronomy within the Research School of Physical Sciences at the Australian National University.

Populariser and publicist

The bright Australian sun shone on Bok as Mount Stromlo entered a new era under his directorship. Within a few months of his arrival he brought Mount Stromlo and Australian astronomy to the attention of the public and parliamentarians. He became the face of Australian astronomy. It all began in October 1957 when the Observatory, according to Mike Bessell, 'announced that it had taken the first photo of the Russian Sputnik'.[2] This precipitated an invitation to Bok from the Australian Parliament. No astronomer before or after Bok has been invited by both Houses of Parliament to address a meeting of parliamentarians, as he was when asked to inform Parliament of the

Figure 6.9: The Russian Sputnik was the first artificial satellite launched, on 4 October 1957. It weighed 83.6 kg and had an orbital period of 96.2 min. It sent shock waves throughout the world and started the Space Age. Photo: NASA.

Figure 6.10: On 8 October 1957, Gottlieb managed to secure this first photograph of Sputnik as it passed over Mount Stromlo, using the Uppsala Schmidt telescope. The wiggles at the start of the exposure were caused by opening the shutter. Together with timed visual observations by Abraham and staff, these data made possible for the first time the determination of the orbit of the satellite. Photo: Mount Stromlo Archives.

implications of the Russian Sputnik in October 1957. They were fearful of its implications for Australia and the world. In a short time, Bok became as well-known to the general public as Australia's Prime Minister, Robert Menzies, who became a firm supporter of his endeavours for astronomy in Australia. According to Bok, 'If I wanted to talk to the Prime Minister, I honestly could take up the phone and call him.'[4]

To bring astronomy to a wider audience, Bok embarked on a series of lectures to schools and to the public: 'I gave three public lectures a month, sort of set an average, one in Canberra or near Canberra, one in a capital city and one in a country centre. This was rather simple because we were travelling about, site testing and looking for new sites, so it was not difficult to get an invitation. I accepted almost any invitation I could get to give a lecture, the Rotary Club and the Lions Club, whatever was going on, but I always said I had a price, and my price was that if I spoke to one of the clubs or anywhere I insisted that in the morning or afternoon I would address the local high school, and this paid off amazingly well in a direction that had never occurred to me, because in a high school audience there is generally a nephew or a niece of a Member of Parliament, and all the Members of Parliament learned about it'.[3] This paid off for Bok and Australian astronomy.

In the ensuing years he 'sold' astronomy and the Milky Way to members of the public and, more importantly, to MPs. When there was a debate about the Anglo-Australian Telescope in Parliament there was great support for it. It was quite an amazing debate as MPs tried to out do one another. According to Bok, 'The members of the Country Party supported the astronomical endeavours as much as did members of the Liberal Party and the Labor Party. I remember, for example, to take one incident, a member of the Country Party getting up, and when he got up to speak the Liberal Party members shouted, Sit down Larry, sit down Larry, all you want is a subsidy for the milk in the Milky Way, and he spoke up immediately and said, You boys on the left side of this Parliament, keep quiet and you boys on the right, keep quiet too. You don't know what astronomy is all about. When I milk my cows – which he hadn't done for thirty years but it sounded fine – When I milk my cows near dawn, I can see the Milky Way, I see the Magellanic Clouds, I know what goes on.'[3]

Bok also offered 'astronomy nights' at Mount Stromlo, which proved very popular with the public. He also gained an Australia-wide audience through the Australian Broadcasting Corporation. On national television, he gave a series of 10 half-hour lectures on astronomy which covered topics from the solar system to the galaxies. Bok had an even larger public audience internationally. He wrote articles in *Sky and Telescope*, *Scientific American* and *Science Monthly*.

At the professional level Bok organised a symposium on the Milky Way at Mount Stromlo as part of the International Astronomical Union (IAU) and International Scientific Radio Union symposium on *The Galaxy and the Magellanic Clouds* in 1963. Sessions were held both in Canberra (on the Milky Way) and Sydney (on the Magellanic Clouds).[9] The symposium was attended by over 50 astronomers, with 20 participants from overseas.[10] It attracted several well-known overseas astronomers such as M. Schmidt, H.C. Arp and Olin

Figure 6.11: In 1963 Bok organised an international conference on the Milky Way Galaxy and the Magellanic Clouds. Photo: Mount Stromlo Archives.

Eggen from Mount Wilson and Palomar observatories and Jan Oort from Leiden. He invited the Minister for Air, D. E. Fairbairn, to open the symposium which showcased Australian astronomy to the international world of astronomy. The discussions centred on the latest developments on the structure of the Milky Way.

The international graduate school

Bok carried on the US influence that had begun with the visit of Kron and Eggen to Mount Stromlo when Woolley was the Director. His goal was to set up an international graduate school along the lines of that run at the Harvard College Observatory. Woolley had set in place the beginnings of a graduate school and Bok took this to completion. According to Whiteoak, 'Bok developed an extensive post-graduate program. He instituted full courses in astronomy, to be given by regular staff. To assist in the future supply of students, he began the successful summer vacation program. This enabled university students from all over Australia to spend summer vacations at the Observatory and gain a "hands-on" initiation into astronomy.'[11]

Bok also initiated a program to recruit international students: 'The presence of a good sprinkling of students with overseas backgrounds serves to keep the Australian-trained scholars on their toes, and gives them a feeling that they are not cut off from world-wide competition. The general aim has been to have about nine or ten Australians being trained as scholars at Mount Stromlo, with another five or six drawn from Great Britain, Western Europe and America, and with at least a couple from either India, Indonesia, Thailand, or from the Pacific area.'[12]

Bok and Priscilla put their heart and soul into the establishment of the international graduate school. From all accounts it was a great success. Whiteoak wrote, 'Bok's attitude towards students, no matter where he was located, was legendary. He was in fact our astronomical "Godfather". We were guided in our PhD work, educated by the lecture courses, advised on our personal problems, and generally trained to become a scientist in Bok's own image'.[11]

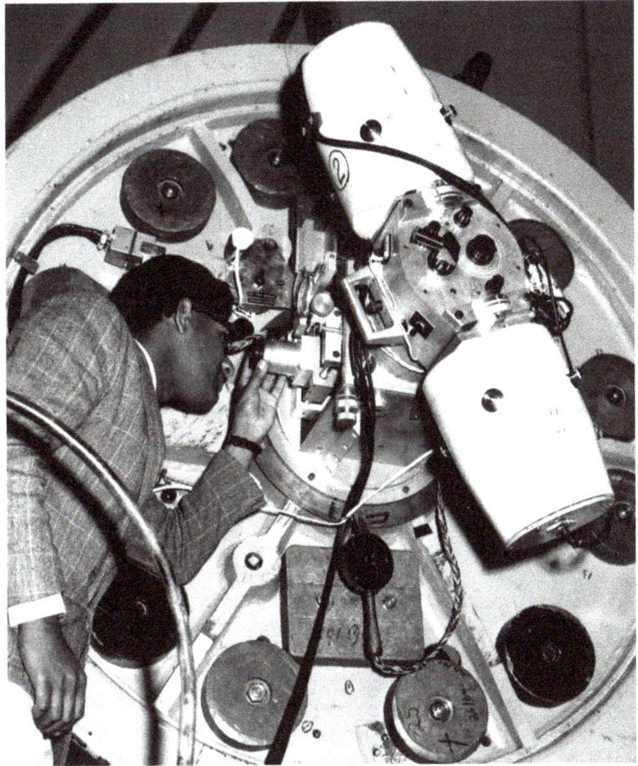

Figure 6.12: Visvanathan (known to colleagues as 'Vis'), with his polarimeter in 1964. He was one of the earliest overseas post-graduate students recruited by Bok and made significant contributions in polarisation studies of galaxies and the Magellanic Clouds. He had a long and fruitful collaboration with Allan Sandage, a protégé of Edwin Hubble. Photo: Mount Stromlo Archives.

Bok's Asian students: Natarajan Visvanathan and Ravi Bhavilai

Natarajan Visvanathan was one of the earliest and most successful student recruited from overseas. He carried out one of the first polarisation studies on the stars in the Magellanic Clouds[13] using the 66 cm reflector at Mount Bingar. After obtaining his PhD with a thesis entitled 'Polarization in the Galaxy and the Large Magellanic Cloud' he went to Mount Wilson and Palomar observatories in California as a Carnegie Fellow in 1962. He collaborated with Allan Sandage on a significant polarisation study of the filaments in the starbursting galaxy M82. They showed that the light from these filaments is polarised. It was a new and influential idea. He then went to Harvard as a Research Associate and Lecturer before spending three years from 1972 to 1975 at the Hale observatories and California Institute of Technology. He became a staff member of the Mount Stromlo Observatory in 1975 and stayed there for the rest of his life.

Ravi Bhavilai from Thailand was another foreign student to graduate with a PhD in the Bok years.[2] Bhavilai became a professor of physics at Chulalongkorn University in Bangkok. Back in Thailand, he was actively involved in promoting studies of Buddhism and science through the Thousand Stars Buddhism and Science Group.

By the end of Bok's tenure 28 students had passed through the Observatory's graduate program and all but five were still actively engaged in astronomy at the time of Bok's death.[11] In his final Annual Report, in 1965, Bok wrote, 'I feel personally that our most important success over the past nine years has been the establishment of a first-class graduate school of astronomy and astrophysics, with comparable emphasis on optical and radio astronomy. We now produce annually between three and five Doctoral Theses, which rank with the best overseas ... Our growing crop of PhDs provides a firm basis for the future of Australian astronomy.'[14]

Bok's success rested very much on the fact that he had made the School a vital part of the Observatory. The students were treated as members of staff and had access to the facilities, the telescopes and advanced instrumentation. Roger Bell, Colin Campbell and John Whiteoak were Bok's first student intake. Alex Rodgers was already at the Observatory, finishing his PhD under the supervision of Woolley; he later became the Director of the Observatory. Several students completed their PhDs under the joint supervision of the Mount Stromlo Observatory and the CSIRO Radiophysics Laboratory staff. These included J.V. Wall, who became the Director of the Royal Greenwich Observatory, R.M. Price, who became Head of the Astronomy Department at Albuquerque, New Mexico, and Ron Ekers, who became the first director of the VLA in New Mexico, the first Director of the CSIRO Australian Telescope National Facility and the President of the International Astronomical Union. He also became a Fellow of the Royal Society of London.

Astronomical programs

One of the reasons cited by Bok for taking up the directorship of Mount Stromlo was that he could devote his time to observing the Milky Way. Despite his heavy workload, which included administrative duties and other pressing matters, such as site testing for a large telescope and the popularisation of astronomy, he and Priscilla made time for a series of magnitude measurements of stars in the southern hemisphere.[15] They had started this project at Boyden Station and completed the work at Mount Stromlo. This was followed by their study of two small fields of blue-giant stars in the Large Magellanic Clouds.[16,17] According to Bessell, 'Bok did a lot of observing at Mount Bingar, Siding Spring and on the 74 inch [1.9 m]. He had Jane Basinski and Priscilla Bok to do his onerous data reduction.'[2] Bok was very good at time management.

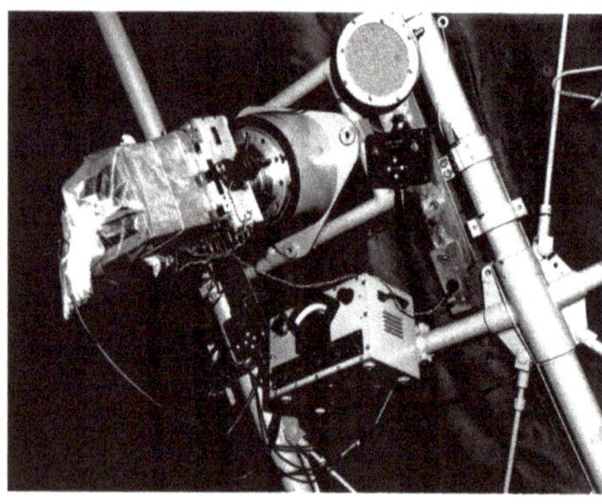

Figure 6.13: The faint star photometer on the 1.9 m telescope. Photo: Mount Stromlo Archives.

Bok was an observational astronomer, so it was natural that he imposed his observational philosophy on everyone at Mount Stromlo. Gascoigne thrived under Bok: 'I learnt about a really faint star photometer, designed and commercially available in the

Figure 6.14: An Atlas of H-Alpha Emission in the Southern Milky Way was produced by Rodgers, Campbell, Whiteoak, Bailey and Hunt. The H-Alpha emission is displayed in red in this Milky Way image. Photo: Bessell, Sutherland, Mount Stromlo Archives.

USA, so I said to Bart, "Look Bart, I really would like to get one of those". It was worth about $10 000, quite a bit of money at that time. Bart, said, "Sure, go ahead." That was wonderful and that photometer really made me.'[18] Gascoigne became an international expert on the Magellanic Clouds. He studied Cepheid stars, which vary in brightness with periods that indicate their intrinsic brightness. By measuring their light variations Gascoigne could compare their intrinsic and apparent brightnesses, and thereby estimate the distance to the Magellanic Clouds.

Some of the other interesting programs that took off under Bok's directorship included a study of the emission regions in the southern Milky Way. They came under the scrutiny of Alex Rodgers, Colin Campbell and John Whiteoak, who published an *Atlas of H-α Emission in the Southern Milky Way* which became a standard reference.[19] Cepheid variable stars also came under investigation. Bessell studied a class of short-period variables which allowed him to resolve a puzzle about their evolutionary status.[20] He became an expert on stellar photometry and spectroscopy.[21]

The cross-fertilisation of ideas and the movement of people from optical astronomy to radio astronomy and vice versa was one of the finest features of Bok's directorship. Although they were students at Mount Stromlo, Ron Ekers and Mark Price spent almost all their time at the Parkes radio telescope doing their PhD research in radio astronomy. Ekers worked on the Parkes interferometer under the direction of John Bolton, one of the pioneers of radio astronomy in Australia. John Whiteoak left Mount Stromlo to join the Radiophysics Laboratory. Whiteoak was a student of Bok: 'Whiteoak wrote a thesis on the southern spiral structure from optical evidence and now is one of the leading radio astronomers in Sydney.'[4] On the other hand, Don Mathewson, a radio astronomer at the Radiophysics Laboratory, took up a position at Mount Stromlo and later became Director of the Observatory.

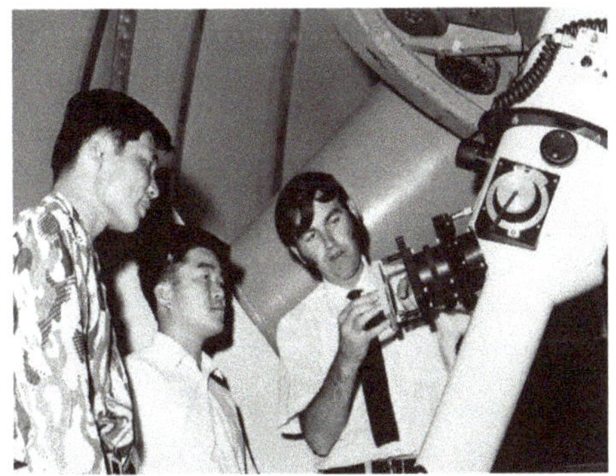

Figure 6.15: Mike Bessell instructing PhD students on the operation of the 1.9 m Cassegrain spectrograph. Photo: Mount Stromlo Archives.

Telescopes and site testing

According to Bok, 'When I arrived there, there were five telescopes on Mount Stromlo, four were not working. Well, the 74 inch [1.9 m] broke down on the night I arrived and had never been properly adjusted, and there were mirror troubles. The Schmidt telescope that was for Uppsala Observatory was in very poor condition, and in boxes, and nobody knew what to do with it. The 50 inch [1.27 m] (the old Great Melbourne telescope) was out of order completely and unusable. The Yale–Columbia telescope was being worked on, but was also in very poor condition. The 30 inch [76 cm] Reynolds telescope was the only one that was working, in a way, but not very good at all. So I arrived at Stromlo with things practically all stopped.'[14] Bok's first impression of the state of the telescopes was hardly flattering. Over the next few years he corrected the problems and introduced the use of digital computers into the Observatory. An IBM 610 was rented for an initial period starting in February 1960, principally for the Time Service. In fact, it was the first computer used by the Australian National University. IBM had introduced the 610 only in 1957. It fitted in a normal office and could be used by only one person. It

Figure 6.16: Inspection of the primary mirror of the 1.9 m telescope. It was found to have astigmatism and it was sent back to the manufacturers for refiguring. It was eventually installed in 1961. Photo: Mount Stromlo Archives.

Figure 6.17: Arthur Hogg, the Acting Director, examining the focus of the 1.27 m telescope. It was originally intended to be used as a Schmidt camera but its mechanical design was found not to be appropriate. New conventional paraboloid-hyperboloid optics were acquired from Grubb Parsons and the telescope was finally commissioned in 1961. Photo: Mount Stromlo Archives.

featured vacuum tubes, a magnetic drum, punched paper tape readers and punchers. User input could also be made from a keyboard; printed output was via an IBM electric typewriter. In his Annual Report for 1965, Bok wrote, 'A permanent connection landline from DSIF 42 at Tidbinbilla, near Canberra, was established in March to make continuous comparisons between a Tracking

Figure 6.18: John Burr with an early photoelectric photometer, in the early 1950s. Photo: Vince Ford Collection. Mount Stromlo Archives.

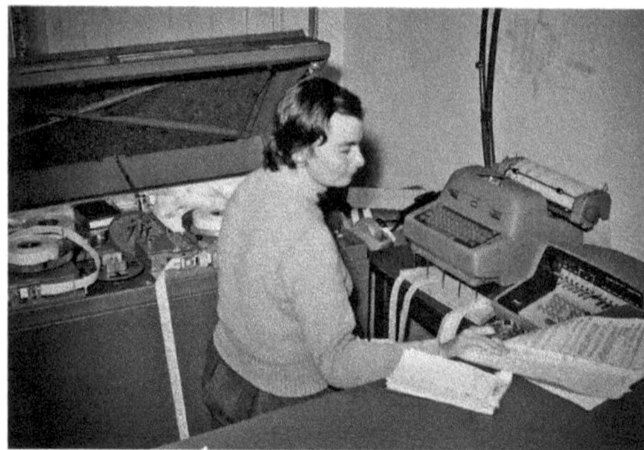

Figure 6.19: Claire Wehner entering data into the IBM 610 in 1960. Photo: Mount Stromlo Archives.

Station rubidium standard and an Observatory standard.' Bok's legacy was that he left all observatories in better shape than he found them.[2]

One of Bok's major tasks was to get a large Coudé spectrograph installed on the 1.9 m. In 1957 he obtained the services of Theodore Dunham Jr, who had built spectrographs for Mount Wilson, to do the job. Dunham is also known for his introduction of the Schmidt camera in spectroscopy, studies of stellar atmospheres and interstellar material, studies of planetary

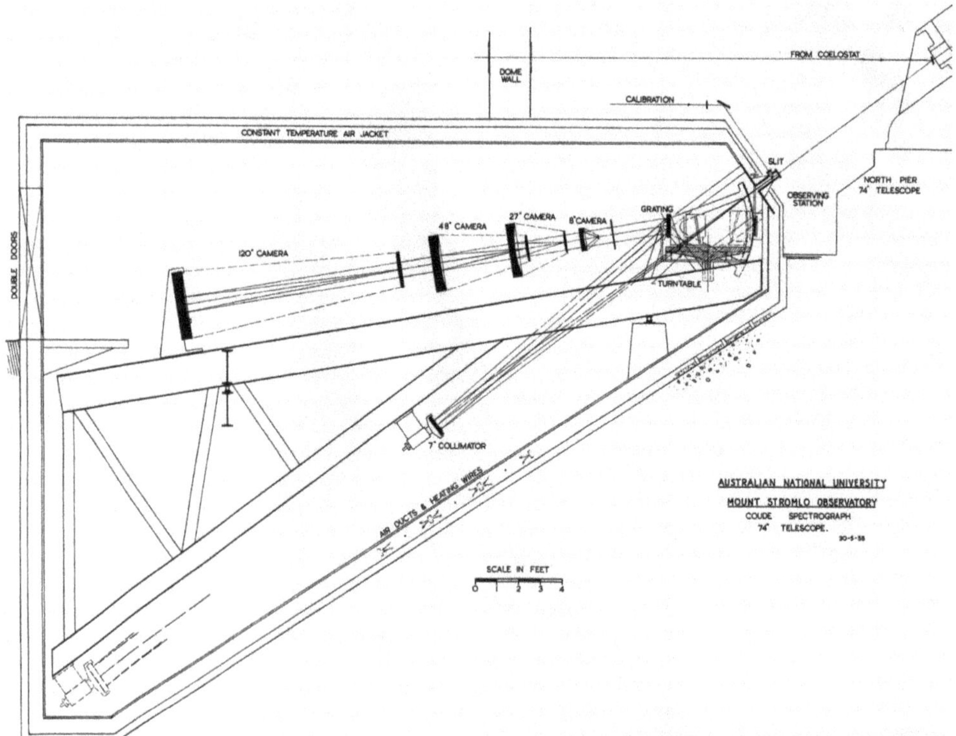

Figure 6.20: The final design for the Coudé spectrograph included thick walls with much of the instrument underground, providing the thermal and optical stability required for high-resolution spectroscopy. Drawing: Mount Stromlo Archives.

Figure 6.21: IAU attendees in the Coudé spectrograph room, then still under construction. Bok organised scientific presentations in the 1.9 m dome as well as a social event at his residence on the mountain. Photo: Shobbrook Collection. Mount Stromlo Archives.

atmospheres, development of photoelectric detectors for spectroscopy, and application of physical methods for research in medicine and surgery.

It was a delicate and intricate matter to install the Coudé. They had to get the light from the telescope to an adjacent room where the spectrograph was housed. A pit had to be blasted for the Coudé. The Coudé room was constructed below ground to the north of the telescope pier. The excavations were quite deep and required the removal of a significant amount of rock. The blasting upset Bok, who thought the telescope would be affected.[2] It was a dangerous exercise but it worked and by 1960 the 1.9 m became a much more useful instrument. According to Bessell, 'The great advance that the Coudé made was to enable Australian astronomers to participate for the first time in high resolution spectroscopy of stars and to derive precise metal abundances, radial velocities and hydrogen line profiles – the era of quantitative analysis of stellar spectra.'[2] It became the workhorse for the Observatory.

One of Bok's other concerns was the light pollution from the growing city. He convinced the university and Prime Minister Menzies of the need to set up a site testing survey to find the best location not only for Stromlo's future telescopes but also for a large Australian telescope. In an address in 1960 to the Royal Astronomical Society of Canada, Bok stated that, 'The other project is an attempt to locate a first-class site anywhere on the Australian continent, which might be used for future major observatory developments, be it either a purely Australian observatory, a British Commonwealth observatory or some major international observatory project.'[22] 'We surveyed the whole continent, Alice Springs, even including Western Australia, the regions north of Perth, but also New South Wales, a number of places.’[14]

Bok appointed Arthur Hogg, his deputy, to take charge of the site testing program. After an extensive search, Mount Bingar and Siding Spring Mountain were shortlisted.[23] Siding Spring was much higher and had better seeing conditions. It also had over 60% of usable nights per year. However, a controversy erupted: Bok favoured Mount Bingar but Hogg, Gascoigne and most of the Observatory astronomers favoured Siding Spring. The issue was resolved by the Australian National University's Vice-Chancellor, Sir Leonard Huxley. He made an executive decision and selected Siding Spring on 12 May 1962.

Bok got to work to build Siding Spring Mountain as the major astronomical site for Australia. Its birth is his legacy to Australian astronomy.[23] Within a few years Siding Spring ‘had grown into a world-class observatory. Not even Bart Bok, most sanguine of directors, could have imagined such a transformation in so short a space of time’.[24]

Figure 6.22: (a) Priscilla Bok and telescope at the Mount Bingar site testing station. Criteria for selecting the site were that it had to be within six to 10 hours of Canberra, have at least 60% clearer nights than Mount Stromlo and be a relatively isolated mountain. (b) Bok observing at Mount Bingar. He favoured this site, partly because of the excellent vineyards in the vicinity. Photos: Shobbrook Collection. Mount Stromlo Archives.

Figure 6.23: An aerial view of the astronomical site at Siding Spring Mountain. By the time this photo was taken, the 1 m and 40 cm telescopes were being installed (on the left), and several accommodation buildings were in place. The 60 cm telescope followed shortly thereafter. Observers slept in bedrooms in the domes while construction work continued. Photo: Mount Stromlo Archives.

Figure 6.24: The opening of the Siding Spring Observatory on 5 April 1965. Sir Leonard Huxley presided, and later commented the establishment of the observatory was one of the highlights of his tenure at ANU. Photo: Mount Stromlo Archives.

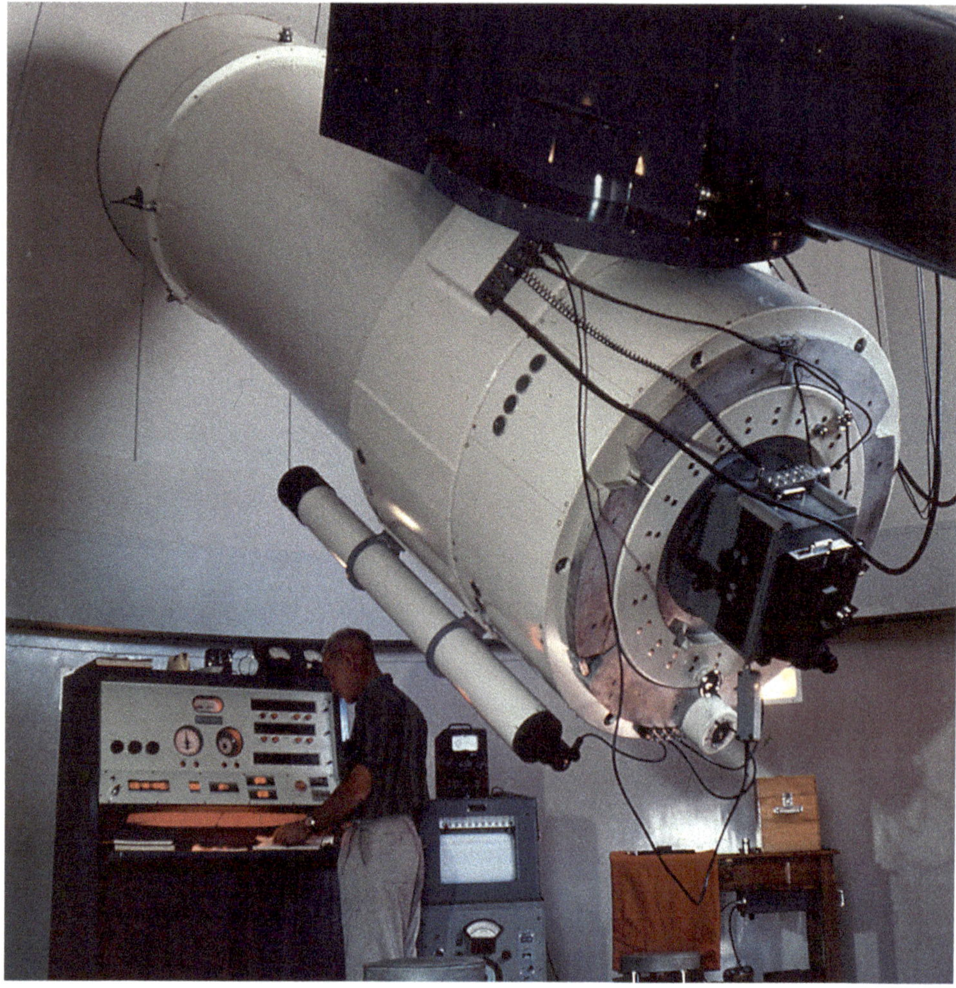

Figure 6.25: The 1 m telescope at Siding Spring, the first up-to-date telescope of advanced design acquired by the Observatory. It was manufactured by Boller & Chivens Co. and provided imaging and spectroscopic capabilities for 45 years. Photo: Mount Stromlo Archives.

Bok and the Anglo-Australian Telescope

Another of Bok's contributions to Australian astronomy was his enthusiastic lobbying for the 3.8 m Anglo-Australian Telescope, to be built at Siding Spring Mountain as a joint venture by the Australian and British governments and astronomers. He even got the MPs excited about it and they had a long debate in Parliament about astronomy in Australia. When the Anglo-Australian Telescope was discussed there was a tremendous burst of enthusiastic support for it as the MPs debated the merits of the size of the telescope. According to Bok, 'I remember, Mr Clyde Cameron, who is now of course one of the distinguished ministers, speaking up and saying – There has been quite a bit of discussion in this Parliament whether we should have a 120 inch [3 m] or 150 inch [3.8 m] reflector, then he said, As everybody knows – for we told them so – the astronomer sits at the prime focus in a cage. If you have a 120 inch reflector, the cage has to be very small and he has got to keep his elbows close to him. For crying out loud, he said, give our Australian boys a little elbow space and let us not talk of a 120-inch when we consider the building of the large telescope.'[4]

Bok's criticism in the 1964 Annual Report regarding the delay in development plans for a 3.8 m reflector was not received well by the ANU administration. The Report was censored and Bok had to rewrite it.

Not only were Parliament and Prime Minister Menzies behind Bok's astronomical endeavours, so were the senior management of the Australian National University. However, not all scientists shared his enthusiasm. The President of the newly formed Australian Academy of Science, Tom Cherry, and the biologists in the Academy were against the proposal. They wanted to build a flower garden on top of Mount Kosciuszko and tried to stall Bok's project. This made Bok see red and he went on the offensive. He was invited to give the Presidential Address to the Australian and New Zealand Association for the Advancement of Science (ANZAAS) meeting and used the opportunity to criticise the Academy. Journalists at the Australian Broadcasting Corporation, sensing a good controversial story, decided to give Bok a prime time-slot and televised his address: 'In that talk, I pointed out very clearly – and the President of the Academy, Tom Cherry, was sitting in the audience looking white and furious – that the project for the big telescope for Australia had been uppermost in the minds of astronomers of Australia, radio and optical, for a long time. They all wanted it. That Parliament had been informed of it and the Parliament had been most sympathetic to it; that the Prime Minister was in favour and wanted to act, and the people of Australia were all behind it. And I said, It is not the function of an Academy to stop it. It's the function of the Academy to accelerate this and make it possible. You've got to go ahead.'[4]

The President and biologist members of the Academy were furious. Bok said, 'the President of the Academy wrote a nasty letter to me. It said, Dear Bart, I think that you should realize that you have offended the establishment with your unfortunate action in favour of your telescope. I think that you should realize that you have outlived your usefulness to our country, and the sooner you go back where you came from, the better it will be for all concerned'.[4] By ' the establishment', Cherry meant the Prime Minister. However, as far as the establishment was concerned things were quite different. At an official luncheon for the King of Thailand at University House in Canberra, the academics had to line up to greet the King and the Prime Minister. According to Bok, 'When he (Menzies) came past me, he never looked at me, but his fist shot out at me, ponk. And he poked me in the belly and he said as he walked on, Bart, you are a bad, bad boy. Then I knew I had not offended the establishment.'[4] Bok knew that he 'should go on fighting the Academy',[3] and he succeeded.

At the opening of the Anglo-Australian Telescope in 1974, Fred Hoyle lavished praise on Bok's determination to see a large telescope built in Australia. The new leadership at the Australian Academy of Science awarded Bok the Selby Fellowship in 1978. The Fellowship is awarded to distinguished

Bok's censored 1964 Annual Report: the 3.8 m reflector and related site testing

The preliminary planning is under way. The Australian Academy of Science has established a large Reflector Committee with a distinguished membership over which the President of the Academy, Sir Thomas Cherry, presides. The Royal Society has a parallel Committee with, as its Chairman, the President of the Royal Society, Sir Howard Florey. In June and July, a distinguished group of British astronomers visited Australia, including Sir Richard Woolley and Professor Redman, and they met with Australian representatives. They consulted on a technical level with a Sub-Committee headed by the Vice-Chancellor of the ANU, Sir Leonard Huxley and with, as members, Professor Gascoigne, Dr Rodgers and Mr Wehner from Mount Stromlo Observatory and Dr J.G. Bolton from the C.S.I.R.O. The joint Technical Committee visited Siding Spring Observatory, the Radio Observatory at Parkes and potential sites at Mount Serle, South Australia and Mount Singleton, Western Australia. Professor Gascoigne reports as follows:

> *'The talks, which were very successful, resulted in detailed financial, administrative and technical proposals, which are to be submitted to the UK Government through the Royal Society and to the Australian Government through the Australian Academy of Science.'*

I regret to report that practically nothing has happened – at least to the best of my knowledge – since the meeting of last July. These delays are most discouraging. Australian astronomers and Australian astronomy stand to lose much unless the two governments are approached promptly and action follows soon. Developments in Chile and elsewhere (see 'Large New Telescope for the Southern Hemisphere' in *Science*, November, 1964; Vol. 146, p. 755) have passed the early planning stage and Australia and Great Britain are falling behind. At a time when Canada on its own orders the prompt construction of a 150 inch [3.8 m] reflector (northern hemisphere), it is disheartening to see Great Britain and Australia failing to proceed with a comparable project on a joint basis.

This seems to be an appropriate place to reflect briefly on the consequences for the future of Mount Stromlo Observatory that arise from the delays that are occurring. We have at Mount Stromlo and at Siding Spring Mountain one really modern telescope, the 40 inch [1 m] reflector. It is excellent for its size, but ridiculously small by world standards. The 74 inch [1.9 m] reflector is a good older telescope and the 50 inch [1.27 m] reflector should be retired within a decade. If the 150 inch reflector does not pass the discussion stage soon, then Mount Stromlo Observatory, as a simple act of scientific survival, will have to undertake the planning on its own of a reflector with an aperture of at least 100 inches [2.54 m], probably greater. Without such an instrument our whole astronomical future looks dim indeed.

Annual Report 1964, Mount Stromlo Observatory, Australian National University.
Dated 20 January 1965.

overseas scientists, enabling them to visit Australia for public tours or seminars to increase public awareness of science and scientific issues. By 1978 Bok had long since resigned his directorship; in March 1966 the Boks had left Mount Stromlo to join the University of Arizona.

References

1. Woolley R. Australian National University Folio 160.
2. Bessell M 2013. Notes to Ragbir Bhathal, 23 January 2013.
3. De Berg H 1996. Bart Bok: The Milky Way (transcript of interview). In *Australian Astronomers: Achievements at the Frontiers of Astronomy.* (R Bhathal) pp. 28–31. National Library of Australia. Canberra.
4. De Vorkin D 1978. Oral history transcript: Dr Bart Bok. American Institute of Physics.
5. Heeschen DS 1983. Bart J. Bok. *Physics Today* **36**(12), 73.
6. Levy DH 1993. *The Man Who Sold the Milky Way.* University of Arizona Press, Tucson.
7. Ewen D 2003. Doc Ewen: the horn, HI and other events in US radio astronomy. http://www.nrao.edu/archives/Ewen/ewen_harvard24and60.shtml.
8. Graham JA, Wade CM, Price RM 1994. Bart Bok: a memoir. *Biographical Memoirs. National Academy of Sciences (US)* **64**, 72–97.
9. Faulkner DJ 1963. Current studies of the Magellanic Clouds. *Sky and Telescope* **26**, 69–72.
10. Bok B 1963. An international discussion of Milky Way research. *Sky and Telescope* **26**, 4–7.
11. Whiteoak JB 1984. Student memories of Bart Bok, an astronomical godfather. *Proceedings of the Astronomical Society of Australia* **5**(4), 608–610.
12. Bok B 1966. Optical astronomy at Mount Stromlo. *Hemisphere* **10**(3), 2–8.
13. Visvanathan N 1965. Polarisation in the Galaxy and the Large Magellanic Cloud. PhD thesis. Australian National University.
14. Bok B 1965. Annual Report. Mount Stromlo Observatory.
15. Bok BJ, Bok PF 1960. Four standard sequences in the southern hemisphere. *Monthly Notices of the Royal Astronomical Society* **121**(6), 531–542.
16. Bok BJ, Bok PF, Basinski JM 1961. Colour-magnitude arrays for two associations in the Large Magellanic Cloud. *Monthly Notices of the Royal Astronomical Society* **123**, 487–496.
17. Bok BJ, Bok PF 1962. Integrated magnitudes and colours of young associations in the Large Magellanic Clouds. *Monthly Notices of the Royal Astronomical Society* **124**, 435–444.
18. Bhathal R 1996. *Australian Astronomers: Achievements at the Frontiers of Astronomy.* National Library of Australia, Canberra.
19. Hyland AR, Faulkner DJ 1989. From the Sun to the universe: the Woolley and Bok directorships at Mount Stromlo. *Proceedings of the Astronomical Society of Australia.* **8**(2), 216–228.
20. Bessell MS 1967. A study of some southern short period variables. PhD thesis. Australian National University.
21. Bhathal R 2009. Stellar astrophysics. *Journal and Proceedings of the Royal Society of New South Wales* **142**, 25–30.
22. Bok B 1960.. A search for new observatory sites in Australia. *Journal of the Royal Astronomical Society of Canada* **54**(6), 257–268.

23. Bessell M 2011. Bok Lecture: Bart Bok – The man who sold the Milky Way.
24. Gascoigne SCB, Proust KM, Robins MO 1990. *The Creation of the Anglo-Australian Observatory.* Cambridge University Press, Cambridge.
25. New Netherland Institute. http//www.nnp.org/nni/Publications/Dutch-American/bokbart.html.

1966–1977

7

A life on the dome floor

Most astronomers are involved in our discipline by compulsion and are overjoyed at being paid to do what we would at least be trying to do, in any case ... it is fun. My opinion is that it has nothing (or very little) to do with a craving for recognition, present or posthumous, but is essentially self-centred with an aim of self-satisfaction.[3]

His scientific achievements include a spectacular paper on how our galaxy formed. This paper, written in 1962 with his colleagues Lynden-Bell and Sandage, is known as ELS and is one of the most influential astrophysics papers ever written.[11] [It had 1499 citations.]

What I [Ben Gascoigne] think he will be remembered for is the controversy over the AAT and the great rift that developed between Stromlo and the rest of Australian astronomy ... It was very damaging; it was years before Stromlo was accepted back into the fold, and it saw the centre of optical astronomy shift considerably from Canberra to Sydney.[5]

Olin Eggen did not apply for the job of Director of Mount Stromlo Observatory. He was invited[1] to fill the role, since there was no one in the Australian astronomical community who had the international recognition and the scientific calibre to be Director of the emerging institution. He continued the US influence begun in the 1950s by him and Gerald Kron and pushed along by Bart Bok. They had brought new ideas and instrumentation to propel Mount Stromlo Observatory into the forefront of international astronomy. Even Woolley, a former Director, remarked that 'I needed fresh contacts to get back into contemporary ideas, and Mount Stromlo needed a man who was soaked in them. It has got one now.'[2] That man, according to Woolley, was Eggen. He was the classic example of a scientific nomad who moved easily not only from one institution to another but from one continent to another. He was a man for all seasons.

Eggen had worked at 'Mount Hamilton in the 1940s, Herstmonceux Castle, Sussex, and Capetown in the 1950s, Pasadena, Mount Wilson, and Palomar in

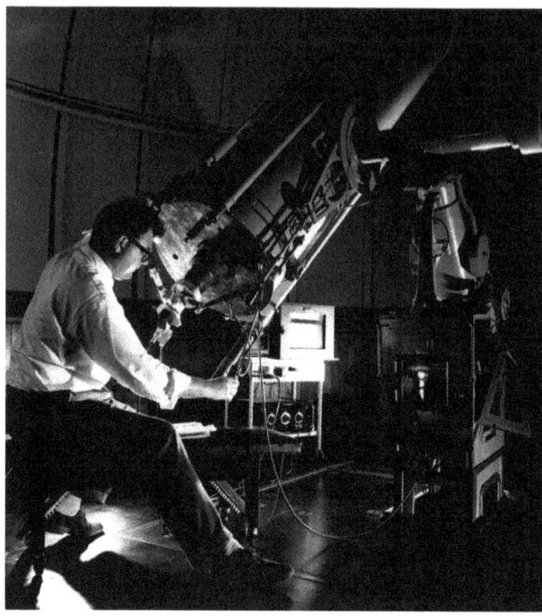

Figure 7.1: Olin Eggen obtained his PhD in astronomy on the light curves of variable stars at the University of Wisconsin-Madison. During the Second World War he served as a Scientific Liaison Officer to the Office of Strategic Services, the precursor of the CIA. He was President of the Astronomical Society of Australia from 1971 to 1972. Photo: Eggen Archives, Steenbock Library, University of Wisconsin-Madison.

the 1960s, Australia in the 1970s, and Chile in the 1980s'.[3] It was at Greenwich that he acquired a taste for things British, such as tweed jackets and Austin Healey sports cars. For some reason he fails to mention in his Memoir that he had a second stint at Greenwich Observatory in the early 1960s. Although he had been friends with Woolley, who had appointed him as the Chief Assistant to the Astronomer Royal at the Royal Greenwhich Observatory (RGO), he was dismissed by Woolley over a row about promotion.[4]

Although Eggen worked at several places it was Australia, he wrote, that 'retains its allure'. Even after he left Australia he spent some time every year at Mount Stromlo Observatory. It became his annual pilgrimage. Astronomy provided a 'very pleasant life ... a life on the dome floor, in the dark'.[3]

The arrival of Eggen as Director of Mount Stromlo Observatory had different effects on the astronomers working there. Some thrived but on others, such as Gascoigne, there was a negative impact. When Eggen came in the 1950s for short observing visits he had worked very closely with Gascoigne and together they had published significant papers. But when he became Director it was quite different. Gascoigne, said, 'Those were halcyon days with Olin when we were thick as thieves, you can see how shocked I was by this turn-about when he came to Mount Stromlo as Director ten years later.'[5]

Gascoigne had been appointed Acting Director, taking over from Arthur Hogg who had died suddenly the night the Boks left the Observatory. According to Gascoigne, 'Eggen arrived in Stromlo in July 1966 to take up the Directorship. From the day he came he made it clear I had no further part in

Planetgate: the missing files

Figure 7.2: The Neptune files, discovered among Eggen's papers after his death. He had been working on a history of Sir George Airy, the Astronomer Royal. Photo: Royal Greenwich Observatory.

Eggen borrowed a shelf-load of historical material from the RGO, including Airy's papers and the Neptune files which contained critical information about the actual discoverer of Neptune.[6] It was a case of 'Planetgate'. When questioned, the librarians said that the papers were 'unavailable' rather than 'missing'. In fact, they were missing for almost 30 years. The whereabouts of these papers 'constituted a mystery … How could documents related to one of the most glorious and important events of astronomy just go missing?'[7] Eggen denied having taken the files but after his death they were found at La Serena Observatory at Cerro Tololo in Chile, among his papers.[6] Over 100 kg of RGO papers were crated by the Chilean authorities to the University of Cambridge library where the RGO archives are now housed. A critical examination of the Neptune files by Sheehan, Kollerstrom and Waff[7] revealed that it was Le Verrier, the French mathematician, and not John Couch Adams, the English mathematician, who discovered Neptune.

running the Observatory. I was given no information, saw no documents, attended no meetings, and was asked for no advice, not even in optical matters. From then on I had the effective status of a junior visiting fellow until I retired in 1980. What saved the day was that in April 1967 the AAT [Anglo-Australian Telescope] was given the go-ahead. The first step was to set up a four-man technical committee, two from each country, and I became a member. This was a half-time job, becoming full-time later on, and was not only a wonderful job but an escape and a lifeline ... The place was effectively run by Rodgers and Mathewson. To me, there was a degree of claustrophobia about the general atmosphere, especially later as the place became more isolated, in the wake of the ANU [Australian National University] stand over the AAT.'[5] Rodgers and Mathewson cut their administrative teeth during the Eggen years and later became directors of the Observatory.

A purpose in life: astronomy at the highest level

Eggen had one purpose in life – to do astronomy at the highest level. He spent very little time on administration, only what was essential to keep the Observatory ticking along. He often slept until 4pm, giving the university administration at most an hour of his time. Most of the administrative chores were carried out by Rodgers and Mathewson. Eggen was more concerned with astronomical research and relished the time he spent on the dome floor. In fact, he set the pace for the Observatory, publishing 99 papers in the 11 years he was Director. The average number of papers per year jumped from about 18 per year in Bok's period to over 40 per year under Eggen.

Eggen moved away from the program of building bigger and bigger telescopes to building the intellectual capital of the Observatory. He allowed a

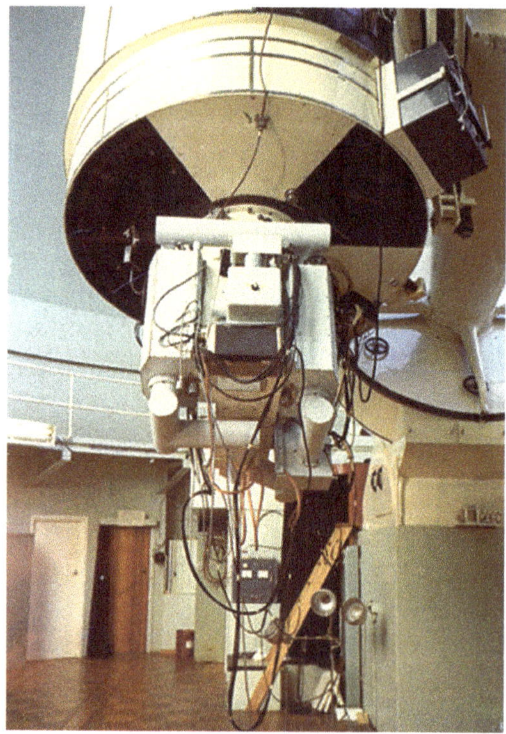

Figure 7.3: Alex Rodgers developed this 33-channel spectral scanner for the 1.9 m telescope. It allowed Stromlo astronomers to accurately measure the energy distributions of distant objects. Energy distributions give essential clues to the compositions and physical parameters of faint stars and galaxies. Photo: Mount Stromlo Archives.

thousand flowers to bloom in the building of new instrumentation, theoretical astrophysics and observational work. He gave the staff free rein to build or acquire new instruments which could open up new fields of research or enhance current areas. For example, an image-tube spectrograph for the 1.9 m was developed by Kent Ford of the California Institute of Technology, a new spectrograph for the same telescope was bought from Boller & Chivens and an echelle grating system for the Coudé spectrograph was developed by Mike Bessell and Harvey Butcher, a postgraduate student at that time. Other instruments included infrared photometers and a 33-channel spectral scanner. Eggen also beefed up the number of technical staff for running the workshop.

Figure 7.4: Don Faulkner was an expert in the theory and computer simulation of stellar evolution at Mount Stromlo. He is widely known for his work with Tom Frame on the history of the Observatory. He served as Acting Director in 1993 and was President of the Astronomical Society of Australia from 1989 to 1991. Photo: Mount Stromlo Archives.

He appointed several astronomers, such as Ken Freeman, Mike Bessell, Don Faulkner, Harry Hyland, John Norris, Agris Kalnajs, Mike Dopita, Bruce Peterson, Peter Wood and Visvanathan, who raised the profile of the Observatory with a series of innovative papers. Freeman in particular flourished during the Eggen years: 'My career got a great boost from working with Allan Sandage here in 1969. Sandage had come here on sabbatical with some particular observational programs in mind and I just by chance worked together on a couple of things ... the contact with Sandage was very good for my career. Sandage became a mentor and a patron and really helped my career greatly through his support, and I think I got tenure fairly swiftly after finishing my Queen Elizabeth Fellowship and I'm sure that Sandage's support was very significant.'[5]

Freeman became an internationally recognised expert on the dynamics of galaxies and a pioneer in demonstrating the importance of unseen matter in galaxies. In 1970, 'I was working on a paper which had to do with the gravitational field of spiral galaxy discs as we observe them. I just took the characteristic light distribution that you see in these galaxies and worked out what kind of rotation you would expect. I compared the expected rotation curves, with what one actually saw in a number of specific galaxies. In one galaxy in particular there was a marked problem in the radius at which the rotation curved turned over – one of the clear predictions of the analytic model is that if you have a certain light distribution you can say that at some radius this rotation curve will stop rising and start to fall – and that was not seen in this particular galaxy. I just made the comment in the discussion of this particular galaxy that if all the data is correct – and I think it was near enough to correct – that means there's got to be a lot of dark matter outside the visible

Dark matter

Studies of the rotation velocities of galaxies were carried out by several astronomers around the world. Freeman himself assigned credit elsewhere for the first unequivocal evidence for dark matter; he said the credit should 'probably go to Mort Roberts' at the US National Radio Astronomy Observatory. The HI data produced by radio telescopes, such as the synthesis array in Westerbork around 1978, that went out to much larger radii than the optical disk 'really clinched the dark matter story', said Freeman. van Albada's paper on the distribution of dark matter in the spiral galaxy NGC 3198 showed that 'the amount of dark matter inside the last point of the rotation curve at, 30 kpc, is at least 4 times larger than the amount of visible matter … The problem of dark matter surrounding spiral galaxies, made evident by the flatness of rotation curves, is one of the most enigmatic questions in present-day astrophysics'.[9]

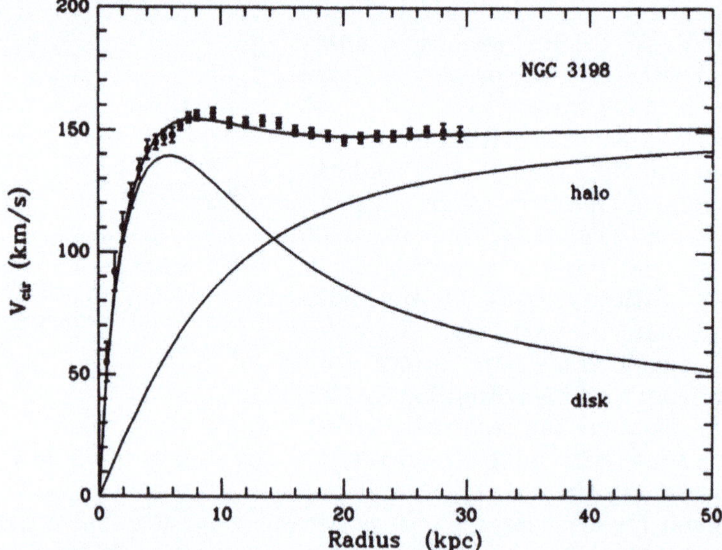

Figure 7.5: Rotation curve for the spiral galaxy NGC 3198. The measurements do not follow the expected change (the curve marked 'disk') of rotational velocity as one moves outward from the centre. Apparently there is a great deal of unseen matter in a 'halo' which causes the deviation. Freeman was among the first (in 1970) to point out that large amounts of halo matter would be required to explain such observations.[9] Image: © American Astronomical Society.

galaxy. I'd really forgotten its significance for dark matter until John Bahcall from the Institute of Advanced Studies in Princeton was writing about dark matter and actually gave me the credit as being the first person in print to say that this was actually a problem.'[5,8]

Galaxy formation

Eggen's work on the motions, ages and compositions of stars in the Milky Way became a research theme for a generation of Stromlo astronomers. He had made a name for himself in the astronomical world before he arrived at Mount Stromlo in 1966, among other reasons being well-known for his

Figure 7.6: Two spiral galaxies (NGC 2207 and IC 2163), each much like our own Milky Way, caught just prior to their merging into one single galaxy. Photo: NASA/ESA Hubble Space Telescope.

influential paper on the formation of the Milky Way Galaxy with Allan Sandage from the Carnegie Institution and Donald Lynden-Bell from Cambridge University.[10] 'This paper, written in 1962 with his colleagues Lynden-Bell and Sandage, is known as ELS and is one of the most influential astrophysics papers ever written'.[11] The authors had taken a top-down approach to the formation of the Galaxy. According to them, the Galaxy formed when a monolithic, rotating cloud of gas collapsed rapidly, over a few hundred million years. As the cloud collapsed inward the protogalaxy began to rotate more rapidly, creating the spiral arms we see today.

In his nightly observing runs, Eggen observed several thousands of stars and classified them into various groupings. He was able to compare their positions and orbits in the Milky Way Galaxy, with other properties such as their stellar ages and chemical compositions. The chemical composition of stars that form in the interstellar medium generally reflects the composition of the gas at that time and place. The chemical compositions of stars of different ages and positions therefore provide important information about the chemical evolution of the Galaxy. Apart from the hydrogen and helium that formed during the Big Bang, the heavy elements (carbon, oxygen, iron etc.) are the products of thermonuclear reactions that occur deep in the cores of stars. Some of these heavy elements are regurgitated into the interstellar medium during supernova explosions. During the long history of the Galaxy the heavy element abundance in the interstellar medium has been evolving. By amassing and analysing this wealth of information Eggen, Freeman and Rodgers[12] put together a picture of the dynamical and chemical evolution of our Galaxy.

One of the spin-offs of Eggen's work was the discovery that widely separated stars can have essentially similar orbits, ages and compositions. The so-called 'moving groups' were believed to have resulted from stars that formed at the same place and time in the Galaxy but that have been

Galaxy formation

For many years, the ELS model of galaxy formation enjoyed widespread acceptance. As with many scientific theories, however, over the years new data have caused revisions. For example, the time-scale of galaxy formation was likely longer than that deduced by Eggen and colleagues. More refined dynamic models show that the protogalaxy would not have collapsed as smoothly as predicted by the simple ELS model; instead the densest parts would have fallen inward much faster than the rarefied regions.[13] John Norris and colleagues at Mount Stromlo later suggested there was a significant decoupling between the formation of the halo and the disk.

Most astronomers today favour a bottom-up scenario for galaxy formation, whereby galaxies form by the aggregation of small clumps of matter over fairly longer periods of time. It is also known now that galaxies change throughout cosmic time, and there are several sources of evidence for an evolutionary scenario for galaxies. For example, in 1978 Harvey Butcher (later Director at Mount Stromlo) and Augustus Oemler Jr, both then at Kitt Peak National Observatory in Tucson, Arizona, found that dense clusters of galaxies located about 6 billion light-years away still contained a large number of galaxies actively forming stars, quite unlike nearby clusters. Evidently, galactic collisions and mergers were more common in the past, triggering bursts of star formation.

A full explanation of galaxy formation and evolution must include the effects of dark matter, which accounts for 90% of the mass of a galaxy. Astronomers have no idea what it is. It does not show up in photographs nor in images taken in any part of the electromagnetic spectrum. Its presence is known only through its gravitational influence on the orbits of stars and gas clouds.

Figure 7.7: Eggen, Lynden-Bell and Sandage. This photo was taken several years after their famous paper on galaxy formation. Photo: John Norris.

Figure 7.8: An all-sky view of the Milky Way Galaxy in the near infrared, which allows us to see through the obscuring dust clouds that affect visible light images. The image demonstrates clearly that the Milky Way is a disk galaxy and that we are located in that disk some way from the centre. Image: Atlas Image mosaic courtesy of 2MASS/UMass/IPAC-Caltech/NASA/NSF.

Figure 7.9: The colliding galaxies NGC 4038 and NGC 4039. Galactic collisions often trigger massive bursts of star formation and lead to the appearance of a new, single galaxy. Photo: NASA/ESA Hubble Space Telescope.

dispersed. Eggen identified some moving groups in the halo of the Milky Way. These halo streams are now recognised as the products of the tidal debris of small galaxies that have been swallowed up by the Milky Way. Since this discovery, several examples of galactic mergers with the Milky Way have been found.

Abundance analyses, Magellanic Clouds and quasars

The great interest shown in the heavy element enrichment processes that took place in the formation and evolution of galaxies led to the active involvement of Bessell, Freeman, Norris and Rodgers in this area of research. Abundance analysis became a core activity during the Eggen years, and continues to the present.

It also was a focus of the PhD theses written at that time. In particular, both Harvey Butcher[14] and Jeremy Mould,[15] future directors of the Observatory, wrote their theses on abundance analysis. Butcher had come to Mount Stromlo from the US since 'I discovered that there were no lecture courses at Mount Stromlo! I was completely fed up sitting in lecture courses and taking exams. And the idea of having four years to do a research project was just unbelievably attractive … Allan Sandage phoned up Olin Eggen and said, "Why don't you take this guy?" And so they did, and I came, and never regretted it.'[16]

Butcher arrived with a thesis in mind: 'I knew before I came to Australia what I wanted to do … The title of my dissertation was *Observational Studies of Nucleosynthesis* which allowed me to study chemical elements in stars of different ages and try and bring those observations into the conceptual framework of stellar evolution and galactic evolution – a chemical evolution of the galaxy.' His PhD supervisors were Bessell and Rodgers.

Don Mathewson had joined Mount Stromlo Observatory in 1965 from CSIRO's Radiophysics Laboratory and was Director of the Observatory after Eggen. He had a radio astronomy background and was one of the first astronomers to cross over from the radio astronomy side to the optical side of the divide in Australian astronomy in the 1960s. During the Eggen years Mathewson and Vince Ford, his Research Officer, used the 60 cm telescope at Siding Spring to study the polarisation of 1800 stars within 5000 light-years of the Sun.[17] By combining their observations with those of other astronomers, they were able to produce a map of stellar polarisation over the whole sky, which showed how the magnetic field of the Milky Way is distributed. The combined observations provided 'conclusive evidence that there is a helical component of the local spiral-arm magnetic field and the well-known "spur" features of the galactic radio emission are "radio" traces of the helical structure'.[18] They extended their work to the Magellanic Clouds, where they found similar evidence for a large-scale, regular magnetic field.[19]

Massive stars end their lives not with a whimper but in a catastrophic explosion, observed as a supernova, a star that for a short time is almost as bright as a whole galaxy. As they explode they eject vast shells of material at supersonic speeds, creating a shock wave which ploughs through the interstellar gas and dust. Its collisions with the interstellar medium heat the gas to very high temperatures and makes it glow. The glowing material marks a

Figure 7.10: Isaac Newton's tomb at Westminster Abbey. Photo: Klaus-Dieter Keller, Germany.

supernova remnant. Supernova remnants came under the scrutiny of Mathewson, who in the 1960s searched for supernova remnants with the Parkes Radio Telescope. He teamed up with Jim Clarke of the University of Sydney to carry out a radio-optical search for supernova remnants. They used the Mills Cross at Molonglo and the 60 cm optical telescope at Siding Spring for their observations. They discovered two remnants in the Small Magellanic Cloud and nine in the Large Cloud.[20,21]

In 1975 Michael Dopita joined the Observatory. He had begun his career in astronomy in the UK when he and a group of young men were interviewed to study for their PhDs at Manchester University by Professor Zdenek Kopal at the tomb of Isaac Newton at Westminster Abbey. It was a fitting start to a brilliant career in astronomy. According to Dopita, 'when I first came to the ANU Don Mathewson had obtained spectroscopic observations of supernova remnants in the Large Magellanic Cloud, which he had identified a few years

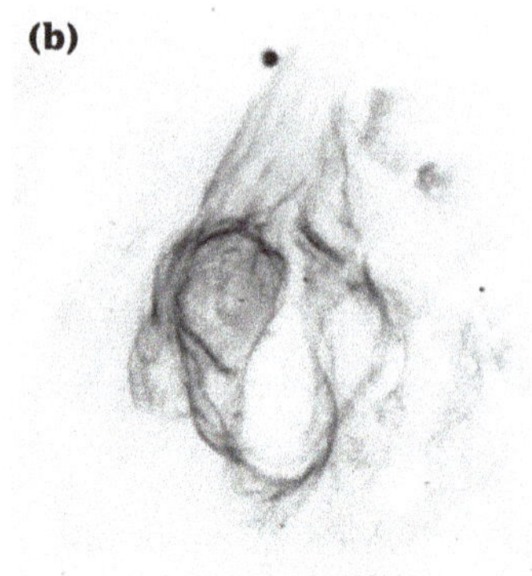

Figure 7.11: (a) Lionel Murphy and his rather large nose made astronomical history, when the N86 supernova remnant was named after him. Photo: ANU Archives. (b) The Lionel Murphy Nebula in supernova remnant N86 in the Large Magellanic Cloud. Photo: © American Astronomical Society.[44]

earlier. I realized that there was no tool to analyse the spectra of these things and so at the time I started to develop a shock modelling code'.[22]

With this analytical tool he was able to show 'that you can use supernova remnants to look at the abundances of various chemical elements in the interstellar medium because in most cases you don't actually see bits of the star which are being blown up, you see the effects of the shocks that the explosion of the star produced in the interstellar medium'.[22,23,24] The supernova remnant N86 in the Large Magellanic Cloud was studied by Dopita, Mathewson and Ford[25] and the resulting paper made quite a splash on the front page of the *Canberra Times*. They had named N86 the Lionel Murphy supernova because of the perceived resemblance of the supernova to Murphy's rather large nose, as depicted in a caricature by the cartoonist Pickering.[1] Murphy, then a Justice of the High Court, was so delighted by this honour and the cartoon that he had it framed and hung in his chambers in place of the Queen's portrait.

At about the same time that Dopita was doing the above analysis he also became interested in the outflows from young stars. He said, 'At this time the so-called Herbig-Haro Objects were thought to be reflection nebulae of some strange kind. But I was able to show that these were actually shocks produced by energetic outflows from young stars.'[22,26] Dopita was the first to show that collimated jets of material can be emitted from stars. It became a very active area of research.

Lionel Murphy

Lionel Murphy had been a member of the Labor Party from a very young age and was appointed the Attorney-General in 1972 by the Whitlam government. His most significant legislative achievement was the *Family Law Act 1975,* which was vigorously opposed by the Roman Catholic Church.[27] It led to the establishment of the Family Law Court of Australia. He was possibly the only Justice of the High Court with degrees in science and law.

Figure 7.12: A Haro-Herbig object lying at the tip of a dust pillar in the Carina Nebula showing opposing outflow jets of material. Photo: NASA/ESA Hubble Space Telescope.

Dopita's work on shock waves led naturally to another field that was opening up and providing information on the early universe: 'I was interested generally in shocks and that led me then to the study of active galactic nuclei.'[22] Essentially all galaxies have been found to harbour massive black holes in their nuclei. These objects also emit collimated jets of material that cause shocks in the surrounding gas, which provides clues to the nature and evolution of black holes.

Gascoigne continued his studies of the Magellanic Clouds. His investigations of the cepheid variable stars in the Clouds allowed him to establish them as good indicators for finding the distance scale of the universe. According to Gascoigne, cepheids make powerful indicators 'with luminosities and intrinsic colours which can be predicted within close limits from a knowledge of the periods and amplitudes only'.[28] He also showed that as distance indicators cepheids are rather sensitive to metal abundance.[29]

Bob Stobie from the Institute of Theoretical Astronomy at Cambridge University spent several years at Mount Stromlo. With Rodgers, he developed numerical techniques to compute the pulsation properties of variable stars. This provided a method of comparing the pulsation properties of the theoretical models with observations.[30,31,32,33] There were initial teething problems

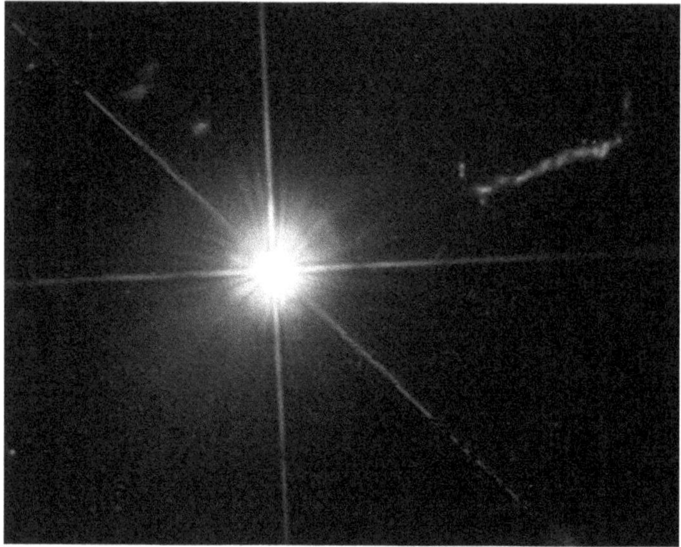

Figure 7.13: The quasar 3C 273, showing its jet (at about two o'clock). The spikes emanating from the central star-like object are artefacts generated by the Hubble Space Telescope's optics. Other faint smudges in the image are very distant galaxies. Photo: NASA, ESA Hubble Space Telescope.

with the models and observations but these were resolved over the years with better observational data.

Agris Kalnajs joined the Observatory in 1973 and strengthened the computational work that was in progress. He carried out several theoretical studies to explore the formation of spirals and bars in flat galaxies.[34,35] An important result was that spiral galaxy rotation curves based only on the optically visible stars and gas cannot reliably reveal the presence of dark matter; the extended rotation curves from radio observations are required to do so.

The discovery of radio sources in the 1950s created the need among astronomers to find optical counterparts for the new objects. Astronomers were baffled by the 1960s discovery of star-like objects at the locations of radio sources 3C 48 and 3C 273 (objects 48 and 273 in the Third Cambridge Catalogue). They were extremely luminous and had thousands of times the radiation output of an ordinary galaxy. Called quasi-stellar radio sources, later quasi-stellar objects (QSOs) or simply quasars, they had very large red shifts and were deemed to be extragalactic in nature.[36,37] John Bolton, then Director of the Parkes Radio Telescope, had a great interest in QSOs and teamed up with Bruce Peterson at Mount Stromlo to make observations of the newly discovered objects. They used the 1.9 m telescope and over a three-year period observed 129 QSOs. A large number of these were new identifications.[38,39,40,41] The quasar program was helped by the arrival of Kent Ford, an expert in image intensifiers. He enhanced the spectroscopic potential of the telescopes and kept the Observatory at the forefront of technological developments in the area of electronic area detectors. Recent studies of quasars show that their tremendous rate of energy production is due to the in-falling of matter into super-massive black holes in the centres of galaxies.

Control of the Anglo-Australian Telescope

Although Bok had many good friends in the House of Representatives and had campaigned vigorously for the construction of the Anglo-Australian Telescope, it was not until September 1969 that a formal agreement was signed between the Australian and British governments. A Joint Policy Committee was established to oversee the project; it consisted of one government member and two scientific members from each nation. Eggen was a member of the Committee by virtue of being Director of Mount Stromlo Observatory. The other members from the Australian side were Ken Jones from the Department of Education and Science and Taffy Bowen, the Chief of CSIRO's Radiophysics Laboratory. The British side was represented by Jim Hosie from the Science Research Council, Herman Bondi from King's College London and Richard Woolley, the Astronomer Royal who had previously been Director of Mount Stromlo Observatory. After the first meeting Bondi was replaced by Fred Hoyle from Cambridge University. Having been sidelined by Eggen, Gascoigne was glad to be a member of the Technical Committee that was set up to oversee the initial development of the project. The other members were Herman Wehner from Mount Stromlo's Technical Workshop and John Pope and Roderick Redman from Britain. A Project Office was later set up to manage the project. Gascoigne and Redman became astronomer-advisers. Gascoigne also served later as the commissioning astronomer for the telescope. A committee chaired by Rodgers provided input on technical matters from the astronomers. Despite some teething problems, the technical side of constructing the telescope and the observatory went ahead fairly smoothly.

Decision-making on scientific control of the telescope and observatory raised a storm, however, and soured relationships between the astronomers. The men who spent hours on the dome floor blissfully observing the grandeur of the universe were at war. Gascoigne wrote, 'At first sight, it may be thought that the problems of management and operations of the Anglo-Australian Telescope would be largely of a technical and logistic nature ... However, the JPC and the AAT [Anglo-Australian Telescope] Board were to find that deciding how to manage the operations would lead to long, drawn-out and at times acrimonious arguments with the upper echelons of the Australian National University before the matter was resolved.'[42] The tussle to control management of the telescope and the observatory continued for almost five years, with several heated meetings.

The wording in the *Anglo-Australian Telescope Agreement Act 1970* was partly to blame for this long controversy. It was open to different interpretations, which were exploited by the various proponents to their advantage. The *Who's Who* of Australian and British politics, astronomy and academia were involved in the rather long and acrimonious dispute. It began with Eggen wanting to control and manage the AAT and the observatory. His view was that the best scientists, not government bureaucrats, should control the telescope's program and that operational control should be local and in

Australia. Especially as the land on which the telescope was sited was owned by the Australian National University.

Eggen teamed up with the ANU Vice-Chancellor, Sir John Crawford, to fight for control of the telescope and the observatory. His gruff personality and desire for scientific control came into play. According to Dopita, 'He wanted it to be run by Mount Stromlo and none of the rest of the community wanted Mount Stromlo to just absorb yet another big facility. So it was a big battle as to how it would be run.'[22] On the British side, Margaret Burbidge, who had succeeded Woolley as Director of Royal Greenwich Observatory and was a member of the AAT Board, strongly favoured an independent management body to control and run the AAT and the observatory. At a most acrimonious meeting held by the Australian and the British parties to the Agreement in April 1972 at La Jolla, a resolution was passed which laid the foundation on which the AAT was to be run and managed. It stated that 'In order to discharge its responsibilities for the control, operation and maintenance of the telescope, the Board considers that it should appoint a Scientific Director and appropriate support staff, both scientific and technical, employed by and answerable only to the Board.'[42] The motion was carried by four votes to two. Burbidge played a key role in getting this resolution passed.

Despite this resolution Eggen carried on the fight. Even Margaret Thatcher, then British Secretary of State for Education, on a visit to Australia in 1972 to meet her counterpart Malcolm Fraser, took the view that total control of the AAT should be vested in the AAT Board. But a change in government brought in Whitlam's Labor Party, which Eggen and Crawford saw as an opportunity to convince the new masters.

Hanbury Brown from the University of Sydney was well-known in international astronomical circles for his highly innovative optical intensity interferometer, and joined the fray. He wrote to Bill Morrison, the Minister for Science in the Whitlam government: 'He attributed much of the trouble with the ANU to the fact that Australia lacked a satisfactory policy for the provision of large-scale scientific instruments available to the whole community and that the ANU had no claim on the AAT since Mount Stromlo neither spoke as a national facility nor aimed to become one ... Eggen was the only university representative on the AAT Board, yet he had not consulted Australian universities on any matters concerning the telescope.'[42] This was the turning-point in the debate. Morrison was finally convinced by senior members of the astronomical community to accept their recommendation for an independent AAT. The Australian government accepted that position in June

Figure 7.14: Hanbury Brown, a key player in the AAT debate. Photo: Australian Academy of Science.

Figure 7.15: Herman Wehner, a member of the Technical Committee set up for the development of the AAT, washing its 3.9 m mirror. Photo: Wehner Collection. Mount Stromlo Archives.

Figure 7.16: The Anglo-Australian Telescope was, both technically and scientifically, one of the most successful telescopes of its generation. Photo: AAO.

1973 after almost five years of wrangling among the astronomers. In protest, Eggen resigned his membership of the AAT Board in August 1973.

The AAT had a pointing accuracy of better than 2 arcseconds, which was remarkable for the 1970s when 1 arcminute was the norm. Astronomers began using it by June 1975.

According to Virginia Trimble[43] from the University of California, Irvine, 'The Anglo-Australian Telescope (AAT) is an outstanding performer, with the third highest total number of citations per paper of any optical telescope in the world regardless of size'. Apart from the observational work in astronomy, the Anglo-Australian Observatory has built astronomical instruments for other organisations.

Having lost the battle for control of the AAT and having soured relationships among members of the Australian astronomical community, Eggen relinquished his directorship on 30 September 1977. Gascoigne has the last word on Eggen's directorship: 'What I think he will be remembered for is the controversy over the AAT and the great rift that developed between Stromlo and the rest of Australian astronomy ... it was years before Stromlo was accepted back into the fold'.[5]

References

1. Frame T, Faulkner D 2003. *Stromlo: An Australian Observatory.* Allen and Unwin, Sydney.
2. Woolley R 1968. Mount Stromlo Observatory *Records of the Australian Academy of Science* **1**(3), 53–57. doi:10.1038/scientificamerican1204-92.
3. Eggen EJ 1993. Notes from a life in the dark. *Annual Review of Astronomy and Astrophysics* **31**, 1–12. doi:10.1146/annurev.aa.31.090193.000245.
4. Kollerstrom N. Neptune's discover: the British case for co-prediction. http://www.dioi.org/kn/nept.
5. Bhathal R 1996. *Australian Astronomers: Achievements at the Frontiers of Astronomy.* National Library of Australia, Canberra.
6. McKie R 2004. Revealed: how Britain put the spin on Neptune. *The Observer*, Sunday 12 December.
7. Sheehan W, Kollerstrom N, Waff CB 2004. The case of the pilfered planet. *Scientific American* **6**, 92–99. doi:10.1038/scientificamerican1204-92.
8. Freeman KC 1970. On the disks of spiral and S0 galaxies. *Astrophysical Journal* **160**, 811–830. doi:10.1086/150474.
9. van Albada TS,, Bahcall JN, Begeman K, Sancisi R 1985. Distribution of dark matter in the spiral galaxy NGC 3198. *Astrophysical Journal* **295**, 305–313. doi:10.1086/163375.
10. Eggen OJ, Lynden-Bell D, Sandage AR 1962. Evidence from the motions of old stars that the Galaxy collapsed. *Astrophysical Journal* **136**, 748–766. doi:10.1086/147433.
11. Freeman K 2000. Olin J Eggen 1919–1998. *Astronomy & Geophysics* **1**, 36.
12. Eggen OJ, Freeman KC, Rodgers AW 1973. Structure of the galaxy. *Reports on Progress in Physics* **36**, 625–694. doi:10.1088/0034-4885/36/6/001.
13. Van den Bergh S, Hesser JE 1999. How the Milky Way formed. *Scientific American* **268**, 72–78.

14. Butcher HR 1974. Observational aspects of nucleosynthesis. PhD thesis. Australian National University.
15. Mould J 1976. The lower main sequence and atmospheres of M dwarfs. PhD thesis. Australian National University.
16. Bhathal R 2011. Interview with Harvey Butcher for the National Project on Significant Australian Astronomers. National Library of Australia, Canberra.
17. Mathewson DS, Ford VL 1970. Polarization observations of 1800 stars. *Memoirs of the Royal Astronomical Society* **74**, 139–182.
18. Mathewson DS 1968. The local galactic magnetic field and the nature of radio spurs. *Astrophysical Journal* **153**, L47–L53. doi:10.1086/180217.
19. Mathewson DS, Ford VL 1970. The magnetic field structure of the Magellanic Clouds. *Astrophysical Journal* **160**, L43–L46. doi:10.1086/180520.
20. Mathewson DS, Clark JN 1972. A supernova remnant in the Small Magellanic Cloud. *Astrophysical Journal* **178**, L105–L107. doi:10.1086/181095.
21. Mathewson DS, Clark JN 1973. Supernova remnants in the Large Magellanic Cloud. *Astrophysical Journal* **180**, 725–738. doi:10.1086/152002.
22. Dopita M 2004. Interview with Ragbir Bhathal for the National Project on Significant Australian Astronomers. National Library of Australia, Canberra.
23. Dopita MA, Mathewson DS, Ford VL 1977. Optical emission from shock waves II: abundances in supernova remnants. *Astrophysical Journal* **214**, 179–188. doi:10.1086/155242.
24. Dopita MA 1977. Optical emission from shock waves II: diagnostic diagrams. *Astrophysical Journal* **Supplement 33**, 437–449.
25. Dopita MA, Mathewson DS, Ford VL 1977. Optical emission from shock waves III: abundances in supernova remnants. *Astrophysical Journal* **214**, 179–188. doi:10.1086/155242.
26. Dopita MA 1978. Optical emission from shocks IV: the Herbig-Haro objects. *Astrophysical Journal* **Supplement 37**, 117–144.
27. Hocking J 1997. *Lionel Murphy: A Political Biography.* Cambridge University Press, Melbourne.
28. Gascoigne SCB 1972. The Magellanic Clouds and the distance scale. *Quarterly Journal of the Royal Astronomical Society* **13**, 274–281.
29. Gascoigne SCB 1974. Metal abundance and the luminosities of Cepheids. *Monthly Notices of the Royal Astronomical Society* **166**, 25P–27P.
30. Stobie RS 1969. Cepheid pulsation I. Numerical techniques and test calculations. *Monthly Notices of the Royal Astronomical Society* **144**, 461–484.
31. Stobie RS 1969. Cepheid pulsation II. Models fitted to evolutionary tracks. *Monthly Notices of the Royal Astronomical Society* **144**, 485–510.
32. Stobie RS 1969. Cepheid pulsation III. Models fitted to a new mass–luminosity relation. *Monthly Notices of the Royal Astronomical Society* **144**, 511–535.
33. Rodgers AW 1970. The masses of pulsating stars. *Monthly Notices of the Royal Astronomical Society* **151**, 133–140.
34. Kalnajs AJ 1976. Dynamics of flat galaxies III. Equilibrium models. *Astrophysical Journal* **205**, 751–761. doi:10.1086/154331.

35. Kalnajs AJ 1977. Dynamics of flat galaxies IV. The integral equation for normal modes in matrix form. *Astrophysical Journal* **212**, 637–644. doi:10.1086/155086.
36. Hazard C, Mackey MB, Shimmins AJ 1963. Investigation of the radio source 3C 273 by the method of lunar occultations. *Nature* **197**, 1037–1039. doi:10.1038/1971037a0.
37. Schmidt M 1963. A star-like object with large red-shift. *Nature* **197**, 1040. doi:10.1038/1971040a0.
38. Shimmins A, Bolton JG, Peterson BA, Wall JV 1971. Identification of southern quasi-stellar objects. *Astrophysical Letters* **8**, 139–143.
39. Peterson BA, Bolton JG 1972. Identification of southern quasi-stellar objects II. *Astrophysical Letters* **10**, 105–108.
40. Peterson BA, Bolton JG 1973. Identification of southern quasi-stellar objects III. *Astrophysical Letters* **13**, 187–192.
41. Peterson BA, Bolton JG, Shimmins AJ 1973. Identification of southern quasi-stellar objects IV. *Astrophysical Letters* **15**, 109–113.
42. Gascoigne SCB, Proust KM, Robins MO 1990. *The Creation of the Anglo-Australian Observatory.* Cambridge University Press, Cambridge.
43. Trimble V, Zaich P, Bosler T 2005. Productivity and impact of optical telescopes. *Publications of the Astronomical Society of the Pacific* **117**(827), 111–118. doi:10.1086/427460.
44. Williams RM, Chu Y-H, Dickel JR, Smith RC, Milne DK, Winkler PF 1999. Supernova remnants in the Magellanic Clouds. II. Supernova remnant breakouts from N11L and N86. *Astrophysical Journal* **514**, 798.

1977–1986

8

An astronomical entrepreneur

Don was the first Australian director – very Australian, energetic, approachable, decisive, a great bustler-around ... Don was largely responsible for taking Australia into the space age.[1]

Mathewson was, I think, in the mould of benign dictators but he certainly also had a lot of connections to Australian astronomy. He also built up for the first time I think credible theoretical capabilities at Stromlo and he was very much trying to drag Australia kicking and screaming into the space age.[49]

Don Mathewson was appointed the fifth Director of Mount Stromlo Observatory in April 1979, having served as Acting Director from September 1977 when Olin Eggen left the institution.

Almost 55 years since the appointment of Duffield as the first Director of the Observatory, an Australian-born astronomer was finally 'the best man available to do the job'. Although the Australian National University authorities were slow in appointing Mathewson as Director rather than Acting Director in the first instance they need not have worried – he lived up to the expectations of a great Director of the Observatory.

He was the right man and he came at the right time to carry the astronomical torch that Duffield had lit in the 1920s. He was well connected to the Australian astronomical community and he mended the bridges that Eggen had broken between Mount Stromlo Observatory and the rest of the astronomical community.

Mathewson began his astronomical career at the CSIRO Radiophysics Division being interviewed by Joseph Pawsey, one of the pioneers of radio astronomy in Australia. Mathewson said he 'had to eat an enormous bag of lamingtons and bananas with Pawsey as he quizzed me on different scientific problems. At the end of the day, Pawsey turned to me and said, well if you want the job you've got the job'.[1]

He joined Chris Christiansen at Fleurs in western Sydney in 1955 and assisted him with building the famous Chris Cross. Three years later he left to do his PhD at Manchester University which had the huge (250 ft, 76 m) Jodrell Bank Radio Telescope run by Bernard Lovell, one of the movers and shakers of British radio astronomy. There Mathewson met Hanbury Brown,

Figure 8.1: Don Mathewson in 1989 with some of his students. Standing, left to right: Emmanuel Vassiliadis, Vince Ford, Markus Buchhorn and Carl Grillmair. Seated, left to right: Angela Samuel, Don Mathewson and Stuart Ryder. Photo: Mount Stromlo Archives.

who later came to Australia to head the Chatterton Astronomy Department at the University of Sydney. Hanbury Brown also constructed a stellar intensity interferometer that could measure the diameters of 32 main sequence stars for the first time.

It was an exciting time for Mathewson. Hanbury Brown became Mathewson's PhD supervisor for his thesis on 'High-resolution studies of the Cygnus Loop, Coma Cluster and Andromeda Nebula', which Mathewson then used as

Figure 8.2: Mathewson (left) with Chris Christiansen, a pioneer of radio astronomy in Australia. Christiansen was one of the first scientists to develop and use the Earth rotation synthesis technique in astronomical studies which embodied most of the essential features of present-day large synthesis telescopes. Christiansen and Pawsey were Mathewson's astronomical heroes. Photo: Mount Stromlo Archives.

Jodrell Bank Radio Telescope

The Jodrell Bank Radio Telescope had an interesting history. According to Hanbury Brown, the antenna had been 'built to detect cosmic rays by radar and unfortunately Patrick Blackett [from the University of London and a 1948 Nobel prize winner for his work on cosmic rays] and he [Lovell] had made a mistake in the calculations.

So they had this instrument which they'd built and had nothing to do with it. Bernard and I decided it was an ideal instrument for looking at the sky and checking the work done by a man called Grote Reber who had published a paper on cosmic static'.[1] The antenna began its life as a radio astronomy telescope instead.

the basis for a couple of significant papers.[2,3] According to Mathewson, 'I don't think I ever spoke to Hanbury Brown when I handed him a copy of my thesis. That doesn't mean to say he wasn't interested. He really sort of left it to me.'[1] It was a good training ground for Mathewson, becoming an independent thinker and researcher quite early in his career.

He subsequently spent some time as a Visiting Professor at Leiden University in the Netherlands where he met Jan Oort, an early pioneer in radio astronomy and a prolific publisher of astronomical papers.[4] While in the Netherlands he gained valuable experience on the Westerbork Synthesis Radio Telescope which consisted of 12 smaller radio telescopes working together to carry out radio interferometry observations. In particular, he used the telescope to carry out a study of M51 (the Whirpool Nebula). The

Figure 8.3: M51, or the Whirlpool Galaxy, is a classic example of a spiral galaxy. Photo: Philip Parkins, ppac9@astrocruise.com.

result verified the Density Wave Theory that gas was compressed at the edge of the spiral arms. Mathewson noted that 'The coincidence of the radio arms with the inside edges of the optical arms is first-class observational evidence for the existence of the gas compression regions predicted by W.W. Roberts to occur on the basis of his non-linear analysis of the density-wave theory'.[5] The map that was published by Mathewson, Van der Kruit and Brouw was used whenever articles were written about Westerbork. The experience Mathewson gained on the Westerbork Telescope would come in handy later on when he proposed ideas about building a radio interferometer in Australia.

Upon becoming Director of the Observatory Mathewson continued building the intellectual capital of the Observatory. During the first year of his directorship 43 papers were published. This rose to 92 per year by the time he completed his term of office. There was also a dramatic increase in PhD students, from 12 when he took over from Eggen to over 30 within a few years. The main lines of research were on stellar astronomy with a focus on the evolution of stars, star clustering, quasars and the determination of the Hubble Constant which remained a vexed question for quite a long time. Theoretical studies were also given a boost during his time as Director, with Mike Dopita, Peter Wood, Peter Quinn, Geoff Bicknell and Bob Gingold taking the lead.

Discoveries and observations

One of Mathewson's major discoveries was the Magellanic Stream. His interest began when he read a paper on a filament that had been found by astronomers in the northern hemisphere: 'I read a paper by Wannier and Wrixon[6] on a filament that they had found in the northern hemisphere. It had a sinusoidal progression, of radial velocity along the length of the filament. I was doodling away on the paper and I just extrapolated the filament with a pencil and I thought to myself, that must go very close to the Magellanic Cloud and it did.' He 'got very excited', he said. 'One of those moments, you know, when your hair bristles and you feel great.'[1]

Because the filament was very strong in neutral hydrogen Mathewson needed a radio telescope to observe it, so he phoned Brian Robinson who was the Director of Research at Radiophysics at the CSIRO: 'Look Brian, can I have a couple of hours on the Parkes telescope? He said, well, look, we're trying out the new correlator. If the engineers don't mind, you can have a couple of hours.'[1] Those couple of hours

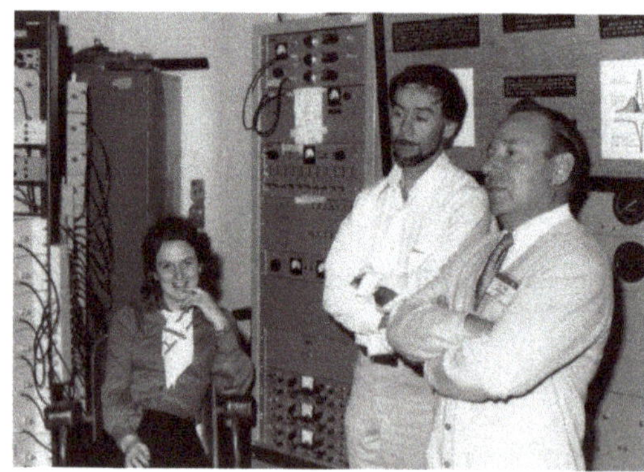

Figure 8.4: Don Mathewson and Martha Cleary watching the Magellanic Stream, using the 18 m dish. Station Manager George Day (at right) is watching the progress. Photo: CSIRO.

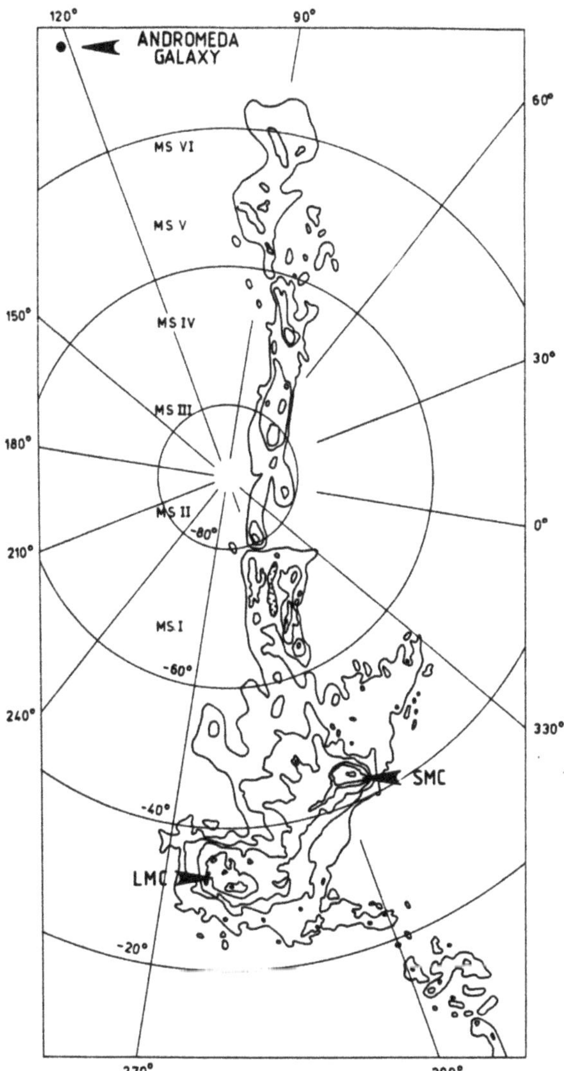

Figure 8.5: The distribution of gaseous hydrogen in the Magellanic Stream. The extent of the optical portions of the Large and Small Magellanic Clouds is shown. Image: Mathewson.[13]

netted the great discovery of Mathewson's life: 'I pointed the telescope to twenty spots along the extrapolated line and at each point to that line it had a big neutral hydrogen signal and at a velocity that was predicted by the sinusoidal run of velocity. I knew then that I'd discovered something really huge, a huge arc of gas 120 degrees wide, going across the sky from the Magellanic Cloud.'[1] He named the filament the Magellanic Stream.[7,8] It formed part of a great circle around the disk of the Milky Way Galaxy.

Having discovered the Magellanic Stream, the next big problem was to explain what caused it. In the next few years several theories, such as tidal interactions between galaxies and ram pressure that pushes the gas out of

Figure 8.6: A composite image of the Magellanic Stream (in pink) produced from observations made by several radio telescopes combined with a visible-light mosaic of the Milky Way from Alex Mellinger. The new observations show that the Magellanic Stream extends an amazing 200° in the southern sky, 400 times the diameter of the full moon. Image: David Nidever *et al.*, NRAO/AUI/NSF and A. Mellinger, LAB survey, Parkes Observatory, Westerbork Observatory and Arecibo Observatory. From Mathewson.[13]

the Magellanic Clouds, were put forward.[9,10] Mathewson favoured the ram pressure model. However, no satisfactory explanation was forthcoming and the debate continued with no resolution into the 1990s.

Reviewing some of the latest literature on his Magellanic Stream, Mathewson said, 'An impasse has been reached after 37 years of hard work by many researchers, and we are now no closer to reaching an understanding of the origin of the Magellanic Stream when I was in the control room of the 210-ft Parkes Radio Telescope pumping air some 40 years ago! Perhaps there is something very fundamental about our Galaxy that we don't know?'[13]

Origin of the Magellanic Stream

One plausible explanation was provided by a new neutral hydrogen survey of the southern sky undertaken by an international team. It was one of the biggest HI surveys undertaken by radio astronomers and used new multibeam equipment fitted on top of the Parkes Radio Telescope.[11] The survey was led by Lister Stavely-Smith at the Australia Telescope National Facility (now at the University of Western Australia) and Rachel Webster from the University of Melbourne. According to Mary Putman, Ken Freeman's PhD student, and her co-astronomers their data revealed that 'a new stream of gas lies in the opposite direction to the trailing Magellanic Stream and leads the motion of the Clouds. The existence of both leading and trailing streams supports a gravitational interaction whereby the streams are torn from the bodies of the Magellanic Clouds by tidal forces'.[12] It appeared that ram pressure was not responsible for the observed phenomena, as suggested by Mathewson.

His deep knowledge of the observational art with both radio and optical telescopes allowed Mathewson to discover that the Small Magellanic Cloud was split by the close passage of the Large Magellanic Cloud. This became quite a controversial issue in astronomical circles. Observations with the Parkes Radio Telescope had previously shown that the Small Magellanic Cloud was not a simple galaxy but showed complex behaviour.[14] The simple dynamic models that were developed by several astronomers to explain the observations failed to provide a good explanation of what was going on in the Magellanic Clouds. According to Matthewson, 'To me it seemed as if it (the Small Magellanic Cloud) was orbiting around the Large Cloud and the tidal interaction had really split it into two'.[1] To show this he, Ford and Visvanathan[15,16] used the 2.3 m optical telescope at Siding Spring Mountain to make observations of several hundred cepheids. The cepheids fell into two lots – a fairly nearby lot and the other more distant at ~20 kiloparsecs. It was controversial 'because, first of all people don't like being told that a galaxy was split into two and it would never reform and it was unstable. It was separating at beyond the escape velocity. So really I was saying that the galaxy was going to disintegrate. Also, the measurement of distant cepheids is always very controversial. As you well know, any distance measurement in the universe is controversial. So people then disagreed with that. But I think they are now agreeing with the fact that it is split into two'.[1] So, astronomers had a Mini Magellanic Cloud to contend with.[17]

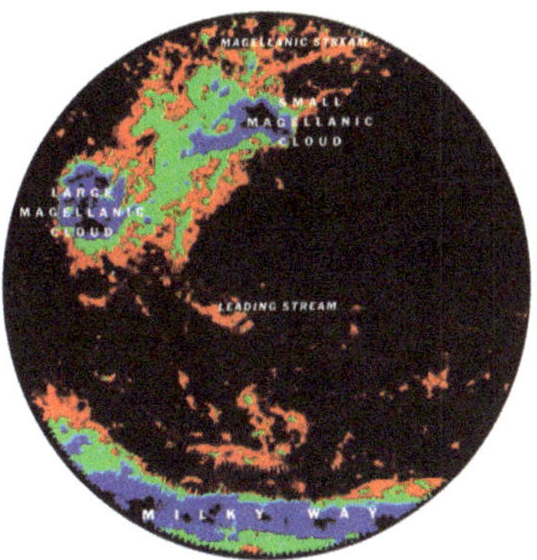

Figure 8.7: The Magellanic Stream in a radio image made by the HI survey multibeam team. Image: Brad Gibson (Swinburne University of Technology/ University of Central Lancashire).

The 2.3 m telescope opened up a major area of research in observational cosmology for Mathewson.[18] Lynden-Bell, Faber, Dressler, Burstein, Davies, Terlevich and Wegner had come up with the proposition that there was a gigantic mass of gas and galaxies called the Great Attractor at a distance corresponding to 4350 km/s. This was attracting all galaxies towards it.[19] Mathewson undertook a five-year project to measure the peculiar velocities of 1355 spiral galaxies using the Tully-Fisher relation. Five PhD students (Angela Samuel, Emmanuel Vassiliadis, Marcus Buchhorn, Carl Grillmair and Stuart Ryder) all worked on the project as part of their first-year program. Rotation velocities were measured using long slit H spectroscopy with the dual-beam spectrograph on the 2.3 m telescope. The results showed that the Great Attractor as postulated by UK and US observers to explain flows, did not exist.[20] The bulk flow of some 600 km/s extends well beyond the hypothetical Great Attractor at 4400 km/s to more than 8000 km/s.[21] At the International Working Party on Large Scale Motions in the Local Universe, which was held at Heron Island in 1995, Mathewson's involvement in the project concluded and he was thanked for his five years of work to solve the problem of the Great Attractor.

The Magellanic Clouds also came under the scrutiny of Mike Bessell. His work on long period variables with Wood and Fox in the Magellanic Clouds led to an important identification with respect to the Asymptotic Giant Branch (AGB) stars and new insights into stellar evolution.[22] AGB stars refer to the last nuclear-burning stage in the evolution of stars after they leave the main sequence. According to Bessell, 'stars in this stage are burning helium and hydrogen in shells surrounding a core of carbon and nitrogen ... One of the interesting things about this time of its evolution is that this kind of burning is very unstable and it can flare up and drop back down again. During the inter pulse periods material can be dredged up to the surface, when we look at the atmospheres of these stars we can see the elements that are being created deep inside these stars. So these stars are actually a primordial source of carbon and nitrogen and also some of the heavier elements such as lead and uranium.'[23]

They found another interesting thing about stellar evolution, being able to work out at 'what mass stars turn into supernovae and which stars turn into white dwarf stars', which was previously unknown. This work was a major breakthrough in AGB evolution. Bessell was also involved in the study of metal or heavy element abundance in stars. He discovered CD-38°245, which was the most deficient star ever formed.[24] It was deficient by a factor of 30 000 with respect to the Sun, which implied that it was formed soon after our Galaxy came into existence. Thus, a study of the abundance of the elements in a star's atmosphere can tell us about the kind of stars that were formed in the very first generation of star formation.

Figure 8.8: Michael Bessell has a global reputation as an expert in the evolution of stars. His research has ranged from studies of stars in the Magellanic Clouds, of the youngest and brightest stars, to brown dwarfs, white dwarfs and the most extremely metal-deficient halo stars. The common thread has been his keen interest in state-of-the-art instrumentation pushed to the limits to obtain answers to astrophysical questions. Photo: Mount Stromlo Archives.

In 1979 Bessell had gone to Kitt Peak Observatory in the US where one of his students, Harvey Butcher (Director at Mount Stromlo from 2007 to 2013), was working. Butcher was 'doing some exciting things with new image detectors. He was also looking at other detectors'.[23] It became obvious to Bessell that the photometric detectors were very poorly calibrated and the northern hemisphere 'standards were really quite terrible'.

In South Africa, Alan Cousins had been carrying out excellent work in setting up standards for photoelectric observations. He was responsible for setting up the Cape-Cousins UBVRI system, which has become the accepted standard system.[25] Most astronomers had been 'brought up to believe that broadband photometry using very wide bands of 800 Angstroms was impossible to do high precision work with. The Europeans, Stromgren and a lot of the Americans assumed that the only way you could do high precision work was to use narrow bands'. Bessell was able to disprove this assumption: 'I started to observe Cousins' standards using both northern hemisphere broadband standards. Cousins' standards were five times more precise than the northern hemisphere broadband standards. So I realised that it was not the broadband that was the problem, it was the lousy work that the northern hemisphere astronomers had been doing.'[23] Butcher encouraged him to write a paper about this, to provide transformations between other photometric systems and to say what he considered to be the best photometric system. The paper attracted very high citations.[26] Bessell went on to investigate in more detail the problems with the northern hemisphere red photometric system. His work 'expanded into trying to understand the bands used for photometric systems so that we could synthesise them with model atmospheric fluxes and do synthetic photometry and from spectral photometric observations of galaxies and quasars and everything else ... And to do that ... you have to reverse engineer standard photometric systems' and so it all kind of tied up over a period of 20–25 years.[27,28,29] He worked in this area up to the time that photoelectric photometry reached its zenith. Today photoelectric photometry has been overtaken by CCDs, which image multiple objects simultaneously; the same calibration problems when using broad and narrow filter bands still come into play.

The age of the universe has been one of the most interesting, intriguing and controversial questions to answer. Visvanathan, who had previously worked with Allan Sandage in the US, took on the challenge of nailing down the figure for the Hubble Constant. His observations of 52 spiral galaxies in the nearby clusters of Virgo, Fornax and Grus and the more distant clusters of Cancer, Peg I and ZW 74–23 gave a figure for the Hubble Constant of 74.3 km s^{-1} Mpc^{-1} which is equivalent to the age of the universe being ~13.5 billion years. This figure is very close to the accepted age of the universe today. It was quite a remarkable result, although it did not get the attention it deserved.

Figure 8.9: Visvanathan tackled the issue of the age of the universe. Photo: Mount Stromlo Archives.

Theoretical studies also played an important role in the astronomical programs at Mount Stromlo. Peter Wood carried out theoretical studies of Mira variables, a class of variable stars in the very late stages of stellar evolution. They expel their outer envelopes with mass loss to form planetary nebulae and a few million years later they become white dwarfs. His studies provided interesting insights into the physical processes involved in the production of mass loss by pulsation and the effect of pulsation on mass flows produced by radiation pressure acting on grains.[30] Since the physical mechanisms which dominate the ejection and subsequent expansion of planetary nebulae were not fully understood, further studies are still being made by astronomers at Mount Stromlo and elsewhere.

Dopita and his colleagues threw further light on the dynamics of mass loss processes by investigating over 60 planetary nebulae in the Magellanic Clouds.[31,32] Geoff Bicknell and Bob Gingold constructed three-dimensional numerical models to investigate the processes of tidal compression of a star by a massive black hole. They found that a star with a solar mass of the Sun passing near a massive black hole is squeezed, that this compression leads to shock heating and nuclear reactions follow with the release of enormous amounts of energy.[33]

As part of the program to determine the true space distribution of quasi-stellar objects (QSOs), Bruce Peterson and a team of astronomers (Ann Savage, David Jauncey and Alan Wright) carried out spectroscopic observations of samples of radio and optically selected QSOs.[34] They were in for a surprise and made a major discovery: 'The QSO identified with PKS 2000-300 has a redshift of 3.78 and is the most distant and most luminous object known.' They opened the celebratory bottle of champagne but unfortunately their record was not to last long, as other QSOs were discovered even further away.

The entrepreneur

Mathewson was an astronomical entrepreneur in the same mould as the great 19th-century Australian astronomer, Henry Chamberlain Russell, the third Government Astronomer of Sydney Observatory.[35] Russell was well connected to the political elite, the NSW bureaucracy and the captains of industry. All these connections played an important role in achieving his scientific goals.

Mathewson came up with the proposal of constructing a Two-Element Synthesis Telescope (TEST) using the existing 64 m and 18 m dishes at Parkes. He reasoned that this combination would yield high-resolution observations of southern hemisphere galaxies quite quickly and would give astronomers valuable experience with the aperture-synthesis techniques that would be required for the Australian Synthesis Telescope (AST), which was under intense discussion within Radiophysics. Their plan was to use the existing 64 m Parkes Radio Telescope in conjunction with a dozen smaller antennae to form an array to be called the AST. While the discussions were going on, the 64 m and 18 m dishes were linked and the system was successfully tested by the astronomers from the CSIRO Radiophysics Division and Mount Stromlo Observatory.[36]

Bob Frater, who had been appointed Chief of the Radiophysics Division in 1981, had other ideas. 'When he came to Radiophysics there was a proposal in the pipeline with a request for funding to the government for the Australian Synthesis Telescope. It was for an amount of money which I think was around $11 million or $12 million, which was grossly inadequate. I suppose

H.C. Russell: Astronomical entrepreneur

Figure 8.10: Henry Chamberlain Russell was the third Government Astronomer of New South Wales. He not only designed and built a large astrograph but also involved the Australian observatories in large international projects. Photo: JD Collection.

Russell was the first Australian-born astronomer to be appointed Director of Sydney Observatory. The two previous directors, William Scott and George Smalley, were appointed by the Astronomer Royal, George Airy, who also set the astronomical agenda for the Observatory. Russell broke away from the mould of a colonial observatory beholden to the preferences of the Astronomer Royal, and kept a healthy balance between independence from and dependence on England. He single-handedly organised the largest Transit of Venus expedition in 1874 in Australia, to determine the distance of the Earth to Sun. He was very successful in that endeavour and his observations were given double weight by Captain Tupman, who collated the observations of British astronomers including those from the colonies. Russell also involved Australian observatories in the International Astrographic Catalogue Project (Carte du Ciel Project), which unfortunately took longer than expected due to economic and political reasons. In fact, the project was completed only in the 1960s, by Harley Wood, the seventh Government Astronomer of Sydney Observatory. In embarking on the project Russell designed an astrograph which was very similar to that designed by the Henry Brothers in Paris for participating observatories, but Russell introduced several improvements . The optics for the camera and its guide telescope were made by Grubb of Dublin and the mounting was made locally by Morts Dock Engineering Co. and Atlas Engineering Co. under the supervision of Russell. He had great faith in the abilities of the local industry and was an early advocate of manufacturing scientific instruments locally.

it was a project that was somewhat unexciting. In any case, the project wasn't funded and we were left with some decisions to make on whether we would actually try again.'[1] He tried again but with a very ambitious project in mind – the Australia Telescope. While reviewing the plans for a new radio telescope Brian Robinson and Frater hit upon the idea for a new configuration for the Australia Telescope – as a long baseline interferometer comprising an

Figure 8.11: The Two-Element Synthesis Telescope at Parkes, comprising the 18 m and 64 m telescopes. Photo: John Sarkissian, Parkes Radio Observatory.

array of antennas extending across the country. This would be unique in the southern hemisphere, and much more competitive than the ideas proposed earlier by Mathewson and several astronomers for building a synthesis telescope. The Australia Telescope was built for Australia's bicentenary. Frater said that 'it's the only major facility in the southern hemisphere and therefore the only one that has access to many of the southern objects. It's in great demand from world astronomers'.[1] Its technical versatility and beauty make it the scientific Taj Mahal of the southern hemisphere.

The Advanced Technology Telescope

One of the lasting legacies of Mathewson's directorship is the 2.3 m Advanced Technology Telescope. He was a man in a hurry and, as Gascoigne said, 'a great bustler around'. Shortly after taking up his directorship, Mathewson embarked on this ambitious telescope, which was mostly designed and built in Australia, in particular by the engineering and technical staff in the Observatory workshops. Like Russell, Mathewson had great faith in the technical abilities of local staff and industry. According to Mathewson, 'Eggen had an application for a 60 inch [1.52 m] photometric telescope at Siding Spring for ten years and it never got funded. I was just thinking about this and one day I thought, well, why don't we build our own alt-azimuth mounted telescope? We've got tremendous expertise at Mount Stromlo Observatory. We've got damn good engineers. We've got tremendous technicians. Why can't we do it ourselves?'[1]

Figure 8.12: The CSIRO Compact Array of Australia Telescope antennas in Narrabri. Photo: CSIRO.

The telescope was to have a novel design. It not only used the alt-azimuth arrangement to control the telescope simultaneously about two axes (horizontally and vertically), but was also mounted in a building that rotated with the telescope and could therefore be much smaller and cheaper.

Mathewson showed his entrepreneurial flair and financial acumen in the purchase of the thin mirror for the telescope. He had become aware of the 2.4 m Cer-Vit mirror blank which had been developed and manufactured by Owens-Illinois of Ohio. Cer-Vit is an extremely low-expansion glass-ceramic material, ideal for use in large high-precision telescopes. The firm's blanks had been used in several large telescopes including the 4 m Anglo-Australian Telescope. Mathewson phoned Henry Cossett of Owens-Illinois in May 1979 and asked, 'Look, have you got a large Cer-Vit blank?' He replied, 'We've got one Cer-Vit blank left. So, I put the hold on it and that was the start of the 2.3.'[1] Mathewson then asked Owens-Illinois to slice the blank in two through the mid-plane. A year later he sold one half of the blank for a price not very different from that which he had paid for the entire blank. A brilliant case of wheeling and dealing!

Figure 8.13: John Hart, Herman Wehner and Gary Hovey were the three engineering brains at Mount Stromlo who made the 2.3 m Advanced Technology Telescope a reality. Photo: Mount Stromlo Archives.

Except for the high-precision drive, which was made by the Swiss company Maag, and the telescope bearings which were made by the German company Rothe Erde, over 95% of the design and construction work of the telescope was carried out by the staff of the Observatory.

Figure 8.14: On 16 May 1984, Prime Minister Bob Hawke formally dedicated the 2.3 m Advanced Technology Telescope at Siding Spring. He recognised that it was a singular achievement in high-technology engineering. Photo: Mount Stromlo Archives.

Figure 8.15: The exterior of the 2.3 m Advanced Technology Telescope at dusk. The housing rotates with the telescope. Each clear evening, the telescope is opened at dusk so that the internal and external temperatures can equilibrate, thereby ensuring the best possible images are obtained. Photo: Bob Cooper, Mount Stromlo Archives.

Figure 8.16: Light from the sky falls on the main mirror at the bottom of the structure, is reflected to a secondary mirror at the upper end of the truss structure, then is redirected back either through a small hole in the primary mirror or to the side through holes in the elevation bearings. Photo: Mount Stromlo Archives.

Figure 8.17: Astronomers Don Faulkner and Vince Ford operating the 2.3 m telescope from its console in the rotating building. It can also be operated over the internet from anywhere in the world. Photo: Mount Stromlo Archives.

John Hart was principally responsible for the telescope design, Herman Wehner for the rotating building and Gary Hovey for the control systems. According to Mathewson, 'I thought of the alt-azimuth mount. John Hart was the mechanical engineer. He designed the telescope – did a magnificent job. Ted Stapinski started off the electronic control systems, followed very strongly by Gary Hovey and Jan van Harmelen. There were software engineers, Hilton Lewis, who at the end of the project was pinched by the Keck Telescope in Hawaii, to become the chief software engineer for those two mammoth telescopes.'[1]

The telescope has a focal ratio of f/2.5. The mirror has a thickness of 22 cm at the edge and 16 cm in the centre. The optical figure is maintained by a sophisticated mechanical mirror-support system.

The telescope was a great success. According to Tom Frame and John Faulkner, 'While each of the three innovative concepts (alt-azimuth, thin mirror and rotating building) had been implemented previously, the ANU's 2.3 m telescope was the first to combine them in the design of a single instrument. It was also the first to employ control computers in such a comprehensive way. Nearly every function of the telescope's operation, its instrumentation, the data collection and the observing environment (including monitoring the weather) became part of an integrated system that placed at the astronomer's fingertips an observational tool of unprecedented sophistication.'[37] The facility was officially opened by the Prime Minister, Bob Hawke, in 1984. It won the Institution of Engineers Australia (Canberra Division) Engineering Excellence Award[38] in 1985 and has been an inspiration for the larger telescopes that came after it was commissioned.

Auxiliary equipment for the telescope was also designed and constructed in the Observatory workshop. Alex Rodgers, Peter Conroy and Gabe Bloxham

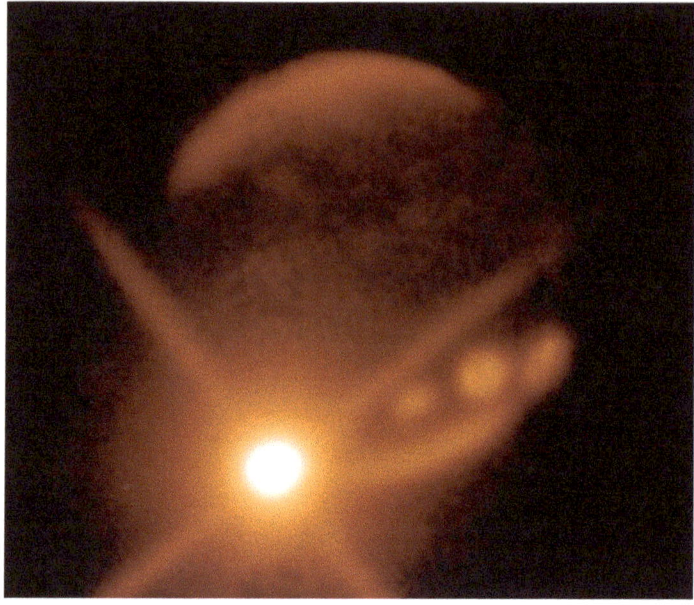

Figure 8.18: The collision of comet Shoemaker-Levy 9 with Jupiter, recorded by the CASPIR infrared imager on the 2.3 m telescope. This particular image appeared in newspapers around the world. Photo: Peter McGregor, Mount Stromlo Archives.

designed and built a dual-beam Nasmyth spectrograph.[39] It allowed observers to make observations at longer and shorter wavelengths simultaneously. Another innovative piece of equipment for infrared work, called CASPIR (Cryogenic Array Spectrograph),[40] was designed by Peter McGregor and came into operation in 1993, just in time to secure spectacular images of the collision of comet Shoemaker-Levy 9 with Jupiter. McGregor has become well-known worldwide for his expertise in astronomical instrumentation.

Mathewson embarked on a very ambitious international project, to take Australia into space astronomy. According to Dopita, 'he was very much trying to drag Australia kicking and screaming into the space age.' It may surprise readers that Australia was the third nation in the world (after the Soviet Union and the US) to launch a satellite into space. However, that short burst into international fame fizzled out by 1980 and the Woomera space program was closed down for lack of will and wisdom by Australia's political class.

The space race and the Cold War between the US and the Soviet Union had remarkable spin-offs for opening up the high-energy end of the electromagnetic spectrum. This gave rise to the deployment of High Energy Observatories such as the International Ultraviolet Explorer (IUE), which was a collaborative project between NASA, the UK Science and Engineering Research Council and the European Space Agency (ESA), and the Einstein X-ray Observatory into space. They were both launched in the late 1970s. The IUE was the first space observatory and was designed to make observations of ultraviolet spectra. According to Kondo, 'a few weeks before the tenth anniversary of launch, 1571 papers in refereed journals were identified as based on IUE results'. The papers ranged from studies of stellar chromospheres, evolution-

Figure 8.19: The International Ultraviolet Explorer, launched on 26 January 1978. Mike Dopita was able to quickly exploit the new technology and energise interest at Mount Stromlo in space-based astronomy. Photo: NASA.

ary processes in interacting binaries, winds from early-type stars, the interstellar medium, supernova 1987A to solar system objects and other topics.[41,42]

Dopita and Ian Tuohy were quick to exploit the new space technology. With Piero Benvenuti from the European Space Agency and Sandro D'Odorico from the University of Padova, Dopita carried out far-ultraviolet spectrophotometry observations in the supernova remnants Cygnus Loop in the Galaxy and N49 and N63 in the Large Magellanic Cloud.[43] They were able to interpret these spectra in the framework of radiating shock-wave models. Using the Einstein Observatory High Resolution Imager, Tuohy and Gordon Garmire from the California Institute of Technology detected a point source of X-ray emission at the centre of the young supernova remnant RCW 103. From their observations they concluded that the 'object is either a hot neutron star or an X-ray emitting pulsar. If the object is a hot neutron star … (it) would represent the first detection of surface radiation from a neutron star'.[44]

Ted Stapinski and Alex Rodgers were also involved in the research and development of a Photon-Counting Array[45,46] which could be used in a space program. During the 1980s, Rodgers and his technical team developed a series of extremely sensitive imaging detectors which were capable of recording the arrival positions of individual photons from distant astronomical objects. Mathewson was not slow to realise how these developments and expertise could be used to get Australia into the space age. The opportunity presented itself in the form of NASA's STARLAB project, which had stalled due to financial cuts to the US space program. Mathewson said, 'I happened to see where NASA wanted a partner in the STARLAB project. Canada was already in it – this was a 1-metre telescope. So I phoned NASA and said, "Look, Australia would like to be a member", and they said, "Well, come on over and tell us how you would like to be involved."'[1] He travelled to NASA's Goddard Space Flight Centre to make Australia's case to join the STARLAB project.

The outcome of the deliberations was that the US would develop the Space Platform and meet all the costs for the first two launches. Canada was to build the telescope while Australia was to be responsible for the instrument package. Mathewson considered that the instrument package was the most sophisticated part of the whole thing, informing NASA that 'The heart of it is the detector, the photon counting array, which had been developed at Mount Stromlo Observatory by, particularly, Ted Stapinski and Alex Rodgers. This was going to be the heart and around it would be an imager and a spectrograph.'[1]

Having made his case in the US, Mathewson now had the job of trying to convince the federal government to release taxpayer funds for the project. But to his dismay, no one was interested. Over the years the Bok touch had been lost by the astronomical community at Mount Stromlo. However, about three months after he had approached the government he received a phone call: 'Phillip Lynch's [a minister in the Liberal government] secretary phoned me and said, "Look, Mr Lynch would like to come up to have a look at your detector this afternoon." Indeed, Phillip Lynch was terribly interested.'[1] The next morning, Lynch asked Margaret Guilfoyle, the Treasurer, to transfer $3 million into the account for STARLAB. That was the beginning of the project.

Figure 8.20: Alex Rodgers adjusting his Photon Counting Array detector system for the Coudé spectrograph of the 1.9 m telescope. Photo: Mount Stromlo Archives.

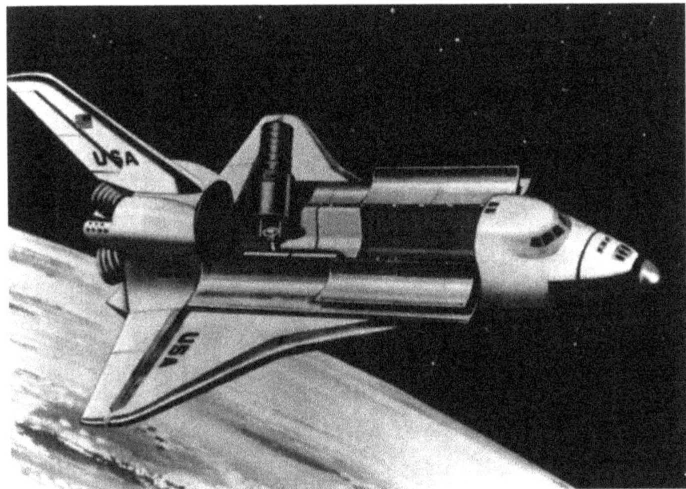

Figure 8.21: The concept of the STARLAB satellite telescope flying on the NASA Space Shuttle. Mathewson grabbed the chance to become involved in the STARLAB project, realising it was a window of opportunity for Australian astronomy and a way to support and challenge the Australian high-tech industry. Auspace Pty Ltd was spun off from the Mount Stromlo engineering lab to take on the project. Photo: NASA.

The project faced many problems. Mathewson used his persuasive powers to get industry involved in his venture. Thus, Hawker de Havilland and SA Matra, the French aerospace company, joined forces to form Australia's first space company, Auspace Pty Ltd.[47] Mathewson said, 'I formed that company and named it, actually with Pip Morgan, who was a Matra agent and now it is a flourishing space company in Australia with Ted Stapinski at the helm. It does other work besides space work, although it does quite a bit of space work. It's a multimillion dollar company.'[1]

A little hut was erected for the STARLAB project at the Observatory and about 15 engineers and technicians were assigned to it. NASA engineers came over to examine the progress of the project. According to Mathewson, 'Through the preliminary design review, the critical design review; they gave us A grade honours for all that, and we went through the cost review.'[1] John Button, the Minister for Industry, Technology and Commerce in the Labor government, gave strong support to the project. All was going smoothly.

Unknown to Mathewson, however, dark clouds were gathering on the horizon: 'Then in March Beggs, the NASA administrator went across to see Roberts, the Canadian Minister for Science, and persuaded Roberts to change from the STARLAB project to the Space Platform. And that was at the eleventh hour, just before it was going to be rubber stamped by the Canadian Parliament and Canada dropped out and that was the finish of the STARLAB project.'[1]

However, about two months later another opportunity presented itself. NASA and the European Space Agency (ESA) approached Mathewson about the possibility of incorporating Mount Stromlo's Photon-Counting Array in their Far Ultraviolet Spectroscopic Explorer (FUSE) or Lyman, which referred to the wavelength range between 0.0900 and 0.1200 microns at which the instrument would be operating. The proposal was quickly taken up by Mathewson

Figure 8.22: The Endeavour Space Telescope in orbit during the Astro-2 science mission on the Space Shuttle. Although the STARLAB project did not happen, the Photon Counting Array detector technology was flown twice on Shuttle missions in the 1990s. Auspace became Australia's premier space technology company and later played a central role in building Australia's FedSat satellite. Photo: NASA.

with funding coming from the Department of Science and Technology. Mathewson's efforts led to a growing interest in government, industry and research circles in space science and technology. For his efforts, Mathewson was elected a Fellow of the Australian Academy of Technological Sciences and Engineering in 1985.

In the same year the Academy published a report, *A Space Policy for Australia*, by the Madigan Committee. This led to the establishment of the Australian Space Office in 1987 with a mandate to support projects of national significance. The FUSE-Lyman Project Office and the staff working on the project were transferred from Mount Stromlo to Auspace Pty Ltd. The emphasis of the project changed and it became more aligned with industry than with pure science.

The project was dealt a severe blow with the withdrawal of funding from the European Space Agency and Australia's involvement with the FUSE-Lyman came to an end. Nevertheless, Mathewson's efforts in space science and technology were not in vain. In the 1996 Budget,[48] the Minister for Science and Technology, McGauran, announced that the first Australian-made satellite would be in orbit in 2001, the centenary of Federation. In December 2002, FEDSAT was launched from Japan. Auspace played a key role in its development. It was the first Australian-built satellite to be launched for almost 30 years. 'It will', McGauran stated, 'conduct scientific experiments and develop practical applications and earth observation.'[48]

Mathewson had a good run as Director of Mount Stromlo Observatory. He also served as Vice-President of the International Astronomical Union from 1992 to 1997. In 1995 he got the IAU Council out to Canberra and taught them sheep shearing at Tidbinbilla! Reflecting on his career as an astronomer, he said, 'there's been nobody ever telling me what to do in my whole life. I've never ever thought of it as going to work ... I guess this is a very spoilt existence. I guess if my achievement in life is making myself happy then that's a terrible thing to think of achievement. I guess one has contributed in some small way. I don't know how. I mean, in a hundred years time will they remember the name Mathewson? I doubt it. And it's good that you disappear from the scene, but there is some little brick up there that people will tread on, that as Newton said, you know, "We all stand on each others' shoulders to see further." Hopefully, I've elevated things by a few inches, but I'm not sure just where and how and why.'

References

1. Bhathal R 1996. *Australian Astronomers: Achievements at the Frontiers of Astronomy.* National Library of Australia, Canberra.
2. Large MI, Mathewson DS, Haslam CGT 1959. A high-resolution survey of the Andromeda Nebula at 408 Mc/s. *Nature* **183**, 1250–1251. doi:10.1038/1831250a0.
3. Large MI, Mathewson DS, Haslam CGT 1959. A high-resolution survey of the Coma Cluster of galaxies at 408 Mc/s. *Nature* **183**, 1663–1664. doi:10.1038/1831663a0.
4. Van de Hulst 1994. Jan Hendrik Oort. *Biographical Memoirs of Fellows of the Royal Society. Royal Society (Great Britain)* **40**, 320–326.
5. Mathewson DS, Van der Kruit PC, Brouw WN 1972. A high resolution radio continuum survey of M 51 and NGC 5195 at 1415 MHz. *Astronomy and Astrophysics* **17**, 468–486.
6. Wannier P, Wrixon GT 1972. An unusual high-velocity hydrogen feature. *Astrophysical Journal* **173**, L119–L123.
7. Mathewson DS, Cleary MN, Murray JD 1974. The Magellanic Stream. *Astrophysical Journal* **190**, 291–296. doi:10.1086/152875.
8. Mathewson DS, Schwarz MP, Murray JD 1977. The Magellanic Stream: the turbulent wake of the Magellanic Clouds in the halo of the Galaxy. *Astrophysical Journal* **217**, L5–L8. doi:10.1086/182527.

9. Murai T, Fujimoto M 1980. The Magellanic Stream and the galaxy with a massive halo. *Publications of the Astronomical Society of Japan* **32**, 581–603.
10. Meurer GR, Bicknell GV, Gingold RA 1985. A drag dominated model of the Magellanic Stream. *Proceedings of the Astronomical Society of Australia* **6**, 195–198.
11. Barnes DG, Staveley-Smith L, de Blok WJG, Oosterloo T, Stewart IM, Wright AE, Banks GD, Bhathal R, Boyce PJ, Calabretta MR, Disney MJ, Drinkwater MJ, Ekers RD, Freeman KC, Gibson BK, Green AJ, Haynes RF, te Lintel Hekkert P, Henning PA, Jerjen H, Juraszek S, Kesteven MJ, Kilborn VA, Knezek PM, Koribalski B, Kraan-Korteweg RC, Malin DF, Marquarding M, Minchin RF, Mould JR, Price RM, Putman ME, Ryder SD, Sadler EM, Schröder A, Stootman F, Webster RL, Wilson WE, Ye T 2001. The HI Parkes All Sky Survey: southern observations, calibration and robust imaging. *Monthly Notices of the Royal Astronomical Society* **322**, 486–498. doi:10.1046/j.1365-8711.2001.04102.x.
12. Putman M, Gibson BK, Staveley-Smith L, Banks G, Barnes DG, Bhathal R, Disney MJ, Ekers RD, Freeman KC, Haynes RF, Henning P, Jerjen H, Kilborn V, Koribalski B, Knezek P, Malin DF, Mould JR, Oosterloo T, Price RM, Ryder SD, Sadler EM, Stewart I, Stootman F, Vaile RA, Webster RL, Wright AE 1998. Tidal disruption of the Magellanic Clouds by the Milky Way. *Nature* **394**, 752–754. doi:10.1038/29466.
13. Mathewson D 2012. Discovery of the Magellanic Stream *Journal of Astronomical History and Heritage* **15**, 100–104.
14. Hindman JV 1967. A high resolution study of the distribution and motions of neutral hydrogen in the Small Cloud of Magellan. *Australian Journal of Physics* **20**, 147–171. doi:10.1071/PH670147.
15. Mathewson DS, Ford VL, Visavanathan N 1986. The structure of the Small Magellanic Cloud. *Astrophysical Journal* **301**, 664–674. doi:10.1086/163932.
16. Mathewson DS, Ford VL, Visanathan N 1988. The structure of the Small Magellanic Cloud. *Astrophysical Journal* **333**, 617–645. doi:10.1086/166772.
17. Mathewson D 1984. The Mini-Magellanic Cloud. *Mercury* **13**(2), 57–59.
18. Mathewson DS, Hart J, Wehner HP, Hovey GR,, van Harmelen J 2013. The 2.3 m new generation telescope at Siding Spring observatory, the Australian National University. *Journal of Astronomical History and Heritage* **16**, 2–28.
19. Lynden-Bell D, Faber SM, Burstein D, Davies RL, Dressler A, Terlevich RJ, Wegner G 1988. Spectroscopy and photometry of elliptical galaxies. V. Galaxy streaming towards a new supergalactic centre *Astrophysical Journal* **326**, 19–49.
20. Mathewson DS, Ford VL, Buchhorn MA 1992. No back-side infall into the Great Attractor. *Astrophysical Journal* **389**, L5–L8. doi:10.1086/186335.
21. Mathewson DS, Ford VL, Buchhorn MA 1992. Southern sky survey of the peculiar velocities of 1355 spiral galaxies. *Astrophysical Journal* **Supplement 81**, 413–421.
22. Wood PR, Bessell MS, Fox MW 1983. Long-period variables in the Magellanic Clouds: supergiants, AGB stars, supernova precursors, planetary nebula precursors, and enrichment of the interstellar medium. *Astrophysical Journal* **272**, 99–115. doi:10.1086/161265.
23. Bhathal R 2009. Stellar astrophysics. *Journal and Proceedings of the Royal Society of New South Wales* **142**, 25–30.

24. Bessell MS, Norris J 1984. The ultra-metal-deficient (Population III?) red giant CD-38°245. *Astrophysical Journal* **285**, 622–636. doi:10.1086/162539.
25. Feast M 2001. Alan William James Cousins (1903–2001). *Astronomy & Geophysics* **42**(4), 4.35–4.36.
26. Bessell MS 1979. UBVRI photometry II: the Cousins VRI system, its temperature and absolute flux calibration, and relevance for two-dim photometry. *Proceedings of the Astronomical Society of the Pacific* **91**, 589–607. doi:10.1086/130542.
27. Bessell MS, Brett JM 1988. JHKLM photometry: standard systems, passbands and intrinsic colours. *Publications of the Astronomical Society of the Pacific* **100**, 1134–1151. doi:10.1086/132281.
28. Bessell MS 1990. UBVRI passbands. *Publications of the Astronomical Society of the Pacific* **102**, 1181–1199. doi:10.1086/132749.
29. Bessell MS 2005. Standard photometric systems. *Annual Review of Astronomy and Astrophysics* **43**, 293–336. doi:10.1146/annurev.astro.41.082801.100251.
30. Wood P 1979. Pulsation and mass loss in Mira variables. *Astrophysical Journal* **227**, 220–231. doi:10.1086/156721.
31. Dopita MA, Ford HC, Lawrence CJ, Webster BL 1985. The kinematics and internal dynamics of planetary nebulae in the Small Magellanic Cloud. *Astrophysical Journal* **296**, 390–398. doi:10.1086/163457.
32. Dopita MA, Ford HC, Webster BL 1985. Extremely energetic planetary nebulae in the Large Magellanic Cloud. *Astrophysical Journal* **297**, 593–598. doi:10.1086/163555.
33. Bicknell GV, Gingold RA 1983. On tidal detonation of stars by massive black holes. *Astrophysical Journal* **273**, 749–760. doi:10.1086/161410.
34. Peterson BA, Savage A, Jauncey DL, Wright AE 1982. . PKS 2000–330: a quasi-stellar radio source with a redshift of 3.78. *Astrophysical Journal* **260**, L 27–L 29.
35. Bhathal R 1991. Henry Chamberlain Russell: astronomer, meteorologist and scientific entrepreneur. *Journal and Proceedings of the Royal Society of New South Wales* **124**, 1–21.
36. Ables JG, Forster JR, Manchester RN, Rayner PT, Whiteoak JB, Mathewson DS, Kalnajs AJ, Peters WL, Wehner H 1987. An HI study of the galaxy NGC 4945 with a two-element synthesis telescope. *Monthly Notices of the Royal Astronomical Society* **226**, 157–171.
37. Frame T, Faulkner D 2003. *Stromlo: An Australian Observatory.* Allen and Unwin, Sydney.
38. Engineers Australia 1985. Power scheme, observatory win awards. 1 November 1985. pp. 26–27.
39. Rodgers A, Conroy P, Bloxham G 1988. A dual-beam Nasmyth spectrograph. *Publications of the Astronomical Society of the Pacific* **100**, 626–634. doi:10.1086/132212.
40. McGregor PJ 1994. CASPIR. *Southern Sky* **7**, 18–24.
41. Kondo Y, Wamsteker W, Boggess A, Grewing M, de Jager C, Lane AL, Linsky JL, Wilson R 1987. *Exploring the Universe with the IUE Satellite.* Reidel, Dordrecht.
42. Kondo Y, Boggess A, Maran SP 1989. Astrophysical contributions of the International Ultraviolet Explorer. *Annual Review of Astronomy and Astrophysics* **27**, 397–420. doi:10.1146/annurev.aa.27.090189.002145.

43. Benvenuti P, Dopita M, D'Odorico S 1980. Far-ultraviolet spectrophotometry of supernova remnants: observations and astrophysical interpretation. *Astrophysical Journal* **238**, 601–613. doi:10.1086/158017.

44. Tuohy I, Garmire G 1980. Discovery of a compact X-ray source at the centre of the supernova remnant RCW 103. *Astrophysical Journal* **239**, L107–L110.

45. Stapinski TE, Rodgers AW, Ellis MJ 1981. A two-dimensional photon counting array. *Publications of the Astronomical Society of the Pacific* **93**, 242–246. doi:10.1086/130814.

46. Rodgers AW, van Harmelen J, King D, Conroy P, Harding P 1988. Large-format photon-counting arrays. *Publications of the Astronomical Society of the Pacific* **100**, 841–852. doi:10.1086/132246.

47. Ford J 1983. Sky's not the limit for Auspace Ltd. *The Australian*, 12 January, p. 2.

48. Woodford J, Dayton L 1996. Australia to develop and launch satellites. Budget '96. *Sydney Morning Herald*, 21 August, p. 14.

49. Dopita M 2004. Interview with Ragbir Bhathal for the National Project on Significant Australian Astronomers. National Library of Australia, Canberra.

1986–1992

9

An instrumentalist and the MACHO project

Over the years, he made an unequalled contribution to the development of the observatories, culminating in the highly successful MACHO project, which has been called one of the great physics experiments of the decade. His role in this project was truly seminal.[2]

Alex Rodgers succeeded Don Mathewson as Acting Director of the Observatory in May 1986, and was appointed the sixth Director in June 1987. It took almost 13 months for Australian National University authorities to grant him the Directorship of the Mount Stromlo and Siding Spring observatories. Ken Freeman, a prolific publisher of outstanding research papers, also threw his hat into the ring. He had a reason for doing this. According to Freeman, 'Alex was a somewhat controversial candidate for the directorship, but I thought he would be good. I was a bit concerned that the committee would not choose him, and that is why I applied. I made it fairly clear that I was not seeking the job and would prefer to see Alex in the job, but that I was willing to do the job if they decided not to choose him.'[1]

Rodgers had the famous Australian larrikin streak, an affinity to the rebellious elements in society and a quirky sense of humour. It was perhaps because of this that he was well liked by everyone who came in contact with him. They enjoyed his rather quirky jokes. Rodgers came from a working-class family and became an intellectual's intellectual through sheer determination, perseverance and ability.

He began his career in astronomy in the 1960s by winning one of the postgraduate scholarships offered by the Observatory during Richard Woolley's directorship. He did his PhD under the supervision of Woolley. He had a great regard for Woolley and even kept a photograph of Woolley on the table in his office. In the 1960s he joined Woolley in 'bringing about the intellectual transformation of Stromlo from an old tradition of descriptive astronomy (with a few notable exceptions) to a new tradition of serious modern astrophysics'.[2]

After completing his PhD he went to the US on a prestigious Carnegie Fellowship, and was appointed a Research Fellow on his return to Mount

Stromlo. He is the only Mount Stromlo astronomer who progressed from being a Research Fellow, Fellow, Senior Fellow, Professorial Fellow and Professor to Acting Director and finally Director of the Observatory – a rather remarkable trajectory.

An instrumentalist: his love of engineering

Rodgers was an exceptional instrumentalist and enjoyed nothing more than spending time in the Observatory's workshops, getting his hands dirty with the technical staff. Engineering was his passion, especially optical engineering. He was a strong advocate for a good workshop and 'argued very strongly that Stromlo's engineering capability was a vital part of its identity and its role in the future'.[2] This has certainly come to pass. He was involved in the design and development of several pieces of equipment for use in the Observatory's research programs. One of his first development projects was the design and building of a spectrophotometer for the 1.27 m telescope (the Great Melbourne Telescope) on which several interesting PhD theses were acquired, for example by astronomers such as Mike Bessell and John Norris, who went on to establish themselves at the international frontiers of astrophysics. He used the 1.9 m as a test-bed for many of his technical innovations and modernised the instrumentation that was used on it. He was responsible for commissioning a new Boller & Chivens spectrograph with a new Carnegie image tube in the late 1960s, thus moving the Observatory away from the old photographic spectroscopic methods. He designed and built the double-beam spectrograph and an imager for the 2.3 m telescope at Siding Spring. He was also responsible for moving the Observatory into the electronic era by designing and constructing photon-counting arrays.[3,4] This technology gave way to charged coupled devices in the 1990s, which were much more efficient and sensitive for astronomical purposes. In fact, he kept the Observatory's instrumentation up-to-date, thus enabling it to remain at the frontiers of international astronomy.

Figure 9.1: Alex Rodgers (1932–1997) was born in Newcastle, NSW. He was well known internationally for his work on normal A-type stars high in the galactic halo, which he argued derived from the merger of a small galaxy with the Milky Way. He was also an instrumentalist who ensured that the scientific capabilities of the Observatory remained at the forefront of technology. Photo: ANU.

He dabbled with Adaptive Optics (AO) when it was first introduced into astronomy and built a functional eight-element system for the 1.9 m telescope. In the 1990s most of the developmental work done on AO by various groups was still

Figure 9.2: Rodgers showing Miss Joan Duffield, daughter of the first Director, W.G. Duffield, around the 1.9 m telescope which he used as a test-bed. Rodgers was sensitive to the history of the Observatory and kept in regular contact with Miss Duffield. Photo: Mount Stromlo Archives.

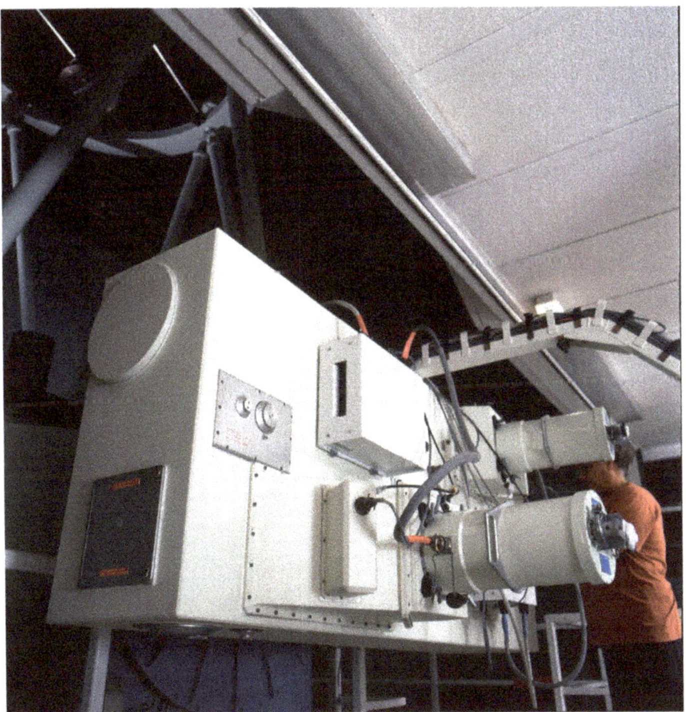

Figure 9.3: The double beam spectrograph mounted at the Nasmuth focus of the 2.3 m Advanced Technology Telescope at Siding Spring. This spectrograph allowed observers to record high-quality spectra of both faint and bright stars. Photo: Mount Stromlo Archives.

Figure 9.4: The 2.3 m telescope was also equipped with a sensitive CCD camera for recording visible light images through a variety of special-purpose filters. Photo: Mount Stromlo Archives.

experimental. Today, the world's largest telescopes use AO to minimise blurring from the Earth's atmosphere and obtain clear and sharp images. Mount Stromlo now hosts one of the world's strongest AO teams. It has won contracts for the design and construction of AO-enabled instruments for both of the international Gemini 8 m telescopes and for the Giant Magellan Telescope.

MACHOs

However, his greatest challenge came when he had to refurbish the Great Melbourne Telescope that had been lying idle for a few years, and convert it into a computer-controlled and wide field imaging telescope for use in answering one of the most intriguing questions in astrophysics – what is dark matter?

For a long time astronomers believed that the mass in galaxies was distributed like the light from them. Thus, the rotation curves should show rotational velocities that would decline as one moved away from the centre of the galaxy. However, much to their surprise, observations of spiral galaxies showed that the rotational velocities remain much higher than predicted out to the most distant measured parts of the galaxies, and in 21 cm radiation from neutral hydrogen even well outside the extent of the visible disk. This was quite a shock.

Mount Stromlo astronomers also found these discrepancies in their observations. John Norris and Mike Hawkins from the Royal Observatory in Edinburgh carried out observations of the motions of stars which extended out to 200 000 light-years into the Galaxy's halo. They found the influence of gravity from unseen material extended to at least half this distance.[5] With his student Claude Carignan, Freeman studied four key galaxies (the Sm galaxy NGC 3109 and the Sd galaxies NGC 7793, NGC 247 and NGC 300 in the Sculptor Group) and found evidence that they had dark halos with comparable mass to their luminous material.[6] As mentioned in Chapter 7, Freeman was cited as one of the first astronomers to suggest that the rotation curves of some spiral galaxies seem to imply the presence of invisible matter.[7]

On a visit to Princeton in 1984, Freeman met Bohdan Paczynski who proposed that it would be possible to test which of the two theories about dark matter (MACHOs or WIMPS) was correct by performing a gravitational lensing experiment.[8] For this to work there has to be almost perfect alignment between a distant star, a dark matter object and an observer. The MACHO would act as a lens and increase the brightness of the distant star by several orders of magnitude. According to Paczynski, 'in any nearby galaxy one star out of a million is strongly microlensed by a "dark" object located in the Galactic halo, if the halo is made up of objects more massive than about 100 million solar masses. Monitoring the brightness of a few million stars in the Magellanic Clouds over a time scale between two hours and two years may lead to a discovery of "dark halo" objects'.[8] He also noted that the 'observational project is not simple'. The best place to try this was in the southern

Figure 9.5: The Great Melbourne Telescope was completely refurbished for the MACHO project, an ambitious search for the large amounts of unseen matter known to exist in the outer parts of the Milky Way Galaxy. The telescope was dedicated to this work from 1992 to 1999. Photo: Mount Stromlo Archives.

WIMPS and MACHOs

The very first observation of large amounts of unseen matter in the universe was made not for galaxies but for clusters of galaxies. The Caltech astronomer, Fritz Zwicky, as early as the 1930s had analysed the relative motions of galaxies in the Coma cluster and found that the galaxies were moving so fast that they should have flown apart and destroyed the cluster. The fact that this did not happen meant that there was much more mass in the cluster, possibly by a factor of more than 100, than could be accounted for by the stars alone.[9] Zwicky named the strange mass *dunkle Materie* (dark matter).

Dark matter is not only intriguing in itself but is also an exciting area of study because it involves several branches of physics. It involves general relativity, high-energy physics, unified field theories, search for gravitational waves, particle physics experiments, such as the search for neutrinos and axions, and almost every branch of astrophysics. The particle physicists have come up with a novel proposal that Weakly Interacting Massive Particles (WIMPS) may play the role of dark matter. If some form of WIMPS are present in our Milky Way Galaxy it should be filled with these almost collisionless particles. The search for WIMPS has generated several groups who are building extremely sensitive detectors to capture the elusive particles. One is the CRESST experiment in the Gran Sasso underground laboratory near Rome. Axions which are low mass bosons postulated in the framework of quantum chromodynamics (the theory of the strong interactions among quarks) have also been postulated to solve the problem of dark matter. In fact, two experiments – one in Livermore, California (US Axion Search) and the other in Kyoto, Japan (CARRACK) – are underway to solve the dark matter problem. Not to be outdone by their particle physicist colleagues, the astronomers came up with the rather colourful name for their particles – the MACHOs. MACHOs stands for Massive Astronomical Compact Halo Objects.

Today, confirming observations of dark matter are made using the technique of gravitational lensing. In 1936, Albert Einstein published a paper in the US journal *Science* titled, 'Lens-like action of a star by the deviation of light in the gravitational field'.[10] In that short paper, Einstein showed that space itself could act like a lens. It could bend the light and magnify it. The bending of light in a gravitational field would mean that if a bright object was located behind an intervening mass, light rays emanating from the object could bend around the intervening mass distribution and converge again, producing either a magnification of the object or the production of several image copies of the object. The intervening mass acts like a lens, albeit a very imperfect one. When Einstein calculated the predicted effects of the lensing of a single distant star he found the effect to be very small. In fact, he noted that the effect would be so small that it would not be measurable. What Einstein did not know was that the effect is measurable if one uses a larger object than a star. Shortly after Einstein's publication, Zwicky wrote a paper in the *Physical Review* in which he showed that one could in fact measure the effect. It works well with large objects such as galaxies. In 1987 the first gravitational lensing of distant quasars by intervening galaxies was observed.

Dark matter and dark energy remain the greatest mysteries of 21st-century physics and astrophysics. 'There are', as Shakespeare's *Hamlet* noted, 'more things in heaven and earth Horatio than are dreamt of in your philosophy'.

hemisphere with the Magellanic Clouds, and Mount Stromlo Observatory was an ideal place to carry out the project.

This idea fired the imagination of US astronomers, Charles Alcock from the Lawrence Livermore National Laboratory (LLNL) and Chris Stubbs from the University of California Centre for Particle Astrophysics (CFPA) at Berkeley. Alcock ran the MACHO project while Stubbs built the detector which was one of the first cameras to use mosaic CCDs. Freeman was instrumental in getting the Observatory involved in the project with Alcock and Stubbs.

This is where Rodgers, as Director of the Observatory, came into the picture. He was the key to the project. According to Alcock, 'Alex went out on a long limb in committing MSSSO (Mount Stromlo and Siding Spring Observatories) to MACHO, at a time when most of the astronomical establishment believed that the project would not succeed. He clearly recognised the potential for MACHO to be very important and he acted on this judgement decisively. He did this at a career stage when most of our colleagues are much more risk averse'.[2]

The MACHO program involved 'the monitoring of the light intensity of millions of stars in the Large Magellanic Clouds, Small Magellanic Clouds and the Galactic bulge fields each night to detect the amplification that may occur through gravitational lensing when a massive object passes close to the sight line to the background star', according to Hart.[11] The program required a 'dedicated telescope of significant aperture and wide field to ensure detection of enough lensing events for statistically useful analysis of the mass spectrum of the lenses and to characterise the structure of the dark halo'.

Figure 9.6: The MACHO project made the cover of the prestigious international science journal *Nature* when its first year of observations were analysed. Photo: *Nature*.

The 1.27 m telescope (Great Melbourne Telescope) was the ideal choice for the project. It was lying idle because it had suffered a catastrophic mechanical failure a few years earlier and there was no money available to fix it, but Rodgers agreed to completely refurbish the telescope and assign it to the project for the next four years. This was later extended for a further period. The responsibility for developing two detector mosaics, each comprising four 2048 × 2048 pixel CCDs, was given to the Centre for Particle Astrophysics.[12] The completed instrument had a field half a degree across – about the size of a full Moon. The provision of computer resources and the development of the software codes needed to handle the enormous amount of data reduction generated by the experiment was given to the Lawrence Livermore group. Freeman, Peterson, Quinn and Rodgers represented the Mount Stromlo team which was made up of a dozen researchers. Observations began in July 1992.

There was great excitement when the first year's observations of microlensing objects were published in *Nature*. In fact, the story made the cover. Writing in the journal, Alcock said, 'We report a candidate for such a microlensing event, detected by monitoring the light curves of 1.8 million stars in the Large Magellanic Cloud for one year. The light curve shows no variation for most of the year of data taking, and an upward excursion lasting over one month, with a maximum increase of ~2 mag. The most probable lens mass, inferred from the duration of the candidate lensing event, is ~0.1 solar mass'.[13] Its mass of about 1/10th that of the Sun made it an excellent MACHO candidate.

The project wound up in December 1999, by which time over 200 000 million individual measurements had been made. The results indicated that about 15 microlensing events had been observed for the Large Magellanic Cloud, two for the Small Magellanic Cloud and several hundred towards the centre of the Milky Way Galaxy. The events seen towards the centre were probably caused by normal stars.

In reviewing the results of 5.7 years of microlensing observations, Alcock said, 'one of our most important conclusions is that a 100% all-MACHO Milky Way halo is ruled out at the 95% confidence level for a wide range of reasonable models'.[14]

Two other groups[15,16] that had taken up the challenge also reported their results in 1993. In the end, the MACHO and similar experiments elsewhere have not solved the problem of dark matter. In fact, they have raised more new questions than provided answers. According to Georg Raffelt, 'the early excitement about the apparent discovery of some or all of the galactic dark matter has given way to a more sceptical assessment – the apparent mass range of the observed events simply does not seem to make sense. Perhaps the least troublesome interpretation is that one is not seeing MACHOs but normal stars as lenses, which is possible if there is an unrecognised population of stars between us and the Large Magellanic Cloud themselves if their distribution is different from what had been thought. Thus, while the observed microlensing events are no doubt real, the question of where and what the lenses are remains for now wide open'.[17]

One of the by-products of the MACHO project was the identification of ~40 000 intrinsic variable stars. Most of these were new discoveries. Several papers were produced on pulsating and eclipsing variables. From the study

Figure 9.7: The Large Magellanic Cloud. The MACHO project used a very large CCD camera on the 1.27 m telescope to observe millions of stars in the direction of the Magellanic Clouds. It tried to detect unseen matter using Einstein's gravitational lensing effect: when objects in the foreground Milky Way passed closely in front of background stars in the Clouds, the effect would cause the background stars to brighten and be seen by the MACHO camera. Photo: Yuri Beletsky, European Southern Observatory.

Figure 9.8: Dark matter in a distant cluster of galaxies. The Einstein gravitational lens effect can also be used to explore unseen matter in the surrounds of external galaxies. Very distant background galaxies are distorted by the lensing caused by the dark matter in a nearby cluster. Here the resulting map of dark matter is coloured blue. Photo. NASA/ESA Hubble.

of these variable stars the researchers concluded that the oldest variable stars in the Large Magellanic Cloud seem to have been born over 4000 million years later than some of their counterparts in the Galaxy.[18] The implications of the MACHO observations of red giant variables were explored by Peter Wood and the MACHO team members.[19] One of the rather surprising discoveries was that the variability of a high number of these giants was due the unexpected presence of binary companions.

Studies of nebulae

The launch of the Hubble Space Telescope in 1990 provided another opportunity for the Mount Stromlo astronomers to enhance their research through international cooperation. The study of planetary nebulae was a major research activity during the Rodger years. When stars with a mass similar to the Sun end their lives they eject material from their surface which forms a gaseous shell around the core. The intense ultraviolet light that is radiated from the core excites the gas in the shell and turns it into a luminous cloud called a planetary nebula. There are many examples of these beautiful planetary nebulae in our Galaxy. However, it is difficult to measure their distances and to study the physics of the ejection process that takes place in these objects in the Galaxy.

The program the astronomers embarked on was to study the planetary nebulae in the Magellanic Clouds, taking advantage of the Hubble Space Telescope's finer resolution. The nebulae promise to provide vital clues to the physics of dying stars. Michael Dopita and his colleagues from the Observatory and from overseas institutions observed ~150 planetary nebulae and obtained some interesting results.[20] They also carried out theoretical modelling of the gaseous shell and of the dying star.[21,22] They found that most remnant stars had masses about two-thirds the mass of the Sun, but some had up to twice this value. There was greater variation in the amount of mass in the shells of the planetary nebulae.

The Orion Nebula is the site of ongoing massive star formation and is the 'Rosetta stone' for the study of Young Stellar Objects (YSOs). This nebula caught the attention of Ralph Sutherland and his colleagues. They made an intensive study of the nebula, obtaining high-resolution images with the Hubble Space Telescope Faint Object Camera and the Planetary Camera.

Figure 9.9: A selection of planetary nebulae imaged by the Hubble Space Telescope. These are nebulae formed during the dying phase of many stars. The outer atmospheres are ejected to form the visible nebula. They often look somewhat like the disk of a distant planet when viewed by eye through a telescope, but in fact have nothing physically to do with planets. Photos: NASA, ESA, Hubble Space Telescope.

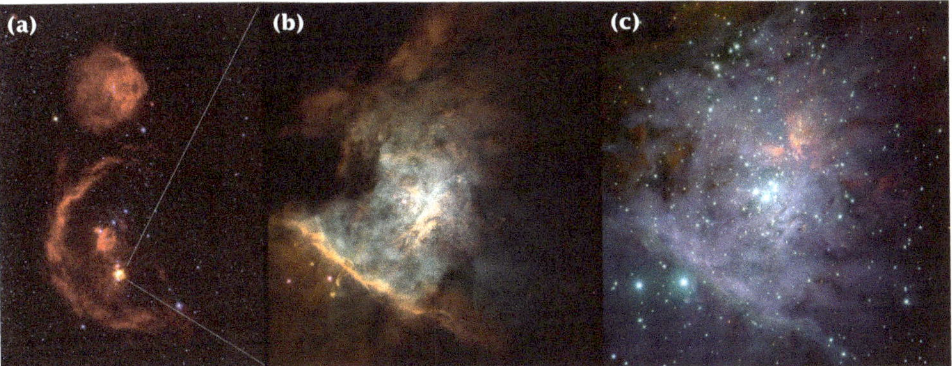

Figure 9.10: (a) The constellation of Orion, the Hunter, showing the stars surrounded by glowing gas. The image extends across the sky for 50 times the diameter of the full moon. The Orion Nebula itself is the bright region at lower centre. (b) A Hubble Space Telescope image of the Orion Nebula in visible light. The many young stars cause the gas to become turbulent. (c) An infrared image of the Orion Nebula taken with the ESO Very Large Telescope. At infrared wavelengths astronomers can see deep into the nebula and study the many young stars making up the star cluster being born. Source: ANU, NASA/ESA, ESO.

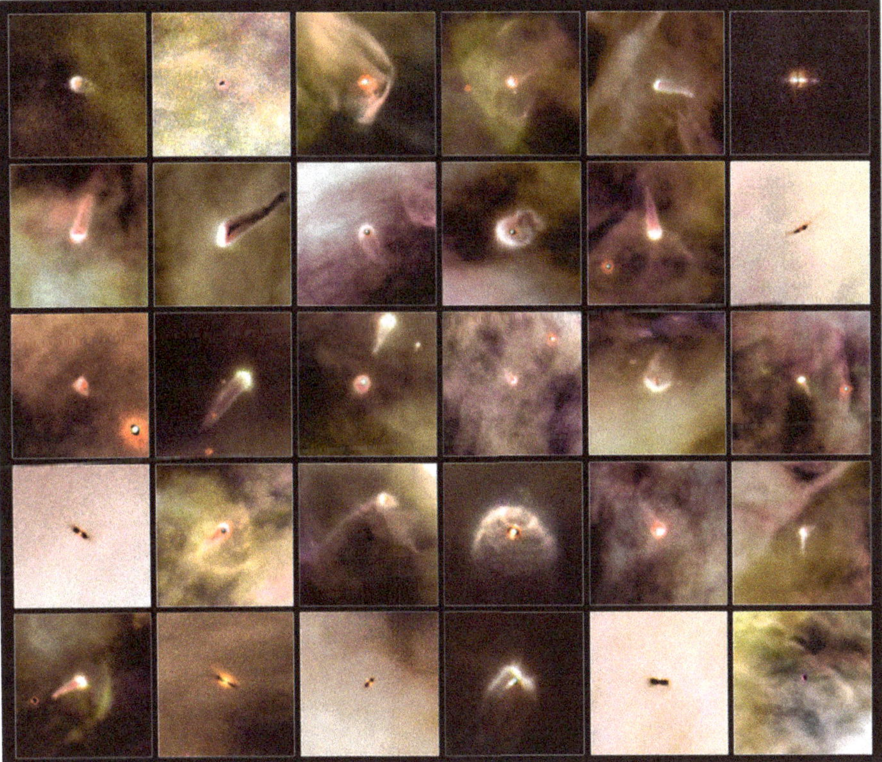

Figure 9.11: A mosaic of young stellar objects in the Orion Nebula, each of which is a star trying to form. Stromlo astronomer Ralph Sutherland and international colleagues studied the physics of star formation in regions like the Orion Nebula where winds of material from the first stars affect the formation of later ones. Photos: NASA, ESA, Hubble Space Telescope. Image © American Astronomical Society.

Figure 9.12: Ralph Sutherland (left) and Mike Dopita (right) proudly displaying their new textbook on the physics of diffuse gas in the universe. Their studies of gaseous nebulae became a major research theme at the Observatory in the 1990s. Photo: Sutherland/Dopita.

Figure 9.13: The Crab Nebula is the remnant of a star that exploded in 1054, an event witnessed by Chinese astronomers. Photo: NASA ESA Hubble Space Telescope.

According to Sutherland, they obtained 'UV images of seven externally illuminated protostellar environments and the first UV spectra that cover the spectral range between 1400 and 3000 Å'.[23] They analysed 43 objects for which the angular resolution had been improved over previous data by more than a factor of 2, and presented an interpretation of the observed morphologies in terms of a dynamical model of external radiation-induced mass loss. That is, the most luminous young stars are so bright their light blows material off the YSOs! They also derived the age of the Orion Nebula as it now appears to be less than 10^5 years and possibly as short as 10^4 years.

Supernova 1987A

The bright explosion that marks the death of a massive star is called a supernova. Historical astronomical records show that, from time to time, a 'new star' suddenly appeared in the sky where nothing was visible the night before. Chinese astronomers called them 'guest stars'. The most famous observation of a supernova goes back to 4 July 1054, when Chinese, Korean and Japanese astronomers recorded the birth of what we now call the Crab Nebula. It remained visible to the naked eye for more than 650 days in the night sky and was visible in daylight for 23 days.

In February 1987, the most exciting discovery of the 20th century saw astronomers rushing to their telescopes to view the 'new' object in the sky.

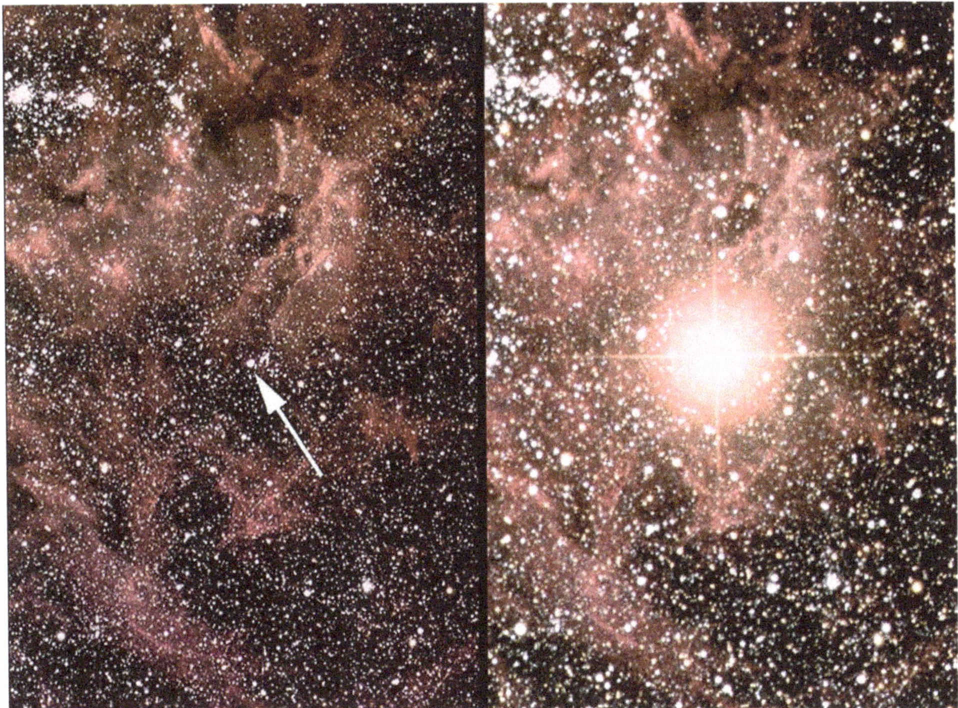

Figure 9.14: A small section of the Large Magellanic Cloud before the supernova exploded (left), and the same area a few days after the explosion (right) of SN 1987A. Photos: AAO.

Figure 9.15: Hubble Space Telescope image of SN 1987A several years after the explosion. The three rings of material surrounding the star became visible several months after the explosion. They are material from the stellar wind from the star before it exploded. The ultraviolet flash from the supernova explosion causes the wind material to glow. Photo: NASA ESA (processed by Noel Carboni).

Supernova 1987A

Astronomers were fortunate to have taken photographs of the star as it was before its explosion. This enabled them to follow the whole process and learn about the origin and nature of supernovae. The event was a great triumph for theory and observation. It was of major importance because it was the closest supernova observed for 383 years – not since 1604 had a supernova been so close as to be visible to the naked eye. It turned out to be an example of a Type II supernova – a type caused by core collapse in a massive star at the end of its luminous life.

One of the most exciting findings was the detection of neutrinos from the explosion, validating earlier theories that had predicted that neutrinos would be manufactured in great numbers during the formation of a neutron star in a supernova explosion. A burst of neutrinos was, in fact, detected a day before the supernova burst into visibility, by neutrino detector laboratories in Japan at the Kamiokande II and in the US at the Irvine-Michigan-Brookhaven experiment. The detection of the neutrinos strongly denoted a Type II supernova, i.e. the death of a massive star and the birth of a neutron star. Neutrinos provide an excellent way to study the physical processes at work in stars that become supernovae.

This new supernova was located in the Large Magellanic Cloud near the region of 30 Doradus. The discovery was made by Ian Sheldon from the University of Toronto on a mountain top in northern Chile while studying a photographic plate of the Large Magellanic Cloud, a galaxy ~170 000 light-years away from us. Designated SN 1987A (because it was the first to be observed that year) it was so bright that it could be seen with the naked eye.

Astronomers at Mount Stromlo lost no time in observing SN 1987A. Peter McGregor made infrared observations[24] while Peter Wood[25] made a study of evolutionary models for the precursor star. A team of astronomers, which included Wood, Paul Nulsen, Mike Bessell and Mike Dopita, used the technique of speckle interferometry on the Anglo-Australian Telescope. At a conference organised in Canberra 500 days after the outburst, Dopita summarised what had been learned from the numerous papers published on SN 1987A.[26] It was found that SN 1987A was initially less luminous than expected because the doomed star was a blue supergiant when it exploded rather than a red supergiant. As the star was relatively small when it exploded, it reached only a tenth of the brightness that a red supergiant would have. Doppler shift measurements indicated an expansion rate of ~17 000 km/s, as expected for a Type II event. The star was estimated to have had a mass of 20 solar masses, a luminosity of 10^5 times and a radius ~50 times that of the Sun.

An Australian Large Optical Telescope

Rodgers was a staunch Australian and when it came to building a Large Optical Telescope he took the view that Australia should build one. The Americans, Russians and the Europeans were well ahead in designing and building large telescopes to make important discoveries for the glory of their

New instruments and discoveries

Continued astronomical discovery requires continued increases in observational capability, whether that means better detectors or larger telescopes. In the 1980s it became clear that detectors would not get much more sensitive, so instruments with primary mirrors in the 8–10 m class would be needed. The first on the drawing board was the 10 m Keck telescope, which used an innovative design with 36 hexagonal submirrors operating together to form a single (segmented) primary mirror.[27] The first Keck telescope began observations on Mauna Kea, Hawaii in 1993. A second one was commissioned in 1996.

Not to be outdone, astronomers from the European Southern Observatory (ESO) built a Very Large Telescope (VLT) at Cerro Paranal in Chile.[28] Its four reflectors, each with a diameter of 8.2 m, form a telescope with an equivalent diameter of 16 m. They could be operated individually or as an interferometer. The first four saw light in April 1998 and the last became operational in September 2000. Another group of nations – the US, UK and Canada – decided to embark on an international facility by building two 8.1 m telescopes. Called Gemini North and Gemini South, they began operations in June 1999 on Mauna Kea and in January 2002 on Cerro Pachon in Chile respectively.

countries. For them it was the Olympic Games of astronomical instrumentation – who could build the biggest and the best telescope.

During the 1970s 4 m class telescopes were built; Australia acquired one in the form of the Anglo-Australian Telescope. It was an excellent telescope and captured some of the prizes in the astronomical competition to open and explore new areas of astronomical research.

By the 1980s astronomers and engineers were beginning to plan for 10 m class telescopes, the primary goal being to increase sensitivity such that objects in the early universe (at high redshift) might be studied. The competition was on to build larger and more powerful telescopes. With all this activity overseas, Australian astronomers reported to the Prime Minister in 1989, in *The Future of Australian Astronomy*,[29] that there was an urgent need for site testing to assess the viability of installing a very large optical/infrared telescope in Australia. They further recommended that, 'in any major astronomy project, the prospects for international collaboration be actively pursued. In the case of the very large optical/infrared telescope, 50 per cent of funding should be sought from overseas'.

Rodgers was steadfast in his view that Australia should build a Large Optical Telescope, comprising a 14 m telescope designed along the same principles as the Keck telescope, and install it on Australian soil as a national facility. He cited the success stories of the 1.9 m at Mount Stromlo, the Parkes Radio Telescope, the Anglo-Australian Telescope and the Australia Telescope to bolster his arguments. Not all the astronomers were convinced. Nevertheless, he initiated testing for a good site in terms of cloud-free conditions and good 'seeing'. By 1992 two sites had emerged for further testing. These were the desert region around Lake Eyre and Freeling Heights with an altitude of 951 m.[30]

Figure 9.16: Rodgers with his model of the 14 m telescope. Photo: Mount Stromlo Archives.

Unfortunately, it soon became apparent that the Australian sites were not suitable for building a large telescope. Australia lacks high mountains and smooth air flow – its winds would cause excessive image blurring compared to the best sites on other continents. It was a great disappointment to Rodgers but he had to accept the hard evidence of the site testings. The model of the 14 m telescope on display in the Observatory foyer disappeared after he stepped down as Director, and was apparently discarded.

Not long after, it emerged that Canada had difficulties in matching its commitment to the Gemini project.[31] This was an international project with the US putting up 50% of the total, Britain paying 25% and Canada 15%. The remaining 10% was being sought from one or more countries in Latin America. According to Bob Bless,[32] 'Being an international collaboration makes it much more difficult' to manage. There are always issues of funding and high politics in such projects. Australia might contribute but it would not be offered a seat on the Gemini Board; the British Board member would handle Australia's interests. This was not acceptable to the Australian government or astronomers.

About the same time, Rodgers' term of office was coming to an end. Rodgers had made valuable contributions to the success of Mount Stromlo Observatory. According to Ken Freeman, 'Over the years, he made an unequalled contribution to the development of the observatories, culminating in the highly successful MACHO project, which has been called one of the great physics experiments of the decade. His role in this project was truly seminal.'

References

1. Freeman K 2012. Email to Ragbir Bhathal. 6 November.
2. Freeman K 1997. Eulogy for Alex W. Rodgers. 15 October. Ken Freeman Collection.
3. Stapinski TE, Rodgers AW, Ellis MJ 1981. A two-dimensional photon counting array. *Publications of the Astronomical Society of the Pacific* **93**, 242–246. doi:10.1086/130814.
4. Rodgers AW, van Harmelen J, King D, Conroy P, Harding P 1988. Large-format photon counting arrays. *Publications of the Astronomical Society of the Pacific* **100**, 841–852. doi:10.1086/132246.
5. Norris JE, Hawkins MRS 1991. Population studies. X. Constraints on the mass and extent of the Galaxy's dark corona. *Astrophysical Journal* **380**, 104–115. doi:10.1086/170566.
6. Carignan C, Freeman KC 1985. Basic parameters of dark halos in late-type spirals. *Astrophysical Journal* **294**, 494–501.
7. Freeman KC 1970. On the disks of spiral and S0 galaxies. *Astrophysical Journal* **160**, 811–830. doi:10.1086/150474.
8. Paczynski B 1986. Gravitational microlensing by the Galactic Halo. *Astrophysical Journal* **304**, 1–5. doi:10.1086/164140.
9. Zwicky F 1937. On the masses of nebulae and of clusters of nebulae. *Astrophysical Journal* **86**, 217–246. doi:10.1086/143864.
10. Krauss LM 2012. *A Universe from Nothing*. Simon and Schuster.
11. Hart J, van Hermelen J, Hovey G, Freeman KC, Peterson BA, Axelrod TS, Quinn PJ, Rodgers AW, Allsman RA, Alcock C, Bennett DP, Cook KH, Griest K, Marshall SL, Pratt MR, Stubbs CW, Sutherland W 1996. The telescope system of the MACHO program. *Publications of the Astronomical Society of the Pacific* **108**, 220–222. doi:10.1086/133713.
12. Stubbs CW, Cook K, Marshall S, Alcock C, Axelrod T, Bennett D, Freeman KC, Griest K, Park H-S, Perlmutter S, Peterson BA, Quinn PJ, Rogers AW, Sutherland WJ 1993. 32-megapixel dual-color CCD imaging system. In *Charge Coupled Devices and Solid State Optical Sensors III. Proceedings of SPIE Conference on Electronic Imaging. 1900.* (Ed. MM Blouke) p. 192.
13. Alcock C, Akerlof CW, Allsman RA, Axelrod TS, Bennett DP, Chan S, Cook KH, Freeman KC, Griest K, Marshall SL, Park H-S, Perlmutter S, Peterson BA, Pratt MR, Quinn PJ, Rodgers AW, Stubbs CW, Sutherland W 1993. Possible gravitational microlensing of a star in the Large Magellanic Cloud. *Nature* **365**, 621–623. doi:10.1038/365621a0.
14. Alcock C, Allsman RA, Alves DR, Axelrod TS, Becker AC, Bennett DP, Cook KH, Dalal N, Drake AJ, Freeman KC, Geha M, Griest K, Lehner MJ, Marshall SL, Minniti D, Nelson CA, Peterson BA, Popowski P, Pratt MR, Quinn PJ, Stubbs CW, Sutherland W, Tomaney AB, Vandehei T, Welch D 2000. The MACHO Project: Microlensing results from 5.7 years of Large Magellanic Cloud observations *Astrophysical Journal* **542**, 281–307. doi:10.1086/309512.
15. Aubourg E, Bareyre P, Brehin S, Gros M, Lachièze-Rey M, Laurent B, Lesquoy E, Magneville C, Milsztajn A, Moscoso L, Queinnec F, Rich J, Spiro M, Vigroux L, Zylberajch S, Ansari R, Cavalier F, Moniez M, Beaulieu J-P, Ferlet R, Grison PH, Vidal-Madjar A, Guibert J, Moreau O, Tajahmady F, Maurice E, Prévôt L, Gry C 1993. Evidence for gravitational microlensing by dark objects in the Galactic halo. *Nature* **365**, 623–625. doi:10.1038/365623a0.

16. Udalski A, Szymanski M, Kaluzny J, Kubiak M, Krezeminski W, Mateo M, Preston GW, Paczynski B 1993. The optical gravitational lensing experiment. Discovery of the first candidate microlensing event in the direction of the Galactic bulge. *Acta Astronomica* **43**, 289–294.
17. Raffelt GG 2001. Dark matter: its nature. In *Encyclopedia of Astronomy and Astrophysics.* (Ed. P Murdin). Nature Publishing Group.
18. Alcock C, Allsman RA, Axelrod TS, Bennett DP, Cook KH, Freeman KC, Griest K, Marshall SL, Peterson BA, Pratt MR, Quinn PJ, Rodgers AW, Stubbs CW, Sutherland W, Welch DL 1996. The MACHO Project LMC variable star inventory II. LMC RR Lyrae stars – pulsational characteristics and indications of global youth of the LMC. *Astrophysical Journal* **111**, 1146–1155. doi:10.1086/117859.
19. Wood PR, Alcock C, Allsman RA, Alves D, Axelrod TS, Becker AC, Bennett DP, Cook KH, Drake AJ, Freeman KC, Griest K, King LJ, Lehner MJ, Marshall SL, Minniti D, Peterson BA, Pratt MR, Quinn PJ, Stubbs CW, Sutherland W, Tomaney A, Vandehei T, Welch DL 1999. MACHO observations of LMC red giants: Mira semi-regular pulsators, and contact and semi-detached binaries. In *Proceedings of IAU Symposium 191: Asymptotic Giant Branch Stars.* (Eds T Le Bertre, A Lebre and C Waelkens) pp. 151–158.
20. Dopita MA 1993. The evolution of the planetary nebulae in the Magellanic Clouds and the Galactic Bulge. In *Proceedings of IAU Symposium 155: Planetary Nebulae.* (Eds R Weinberger and A Acker) pp. 433–441. Kluwer Academic Press, Dordrecht.
21. Dopita MA, Meatheringham SJ 1991. Photoionization modelling of Magellanic Cloud planetary nebulae. *Astrophysical Journal* **367**, 115–125. doi:10.1086/169607.
22. Vassiliadis E, Wood PR 1993. Evolution of low- and intermediate-mass stars to the end of the asymptotic giant branch with mass loss. *Astrophysical Journal* **413**, 641–657. doi:10.1086/173033.
23. Bally J, Sutherland RS, Devine D, Johnstone D 1998. Externally illuminated young stellar environments in the Orion Nebula: Hubble Space Telescope Planetary Camera and ultraviolet observations. *Astrophysical Journal* **116**, 293–321.
24. McGregor PJ 1988. Observations of SN 1987A in the near-infrared. *Proceedings of the Astronomical Society of Australia* **7**, 450–461.
25. Wood PR 1988. The progenitor star and its evolution. *Proceedings of the Astronomical Society of Australia* **7**, 386–389.
26. Dopita M 1988. SN 1987A: an overview. *Proceedings of the Astronomical Society of Australia* **7**, 344–351.
27. Sinnott RW 1990. The Keck Telescope's giant eye. *Sky and Telescope* **80**, 15–22.
28. Tarenghi M 1998. Eyewitness view: first sight for glass giant. *Sky and Telescope* **96**(5), 46–51.
29. The Future of Australian Astronomy 1989. A report to the Prime Minister by the Australian Science and Technology Council. AGPS, Canberra.
30. Frame T, Faulkner D 2003. *Stromlo: An Australian Observatory.* Allen and Unwin, Sydney.
31. Dickson D, MacIlwain C 1993. Gemini telescopes in suspended animation. *Nature* **364**, 566. doi:10.1038/364566b0.
32. Mervis J 1992. Gemini telescope project shifts into high gear. *Nature* **357**, 430. doi:10.1038/357430a0.

1993–2002

10

Masters of the universe

The distance scale path has been a long and tortuous one, but with the imminent launch of the HST (Hubble Space Telescope) there seems to be good reason to believe that the end is finally in sight [on the Hubble Constant].[13]

Although we usually liked to push the fledglings out of the nest, Brian was so extraordinary he won one of the competitive postdoc jobs at the Centre for Astrophysics. This gave him the chance to step out as an independent worker.[11]

Earlier surveys have been eclipsed by the Two Degree Field Galaxy Redshift Survey (2dFGRS). In four short years, the 2dFGRS obtained 246,677 unique galaxy redshifts, covering some 5 percent of the sky while reaching out to a distance of 3 billion light-years.[52]

Jeremy Mould returned to Mount Stromlo as a prodigal son in December 1993. After finishing his PhD at Mount Stromlo he became a scientific nomad, moving from one astronomical institution to another to study the universe at large. His nomadic life took him to the Royal Greenwich Observatory, Mount Wilson and Las Campanas observatories, the Kitt Peak National Observatory and the California Institute of Technology, where he was appointed a professor. It would seem at that point that his travels had come to an end. But this was not the case. So what was the reason that made him return to Mount Stromlo? 'Two things: one that I had always been keen to return to Australia in which astronomy is one of the major sciences, and the second was the enormous challenge that is represented by the needs of optical and infrared astronomy in this country in the next few years. We really stand at the moment where excellent future plans can be made but they're entirely conditional on funding and the support and sponsorship primarily of government. So, I am very keen to lead that effort hopefully to success.'[1]

He came back to Australia with a string of valuable contacts in the major overseas centres of astronomy. These were to be especially useful as he pursued a new program for Mount Stromlo. His vision for the next five years was concerned 'with building new facilities which will serve Australian

Figure 10.1: Jeremy Mould when he was Director of the US National Optical Astronomy Observatory, just after he left the directorship at Mount Stromlo Observatory. Photo: NSF/NOAO.

astronomy for many years to come. Examples of this are the move to join the European Southern Observatories so that we can take part in building telescopes of the new generation of 10-metre aperture. We plan to involve ourselves and to lead, together with the University of New South Wales, site testing for astronomical facilities in Antarctica and we plan to sharpen up our astronomical instrumentation on major telescopes – the 2.3 metre and the 1.02 metre (40 inch) telescope at Siding Spring. There's a list of new initiatives which I hope at the end of five years, resources permitting, will put this place in a much better position than it is today.'[1]

The site testings which Rodgers had initiated showed quite conclusively that Australian sites were not suitable for building the Large Optical Telescope. Australia had to find a solution by cooperating with other nations, such as those in Europe, by getting a share of the large telescopes that they were building in the southern hemisphere. Mould joined Mount Stromlo at the right time. Soon after he arrived, he became involved with the Australian Academy of Science's National Committee for Astronomy (NCA) which was undertaking a 'discipline review'.[2] Mould was given the responsibility of chairing the committee on Future Facilities.[2] The Review Committee recommended that top priority be given to obtaining 'significant access to a large optical/infrared telescope'. Furthermore, 'Australia should immediately accept the European Southern Observatory's (ESO) invitation to join ESO and participate in the world's premier astronomy project, the Very Large Telescope (VLT)'.

The path to obtaining access to the VLT was not going to be easy. As usual, high politics intervened. The $28 million ESO bid from the astronomers received top ranking from two expert committees set up by the government. But in a pre-election announcement in December 1995, the Prime Minister, Paul Keating, on the advice of his Science Minster, Peter Cook, overturned the Committee's advice. Then, Senator Peter McGauran, the Minister for Science in the incoming Liberal–National Coalition led by John Howard, killed the proposal.

Mould was not deterred by this failure. Through the Large Telescope Working Party, he continued negotiations with ESO. This was to bear fruit. In June 1997 the ESO Council approved a protocol which allowed non-member states such as Australia to provide enhancements to ESO facilities in return for guaranteed access and rights to use these facilities, always as non-members.[3]

Figure 10.2: The Gemini South telescope. Mount Stromlo was responsible for building an Adaptive Optics Imager for this telescope and the Near-infrared Integral Field Spectrograph for the Gemini North telescope in Hawaii. Mould placed the Observatory's technical skills on the international market with success. Photo: NSF.

Eight months later David Kemp, the Education Minister (with responsibility for the Australian Research Council), announced the government's support for funding to give Australian astronomers 4.76% observing time on the twin 8.1 m telescopes[4] of the Gemini Observatory. Gary Da Costa was appointed as Project Scientist while Lawrence Cram from the University of Sydney became the first Australian member of the Gemini Board. An Australian Gemini Office was also set up at Mount Stromlo.

Instrumentation projects followed. In collaboration with the Institute for Astronomy, University of Hawaii, Mount Stromlo won a contract to design and build a Near-infrared Integral-Field Spectrograph for the Gemini North Telescope. The $4.5 million project was led by Peter McGregor[5] (Project Scientist) and Jan van Harmelen (Project Manager).

Matthew Colless was a co-Principal Investigator of a collaborative venture by Stromlo, the Anglo-Australian Observatory and the University of New South Wales to develop instrumentation for the ESO's VLT.[6] A project team led by Peter McGregor, Jan van Harmelen and John Hart was awarded a $6.3 million contract for an Adaptive Optics Imager for the 8.1 m Gemini South Telescope in Chile.[7]

Mould's ideas about the feasibility of generating income from the commercialisation of astronomical instrumentation were being vindicated. While previous Directors had focused on building instrumentation for the Observatory's telescopes, Mould went beyond that by involving the Observatory in international instrumentation ventures. He placed the Observatory's technical skills on the international open market.

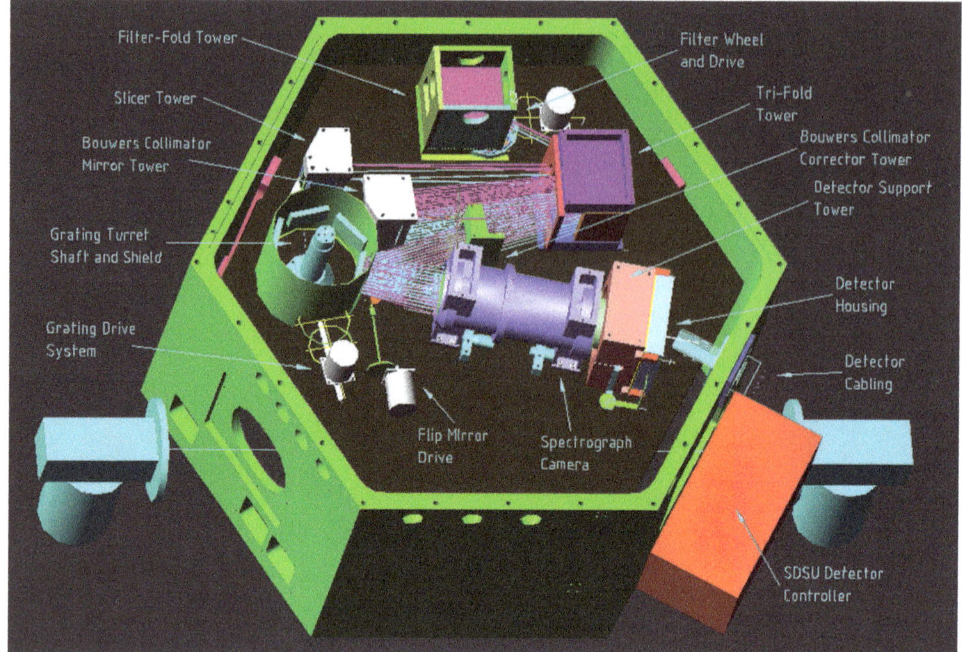

Figure 10.3: Computer drawing of the open Near-infrared Integral-Field Spectrograph (NIFS) showing the integral field unit (top), grating turret (left) and detector housing (lower right). Image: Mount Stromlo Archives.

Theoretical astronomy was not forgotten during the Mould years. To strengthen this area and create synergies between the mathematicians from the Australian National University's School of Mathematical Sciences and Mount Stromlo theoretical astronomers, Mould established the Astrophysical Theory Centre. However, due to personality differences among participating staff, the marriage was short-lived.

Another joint venture that Mould was involved in during the 1990s was the Joint Australian Centre for Astrophysical Research in Antarctica (JACARA). Mount Stromlo astronomers Mike Dopita, Ralph Sutherland, Gary Hovey, Peter Conroy, Peter Wood, Brett McLinden and Mark Jarnyk were involved in this venture. Antarctica offers several advantages for astronomical observations, such as a long interval of total darkness, extremely cold conditions and a very dry atmosphere. It is therefore an exceedingly good place for making infrared observations. According to Michael Burton from the University of New South Wales, 'the possibilities for considerably furthering our understanding of the universe are sufficiently high as to merit pursuit of the goal of establishing an astronomical observatory on the Antarctic Plateau. The opportunities for infrared and sub-millimetre astronomy are exceptional.'[8] However, in 2001 Mount Stromlo withdrew from the venture due to its involvement in too many projects. It had to set priorities.

Mould was involved in bringing astronomy to the public, but not on the same scale as Bart Bok. A successful hands-on Mount Stromlo Observatory Visitor Centre was set up at the Observatory. In 2000 it attracted over 27 000 school children and members of the public who were curious about the

Figure 10.4: The Automated Astrophysical Site Testing Observatory at the South Pole became operational in January 1997. It was formally opened by Senator Robert Hill in a ceremony at the Amundsen-Scott South Pole base. The facility aimed to test the site's suitability for astronomical observations. Mount Stromlo designed an instrument to measure atmospheric turbulence above the site and the University of New South Wales built a number of infrared and ultraviolet instruments to measure how dark and transparent the Antarctic skies are. Photo: Ralph Sutherland.

Figure 10.5: Mount Stromlo astronomer Ralph Sutherland accompanied the ANU–UNSW site testing team to the South Pole. Photo: Ralph Sutherland.

Figure 10.6: Jeremy Mould oversaw the construction of a modern Visitors' Centre at Mount Stromlo to showcase the research at the Observatory and explain current developments in astronomy generally. Photo: Mount Stromlo Archives.

Figure 10.7: An interactive exhibit in the Visitors' Centre, on the colours of white light. Photo: Mount Stromlo Archives.

workings of the mysterious universe. However, due to financial constraints it was closed a few years later.

The Hubble Constant

The Mould years saw three major cosmological projects that had far-reaching implications for astronomy, and which placed the Observatory at the forefront of international astronomy and the related controversies in the US. During his time in the US Mould became involved in the Hubble Constant wars. Allan Sandage was the lord and master of the Hubble Constant for several decades, citing a figure of 50 km per second per Megaparsecond (km s^{-1} Mpc^{-1}).[9] His greatest critic was de Vaucouleurs, who had previously worked at Mount Stromlo. According to de Vaucouleurs' observations, the magnitude should be ~100 km s^{-1} Mpc^{-1}. They were both wrong.

In the 1970s younger astronomers began to challenge the earlier calibrations of the Hubble Constant. Brent Tully and Richard Fisher, graduates from the University of Maryland, found a new method to measure the true luminosities of spiral galaxies and thus their distances. It came to be known as the Tully-Fisher relation. They showed that there was a correlation between the rotation rate of a galaxy and its luminosity. Using a radio telescope aimed at the Virgo cluster, their observations showed that the Hubble Constant was more in keeping with de Vaucouleurs' figure of 100 rather than Sandage's value of 50. Another challenge came from Marc Aaronson, a graduate student from Harvard University. His PhD thesis was a study of the infrared properties of galaxies, and as part of the work he had built an infrared photometer which he took to the University of Arizona's Steward Observatory. There he teamed up with John Huchra and Mould, who was across the street from Steward Observatory at the Kitt Peak headquarters, for postdoctoral work. According to Mould, 'I met Marc Aaronson with whom I started on two research projects that have lasted career long, one on the evolution of intermediate mass stars which turn into carbon stars – at the time nobody had quite made the link between carbon stars and stellar evolution – and Marc and I were able to at least provide the observational impetus. The second was attempting to measure galaxy distances using one of the classic standard candle relationships of astronomy, between luminosity and rotational velocity of galaxies and it was my idea to attempt to do this using infrared wavelengths rather than blue wavelengths that had been used before and Marc was a top infrared observer of galaxies.'[1]

Using Aaronson's infrared photometer on the 91 cm telescope at Kitt Peak, they[10] measured the infrared flux from about 20 spirals in the Virgo and Ursa Major clusters (the same as those that Tully and Fisher used) and found the Hubble Constant to be ~65 km s^{-1} Mpc^{-1}. This did not make them popular with Sandage. According to Dennis Overbye, Sandage didn't trust the Tully-Fisher relation, and sometimes referred to it as 'Fishy-Tuller'.[9]

The measurement of the Hubble Constant became Mould's primary area of interest. According to him, 'It became clear that the Hubble Space Telescope was going to be very significant for measuring the Hubble constant'.[1] Astronomers were banking on the fact that its higher resolution would allow them to solve the problem of the Hubble Constant. However, the Challenger disaster delayed the launch of the Hubble Space Telescope. There was another delay when NASA scientists and engineers found the mirror suffered from spherical aberration. 'The Hubble Telescope had a classic spherical aberration in its mirror', Mould said. 'The result is that light from the outer zones of the mirror is focussed in a different place from light from the inner zones of the mirror, leading to fuzzy images as opposed to sharp images, which is one of the two reasons why the Hubble is there, that it can produce much sharper images, in fact ten times sharper images than we can obtain from the ground where we're affected by our atmosphere.'[1] The fault had arisen because of an error in the test set-up at the NASA contractor for the primary mirror. The mirror shape was perfect, but perfectly wrong. The fault could be fixed in orbit by adding a lens. Mould received a phone call at his Caltech office: 'I received a phone call from John Trauger at JPL and he said, Would you like to join the Wide Field/Planetary Camera II [WFPCII] team? We know how to fix the aberration but we need somebody who has some scientific goals which require that the telescope be fixed and since you are leading the key project to measure H_0 [Hubble Constant] we'd like to have you on the WFPCII team.'[1]

Mould was the leader of the team with Wendy Freedman of the Carnegie Observatories and Rob Kennicutt from the University of Arizona, who had Hubble observing time to measure the Hubble Constant to an accuracy of ±10%. Freedman worked in the same institution as Sandage and was three decades younger. According to Kirshner, 'Observational cosmology is not exactly a contact sport, but it helps to be tough and competitive if you are working in the same field as Allan Sandage, and at the same institution, and especially if you get a different answer.'[11] Freedman's group did get a different answer, and Freedman was willing to answer any criticisms from Sandage or other astronomers. To be absolutely sure there was no mistake, they cross-checked their work by using different methods to measure the distances to the galaxies to derive the value for the Hubble Constant. Distance measurements using Cepheid variables were an established method for measuring distances to distant galaxies. However, these have a practical limit of only a few Mpc (Mega-parsecs). According to Freedman, 'Although dedicated campaigns with ground-based 4 m class telescopes have yielded candidate Cepheids in more distant galaxies, these measurements have proven to be difficult. As a result, precise ground-based Cepheid distances remain limited to galaxies within 4 Mpc, more than an order of magnitude short of the distance needed to reliably sample the Hubble flow. Bridging this "twilight zone" requires the application of one or more "secondary" distance indicators, which are calibrated locally then used to extend the scale to the 100 Mpc range.'[12]

The strategy Mould, Freedman and Kennicutt adopted was to tie a Hubble Space Telescope-based local Cepheid scale to several (largely ground-based) secondary indicators to provide a rigorous determination of the Hubble Constant and its associated uncertainty. Earlier Aaronson and Mould[13] had defined several criteria which the ideal distant indicator should satisfy: (1) it should exhibit a small, quantifiable dispersion – i.e. it is a standard candle,

(2) it should be measurable in enough galaxies so that it can be calibrated locally, (3) it should have a well-defined physical basis, and (4) it should be luminous enough to be useful at large distances, preferably into the region of the Hubble flow. They used five indicators that satisfied the above criteria: the infrared Tully-Fisher relation, the planetary nebula luminosity function, surface brightness fluctuations of galactic spheroids, the expanding photosphere method for Type II supernovae, and the peak luminosity of Type Ia supernovae for obtaining improved Cepheid-based calibrations. Their measurement of Cepheids in the Virgo cluster spiral galaxy, M 100, gave a value of ~80 km s^{-1} Mpc^{-1} for the Hubble Constant. The distance to the cluster was found to be 17.1 ± 1.8 Mpc.[14] Three years later into the project they obtained a new value of ~73 for the Hubble Constant.

However, this raised a problem. The Hubble Constant of 73 implied that the age of the universe was ~13 billion years. This was in conflict with the age of the oldest stars in globular clusters which had an age of ~15 billion years.[15,16] The universe could not be younger than the oldest objects in it. The astronomers scurried back to their telescopes and computers to obtain a more accurate age for the globular clusters, revising their ages downwards as new observations and computations were carried out.[17,18] The globular clusters were found to be ~11.5 billion years old with an uncertainty of 20%.[19]

Five years after it began, the Hubble Space Telescope Key Project on the distance scale came to an end, with Mould's team producing a value for the Hubble Constant of 72 (± 10%).[20] The project had achieved its objective of getting a value for the Hubble Constant within 10% accuracy. Mould, Freedman and Kennicutt became the new kingpins of the Hubble Constant. Unfortunately, Aaronson was not with them to enjoy the new result. He had died in a freak telescope accident at the Kitt Peak 4 m Mayall telescope.

The accelerating universe

The second cosmological project was undertaken by Brian Schmidt, who joined Mount Stromlo in 1995. Born in the small country town of Missoula, in Montana in the northern US, Schmidt had spent most of his adolescent life in Alaska. Four of the teachers at his high school had PhDs; this made 'going to university ... quite a step down' from school. His undergraduate studies were done at the University of Arizona, the astronomical home of Bart Bok, a former director of Mount Stromlo. It was here that he was first exposed to the study of supernovae. For his undergraduate research project, 'I got embedded in with a guy by the name of John McGraw and interestingly enough, one of the things I got involved in doing was looking for supernovae'.[21,22] Some of the astronomers there at the same time were Peter Strittmatter, Simon White, Rob Kennicutt, Dave Arnet, Craig Hogan, Jim Liebert and Frank Low. Liebert was working on white dwarfs and Low was carrying out his well-known studies in infrared astronomy.

On completing his undergraduate studies, Schmidt had to decide where to go for his PhD studies. He could have gone to the University of California, Santa Cruz, where Sandy Faber, Mike Bolte, Stan Woosley, George Blumenthal, Peter Bodenheimer and David Koo were extremely active in studies

Figure 10.8: Supernova 1994D, which burst forth in the outer parts of the galaxy NGC 4526 in 1994. Such exploding stars at their peak can rival their host galaxy in brightness, and can be seen across much of the distant universe. Photo: NASA/ESA Hubble.

which interested him. Instead he chose to go to Harvard University where Robert (Bob) Kirshner was actively engaged in research on supernovae. However, he decided that he was not too interested in the supernova topics that Kirshner was suggesting. He had his own ideas. He informed Kirshner that he would like to measure the Hubble Constant using supernovae, not just any supernovae but SN II. Unlike Supernovae Ia which are thermonuclear explosions in white dwarfs, Type II supernovae result from the collapse of a massive star. When the outside of an SN II is ejected, it is still mostly hydrogen. The properties of the expanding, cooling atmosphere can be computed in detail and this was done by Ron Eastman, a postgraduate student in Kirshner's group.[23] By repeated measurements of the temperature, speed and brightness of the supernovae atmosphere, it is possible to figure out how large the atmosphere is and compute the distance to the explosion.

Schmidt teamed up with Eastman, who was writing a very fancy computer code which modelled what was happening in the supernova from a physical basis. He worked with Eastman and Kirshner to model these supernovae. The model showed how many watts the supernovae put out so that the researchers could find out how bright they appeared on Earth and compare that to how many watts intrinsically they were. Then, using the fact that light gets fainter with distance squared, they could measure the distance directly to objects outside the local universe. Using the expanding photosphere method, Schmidt measured 14 supernovae for his PhD thesis. This enabled him 'to measure a value of the Hubble Constant which was independent of any other means and could be compared directly with Cepheids, and it turns out the value I got, 73 plus or minus eight kilometres per second per Megaparsec', was very close to the now accepted value of ~72 km s^{-1} Mpc^{-1}.[24]

It is interesting to note that some of the galaxies with SN II data and expanding photosphere distances were also galaxies in Wendy Freedman's and Mould's Key Project sample. The results agreed very well. Schmidt's PhD was significant and it provided another confirmation of the Key Project Team's measurements of the Hubble Constant with the Hubble Space Telescope of 72 km s^{-1} Mpc^{-1}. According to Schmidt, the work he did on the Hubble Constant 'was a good grounding in what came later, the accelerating universe'.[21]

On a visit to Harvard, Mario Hamuy from Chile not only showed Schmidt and his colleagues new supernova data but also showed them very clearly that Type Ia supernovae could be used to measure accurate distances. According to Schmidt, 'In 1991, there was the idea that Type Ia supernovae were perfect standard candles. I was very sceptical about using Type Ia supernova for cosmology because we didn't know very much about them.' In 1991 two unusual supernovae, SN 1991T and SN 1991bg, were discovered. They strengthened the case that there were real differences among SN Ia and their discovery cast doubt as to whether SN Ia could be used as standard candles.[25] Supernova 1991bg was intrinsically faint and appeared to be 10 times fainter than an earlier SN Ia in the same galaxy. It rose and fell much more quickly. Supernova 1991T was exactly the opposite; it was much brighter and it rose and fell more slowly. This intrigued Mark Phillips, well known internationally for his work on SN 1986G and SN 1987A, the 1990 Calan/Tololo Supernova Search and his observational studies of all classes of supernovae. By plotting the luminosity of several supernovae he found that the rate at which supernovae rise and fall is a very good indicator of how bright they are intrinsically.[26] According to Schmidt, 'that was the relationship with a new independent dataset that was shown to me in 1994. They found that Phillips' relationship made the supernova behave very nicely, you can measure distances to about eight per cent accuracy which may not sound brilliant on Earth, but in astronomical terms when we are used to everything being about thirty per cent, that was good. Eight per cent was outstanding.'

Hamuy had taken Phillips' idea and used it to solve the puzzling luminosity differences in the supernovae, publishing the results in a paper of which he was the first author.[27] He showed that Phillips was correct. The slowly declining supernovae are the bright ones and the fast decliners are the faint ones. Measuring how fast a Type Ia supernova fades after it reaches maximum brightness means that astronomers are not likely to make any mistake in assigning it the wrong distance. The scene was set for using Type Ia supernovae as distance indicators.

In the same month, Schmidt had learned that Saul Perlmutter and his group at the University of California, Berkeley, had found seven distant objects that enabled them to trace back the expansion history of the universe. They had begun a serious study of supernovae in the late 1980s with a combination of Rich Muller from the Physics Department and the Lawrence Berkeley Laboratory, including Carl Pennypacker.[28] Alex Filippenko from the Astronomy Department joined them in 1994. Perlmutter, who had a more forceful approach, joined later and became the leader of the group although he was quite a junior academic. Schmidt's view was, 'if those guys at Berkeley can find supernovae, we sure as hell can. We had all the supernovae expertise, let's get out and do that'.[21]

Figure 10.9: The famous High-Z team, summer 2001. Front row, left to right: Saurabh Jha, Adam Riess, Brian Schmidt, Robert Kirshner. Back row, left to right: John Tonry, Nicholas, Suntzeff, Bruno Leibundgut, Alex Filippenko, Mario Hamuy. Photo: Brian Schmidt.

That was the genesis of the High-Z Supernova Team, set up at about the time Schmidt arrived at Mount Stromlo Observatory. He was 27, a remarkably young age at which to lead an international team that would explore the history and ultimate fate of the universe: 'I knew I wanted to go back and measure the past history of the universe, but this was going to require telescope time which I did not have access to in Australia.'[21] He knew he needed the world's largest telescope, the Keck telescope, access to the Hubble Space Telescope and a huge amount of time on the wide field imager on the Cerro Tololo 4 m. 'And so I went through the people I knew in supernovae which was the Chileans, who helped determine the relationship that allows us to use them. Alex Filippenko from Perlmutter's group at Berkeley joined us in 1995 when the power of the Keck telescope became apparent. Filippenko had access to Keck. Chris Stubbs, who worked on the MACHO experiments here and was someone who was interested in large data sets was invited. Bob Kirshner, my supervisor for his general expertise and Bruno Leibundgut who had access to European facilities.'[21] The team started with about 15 people, and eventually grew to 20.

In 1995 they found their first supernova –SN 1995K, which turned out to be extremely interesting. It was 40% fainter than he had expected. 'This single object seemed to show that the universe was speeding up,'[21] Schmidt said, quite excitedly. But to provide concrete evidence they needed more objects.

1998 was Schmidt's year. It began with publishing several papers which tackled the problem of the state of the universe, experimental techniques to study the universe and the equation of state of the universe.[29,30,31,32] 'At the end of 1997, Peter Garnavich had put together four new objects (five in total), and these showed that the universe was not slowing down quickly. It was a month later, when Adam Riess (along with Schmidt and Garnavich – who were the three postdocs in the group) assembled fourteen objects, that we saw the signal of acceleration',[21] Schmidt said. In 1998 they showed that the universe wasn't slowing down at all. Indeed, it seemed to be speeding up. 'It was a big thing because here we have observations that the universe is speeding up. What does it mean? It means that the universe has to be full of an energy which pushes on the universe rather than pulling it. We were telling the world that seventy per cent of the universe is made up of a material that you did not know existed. And we did this with the Perlmutter group who it turned out had independently made the discovery. Our papers came out ahead of theirs. They hit the headlines before we did. Anyway, the important thing is that the two groups were independent.'[21]

Figure 10.10: Brian Schmidt (shown here), with Saul Perlmutter and Adam Riess, discovered that the expansion of the universe is accelerating. It was one of the greatest astronomical discoveries of the 20th century. Photo: *The Australian.*

It was a remarkable discovery, all the more so because although there was not much love lost between the two groups when they began this quest they ended up agreeing that the universe is accelerating. The discovery was *Science* magazine's discovery of the year for 1998. Einstein's blunder was not a blunder after all: earlier in the 20th century Einstein had added a cosmological constant to his General Theory of Relativity to balance the motion of the universe so that it would be stationary, following the paradigm of a stationary universe that prevailed at that time. Did the accelerating universe confirm Einstein's theory? According to Schmidt, 'In some sense it does. I'm not sure whether it confirms it but it certainly is pointing towards the

Reactions to the accelerating universe

Brian Schmidt:

> *My own reaction is somewhere between amazement and horror. Amazement because I just did not expect this result and horror in knowing that it will likely be disbelieved by a majority of astronomers – who like myself are extremely sceptical of the unexpected.*

Adam Riess:

> *The results are very surprising, shocking even. I have avoided telling anyone about them for a few reasons. I wanted to do a few cross checks (I have) ... Approach these results not with your heart or head but with your eyes. We are observers after all.*

David Spergel (Princeton University):

> *The implications are so profound that they really need to assemble a more solid case.*

http://hubblesite.org/hubble_discoveries/dark_energy/

Figure 10.11: The discovery that the universe is accelerating hit the front page of *Science* as the 'Science Breakthrough of the Year' for 1998. A shocked Einstein looks on in amazement.[53] Photo: © 1998 American Association for the Advancement of Science.

cosmological constant. But it is hard to understand why the cosmological constant is so small and not zero.'[21]

Supernovae provided the evidence for the acceleration of the universal expansion, but when combined with measurements of the cosmic microwave background (CMB) an astonishing picture of the universe emerges. The measurements allowed the astronomers to pin down how much dark energy (Ω) and dark matter (Ω_m) the universe contains. The supernova data gave a value of $\Omega_m - \Omega$ and the cosmic microwave background gave a value of $\Omega_m + \Omega$. In their paper on the constraints in cosmological models, the High-Z Supernova team combined the data from supernova with those of the CMB[30,33] and to their amazement found $\Omega_m = 0.3$ and $\Omega = 0.7$. In effect, the results were telling the astronomers that in the early universe matter dominated the dynamics, but at some point in cosmic time dark energy took over to give us the accelerating universe in which we live today. But of course, astronomers still don't have a clue what the dark energy is that is causing the acceleration of the universe.

Has the problem of the cosmological parameters been completely solved or are there other problems to be addressed? According to Schmidt, 'I'm fairly heretical on this. And I believe we've done most of it. The theorists are aching to show that it isn't the cosmological constant. My view is we've shown that it's close to the cosmological constant. I think we have done most of what we can do and it's time to move on to other problems which are more interesting.'[21]

Schmidt has turned his attention to the SkyMapper project, with which he hopes to construct a comprehensive digital map of the southern sky. It will be used, 'to find very rare objects and help us answer a whole range of different science issues'.[21] He wants to find out how the universe was turned on, at what age this happened and the process behind it. Another interest 'is to look at the first stars in our galaxy. We can pinpoint the stars that have almost nothing other than hydrogen and helium in them. And right now this university [the Australian National University] is leading the world in this area of research'.[21]

Mapping the universe: the 2dF Galaxy Redshift Survey

The third cosmological project was undertaken by Matthew Colless. Colless began his astronomical career at the Institute of Astronomy at Cambridge University on a Shell Australia Science and Engineering Scholarship. It was there that the seeds of the Two Degree Field Galaxy Redshift Survey (2dF Galaxy Redshift Survey) were sown. Colless began his PhD thesis working with Craig Mackay, one of the first to recognise the importance of the Charge-Coupled Device (CCD) for astronomy. However, their Heath Robinson multi-object spectrograph did not produce any data for Colless – not because it didn't work, but because they had terrible weather.

After almost a year without any data Mackay wisely suggested that Colless should change to a new topic. Thus, he began work with Paul Hewett, with a lot of advice from George Efstathiou, on the dynamics of clusters of galaxies. 'The focus', he said, 'was very much to measure the amount of dark matter in clusters of galaxies.'[34] He was able to show that clusters of galaxies were not dynamically relaxed systems. 'There was clear evidence for substructure and strong support for the hierarchical formation of structure in the universe that was being then promoted by Simon White and George Efstathiou and Carlos Frenk.'[35] His PhD work gave Colless a chance to study both the cosmology of dark matter and the properties of galaxies. He looked at the luminosity function of the galaxies, which had been 'a thread through my scientific work'. The work he did for his PhD was also important: 'It was the first application in a very clear and effective way of using the multi-object spectroscopic technique, another major thread in my whole career. I have been fortunate

Figure 10.12: Matthew Colless worked with an international team to carry out the 2dF Galaxy Redshift Survey, which reached out to a distance of 3 billion light years. Photo: Australian Astronomical Observatory.

enough to pick areas to work that have blossomed. Cosmology and the high-redshift universe were the preserve of a small fraction of astronomers when I started, yet today this is one of the biggest fields of astronomy. I have been incredibly fortunate to be part of that Golden Age of cosmology.'[35]

Richard Ellis, on a visit to Cambridge, offered Colless a postdoctoral fellowship at the University of Durham even before he had finished his PhD. Before settling down for a three-year stint at Durham Colless spent a six-month period at Kitt Peak National Observatory, which was important for his career development. He began an important collaboration with Roger Davies (Oxford University), Dave Burstein (Arizona State University) and Gary Wegner (Dartmouth), who were beginning their studies on peculiar velocities of galaxies.

'What really made me was the three-year period that I had with Richard Ellis',[35] said Colless. Ellis had been involved with Keith Taylor at the Anglo-Australian Observatory in building an instrument called the Low Dispersion Survey Spectrograph (LDSS) and had hired Colless for the project. 'We were able to show convincingly that there really had been significant evolution of the galaxy population out to redshifts of 0.3 to 0.5.'[36] It was a remarkable result at that time. According to Colless, 'It got a lot of publicity and produced a couple of papers that are still highly cited. That was undoubtedly the breakthrough episode of my career. Richard Ellis deserves an enormous amount of credit for that work and certainly helped make me have those early achievements that made a big difference to my career.[35]

Colless is immensely proud of his work on the structure and dynamics of the Coma Cluster, which began at this stage of his career and continued when he moved back to Australia. He described it as being 'more synthetic' than his other work, which had been concerned with pushing instrumental and observational boundaries, doing big surveys and interpreting those within a fairly limited conceptual framework. The idea of the study was to try to get 'every bit of information about it that we could and then try to synthesize an understanding of how that cluster was formed'.[35] It was triggered by the very beautiful results from the ROSAT satellite (Rontgensatellit; in German X-rays are called Rontgenstrahlen, in honour of Wilhelm Röntgen, the recipient of the first Nobel Prize for Physics in 1901 for his discovery of X-rays) showing the X-ray structure in the Coma Cluster. He and Gus Oemler used the Hydra multi-object spectrograph at Kitt Peak to take several hundred redshifts for the Coma Cluster, and he later obtained a lot more with the 2dF at the Anglo-Australian Telescope (AAT). 'We found that it hadn't formed in a simple monolithic way. In fact, it had formed by accreting other structures which had been partially absorbed and it was still in the process of absorbing another structure even today. We were able to put forward a consistent life history of the Coma Cluster and of the dominant galaxy within it.'[37]

In the Autofib Redshift Survey, his group focused very much on the galaxy luminosity function. 'Up until then, people – including ourselves – had been doing single magnitude limit samples. And a single magnitude sample is a rotten way of trying to understand a wide range of magnitudes.'[35]. So they developed a new strategy. 'What we decided instead to do – and this is a strategy that's now quite common – is a sort of "layer cake" effect, where you have a very wide survey nearby. We used the Durham Anglo-Australian Redshift Survey of Ellis, Peterson and Shanks, and we combined that with our own LDSS surveys, which were the deep end, and then filled that in with the Autofib survey, which was intermediate both in depth and in sky coverage. And in that way we actually built up a sample that had good coverage over a range of luminosities and allowed us to study the luminosity function over a good range, from giant galaxies down to nearly dwarf galaxies at a range of different redshifts.'[38] This was a unique study and a vital component in studies of galaxy evolution, and is still widely cited.

In 1993 Colless returned to Australia as a Research Fellow at the Mount Stromlo and Siding Spring observatories of the Australian National University. By this time he had acquired the expertise necessary to chart the far corners of the universe, even as his hero Captain Cook had in an earlier century charted the far corners of Earth. 'I came back to Australia with the very clear intent of setting up a project team here to match the one that was forming in the UK and running the Two Degree Field program.'[35] Richard Ellis and Colless became the co-leaders of the 2dF Galaxy Redshift Survey, 'with the explicit goals of studying both cosmology and the galaxy population with a precision and with a sample vastly superior to anything that had been done before'.[35] They planned to do, on a much grander scale, the work that had been pioneered by Margaret Geller, John Huchra, Mark Davies and others at the Centre for Astrophysics in the US.[39,40]

But they had to convince the Anglo-Australian Observatory and the user community that their project was an important piece of science and that they needed very large numbers of nights to carry it out successfully. This was a break with tradition. Typically, astronomers using the AAT had been allocated a few nights on the telescope to carry out their projects and produce one or two papers. 'Proposing that we needed hundreds of nights was an enormous break with tradition and everyone was saying that this would destroy the scientific productivity of the telescope and the world would come to an end and the sky would fall and nothing would come of it, and so on,'[35] said Colless. It was a long and difficult struggle to convince people that this was really the way forward, particularly for the AAT; that it was coming into an era of 8 m telescopes, that it had to very much play to its strengths of wide-field spectroscopy, of the sort of cosmology that the UK community in particular had been doing so well for such a long time. Colless and his team won the argument – eventually.

The 2dF spectrograph was being built when Colless came back to Australia, and through his connections with the LDSS and Autofib surveys he already knew Keith Taylor at the Anglo-Australian Observatory. 'Taylor and Richard Ellis were very much the brains behind the idea of having a two-degree field',[35] he said. Charles Wynne was another person who was critical of the 2dF spectrograph, pointing out that the AAT was configured in such a way that it would be possible to build for it a corrector that gave a very wide field of view – a 2° field of view – which was unheard of on 4 m telescopes. 'Keith, Richard and Charles had got together and come up with this idea of using Charles' wide field corrector, with the knowledge that Keith had of fibre spectroscopy and robots', according to Colless.[41] They built a system

Figure 10.13: The 2dF spectrograph mounted on the top end of the 3.9 m telescope at Siding Spring Observatory. The instrument allows spectra to be taken of some 200 galaxies at a time. Photo: Australian Astronomical Observatory.

that was very much targeted at both galaxy population studies, the sort of work that Ellis and Colless had been doing and that Geller and Huchra had been doing in the US, but on a larger scale.

The 2dF Galaxy Redshift Survey produced more than 50 papers, of which more than 20 were in the highly cited category. Colless and Ellis had been vindicated. The main goals of the survey ranged from measuring the amount

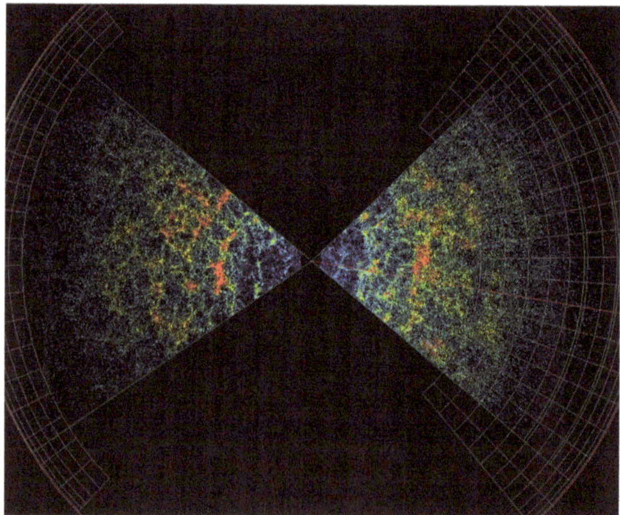

Figure 10.14: The results of the 2dF Galaxy Redshift Survey. The estimated distances to a large number of galaxies are plotted over the parts of the sky visible from Siding Spring. The clumpy distribution of galaxies is immediately evident. Image: Australian Astronomical Observatory.

of dark matter in the universe by looking at the galaxy distribution, to specific studies of the galaxy population with this very large sample.

A highly cited paper in *Nature* in 2001 gave their first results on large-scale structure in the universe.[42] They used the power spectrum for statistically characterising the galaxy distribution, which they found to be a very complex distribution with a web-like structure. By looking at the shape of the power spectrum they were able to tell what sort of dark matter it is: 'It turns out the shape of the power spectrum is precisely the shape for a cold dark matter dominated universe as predicted by theoreticians in the 1980s.'[43] The galaxy survey, according to Colless, also gave a 'second nearly independent way of getting an estimate of dark matter'. They have since updated and refined their 2001 paper and now have a total mass density of the universe which 'is 23% of the critical density, and we have that measurement to slightly better than 10% precision'.[44]

There were two other outcomes of the survey. One was the detection of baryonic acoustic oscillations: 'It is a tool that we are using to try to measure the geometry of the universe at different times to get a handle on the dark energy, because the dark energy affects the geometry,'[35] said Colless. The second was that they were able to set an upper limit to the mass of the neutrino.

The 2dF Galaxy Redshift Survey was Colless' greatest achievement: 'The single biggest thing I've done is the 2dF Galaxy Redshift Survey. It is, without doubt, the thing that I've done that has had the most impact, both in importance and across the field of astronomy.'[35] In 2004 Colless left Mount Stromlo to take up the directorship of the Anglo-Australian Observatory.

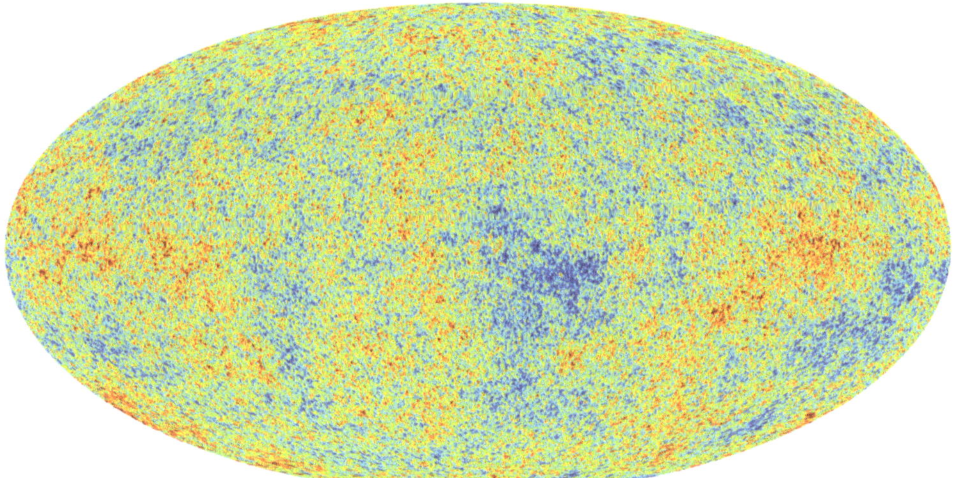

Figure 10.15: The WMAP data has recently been superseded by data from the Planck satellite, shown here. The image shows a detailed map of the oldest light in the universe, emitted when the universe was only 380 000 years old. The temperature fluctuations (shown as colour differences) correspond to the seeds that grew to become the galaxies we see today. The data from these satellites confirmed that the geometry of the universe is flat and the age of the universe is 13.8 thousand million years. Image: ESA.

The outcome of the three cosmology projects was complementary and gave a new view of our universe. In 2001, the Wilkinson Microwave Anisotropy Probe (WMAP) was launched, to measure the CMB. By February 2003 the first data were released, confirming that the geometry of the universe on a large scale is flat – to be expected if inflation is correct. Furthermore, when the WMAP measurements were combined with the ground-based evidence, the best age of the universe[45] was found to be ~13.7 ± 0.2 billion years, the long-debated Hubble Constant was found to be 71 ± 4 km s^{-1} Mpc^{-1} and the composition of the universe was found to be $\Omega_m = 0.27$ and $\Omega = 0.73$.

The measurements taken by the Mount Stromlo researchers and their international colleagues, together with the cosmic background results, show a universe mixed with dark matter and dark energy. For the first 7 billion years of its existence the universe slowed down due to dark matter, then it shifted to acceleration when dark energy drove the cosmic expansion.

Stars and globular clusters

An alternative way to do cosmology is to look at stars in nearby galaxies, including in the Milky Way, and try to understand what they tell us about the early history of the universe – near-field cosmology, it is often called.

John Norris and Mike Bessell began to study stars with very low metal abundances. These stars are among the earliest that were formed in the Milky Way. They were an important area of study since their chemical composition reflects the composition of the gas from which they were formed, thus they provide crucial clues to the star formation history and the synthesis of chemical elements in the early universe.

According to Christlieb (Hamburger Sternwarte, Germany) and Bessell, 'They are the local relics of epochs observable only at high redshifts;[46] if totally metal-free ("Population III") stars could be found, this would allow the direct study of the pristine gas from the Big Bang. Earlier searches for such stars found none with an iron abundance less the 1/10 000 that of the Sun, leading to the suggestion that low mass stars could form from clouds above a critical iron abundance.'[47]

In was quite an amazing discovery when they found a low-mass star (HE0107–5240) with an iron abundance as low as 1/200 000 of the solar abundance of iron. This discovery suggested that the first-generation stars could contain long-lived low-mass objects which might still be discoverable. According to Bessell, 'interestingly enough these stars, even though they were low in iron, they were very high in carbon, oxygen, nitrogen and sodium'. There seems to be an explanation for this. 'So what we believe is HE0107-5240 was formed from the supernova ejecta of a very massive 25 solar mass first generation supergiant formed soon after the Big Bang. It produced a lot of carbon and nitrogen through normal chemical evolution of hydrogen and helium so it had an outside envelope of carbon, nitrogen and oxygen. This was ejected partly with stellar winds and then the star underwent a massive supernova explosion but there was so much envelope on top of the star that the supernova did not completely break out. A lot of the material got remixed and reprocessed but only a little was ejected. Most of it fell back into a black hole. So you have a 25 solar mass star which ends up as a 20 solar mass black hole. A lot of the carbon, nitrogen and oxygen was expelled in the interstellar medium but only a very small percentage of iron escaped.'

Globular clusters of stars have long been an active area of study at Mount Stromlo. They are the oldest known composite objects in our Galaxy. Their chemical and kinematic properties and their spatial distribution provide information on the formation and early evolution of the Galaxy and their relative nearness to the Sun allows detailed observations to be made of their structure. According to Freeman, 'The chemical inhomogeneities observed in individual clusters present intriguing problems in the evolution of low mass stars and the chemical evolution of the clusters themselves.'[48]

Along with 47 Tucanae, Omega Centauri is one of the most well-known southern globular clusters. Omega Centauri has been studied for over 20 years since it was shown, in the 1970s, to possess a wide range of chemical abundances. In 1973, Russell Cannon and Bob Stobie[49] carried out a photometric study of 126 bright stars in Omega Centauri, which showed that Omega Centauri had an intrinsically wide giant branch. However, they noted that it was premature to speculate on the possible explanations for the width of the Omega Centauri giant branch. Freeman and Rodgers[50] were the first to measure a metallicity spread directly among RR Lyrae stars in Omega Centauri. Their observations revealed diversity in the chemical composition of these stars. Arising from this, several papers were published on CN and metallicities in Omega Centauri, leading to the papers written by Norris, Freeman and other astronomers in the mid 1990s. According to Norris and Da Costa, 'Omega Centauri has the distinction among Galactic globular clusters of showing not only

Figure 10.16: HE0107–5240 was discovered by Norris and Bessell to have less than 1/200 000 the heavy element content of the Sun, making it among the earliest stars to have formed in the Milky Way. Photo: European Southern Observatory.

Figure 10.17: Ken Freeman (centre) with the President of the American Institute of Physics (left) and Robert Gehrz, the President of the American Astronomical Society. Photo: American Astronomical Society.

evidence for evolutionary phenomena (as do all such systems) but also the clear signature of primordial chemical enrichment. As such, it offers one the possibility of insight into the manner in which a relatively isolated system enriches itself and of the stars responsible for the enrichment.'[51] They studied 40 red giants in the chemically inhomogeneous globular cluster with a view to 'disentangling the primordial and evolutionary phenomena, and thus to gaining insight into the manner in which primordial enrichment occurred'. Among other things, they found that chemical enrichment in Omega Centauri had occurred over an extended period of perhaps greater than one Gigayear.

International honours for Mount Stromlo astronomers

Mould ended his directorship of Mount Stromlo Observatory in 2002 on a high note, with many of the Observatory astronomers being awarded international honours during his tenure.

Ken Freeman had been elected a Fellow of the Royal Society of London in 1998 for his work on the formation of galaxies and globular clusters and dark matter in galaxies. The Royal Society is one of the oldest scientific societies in the world and some of the world's greatest scientists have been Fellows, such

Figure 10.18: Citation Laureates. Front row, left to right: Jeremy Mould, Ken Freeman and Bruce Peterson. Back row, left to right: Michael Dopita, Mike Bessell and Matthew Colless. Photo: Mount Stromlo Archives.

as Isaac Newton, Albert Einstein and Niels Bohr. Other honours followed for Freeman, including the prestigious Dannie Heineman Prize in 1999.

In 2000 Schmidt was the inaugural winner of Australia's Malcolm McIntosh Award for Achievement in the Physical Sciences. And in 2001, six of Mount Stromlo Observatory's astronomers (out of 33 Australian scientists total) were selected by the Institute of Scientific Information as Citation Laureates: Mike Bessell, Matthew Colless, Mike Dopita, Ken Freeman, Jeremy Mould and Bruce Peterson.

Mould was a diplomat *par excellence* and got things done. He had achieved most of his planned programs at the Observatory and placed it 'in a much better position than it is today'. On the occasion of the 75th anniversary of the Observatory, Mould said that the Observatory had been successful because 'Maximising the responsibility we put in the researchers' hands is a winning strategy ... It must be the demonstrated excellent performance that sustains a School which continues to plot its course.'

John Norris took over as Interim Director until the arrival of Penny Sackett, eighth Director of the Observatory, on 22 July 2002.

References

1. Bhathal R 1996. *Australian Astronomers: Achievements at the Frontiers of Astronomy*. National Library of Australia, Canberra.
2. National Board of Employment, Education and Training, Australian Research Council. Discipline Research Strategies. 1995. *Australian Astronomy: Beyond 2000*. AGPS. Canberra.
3. European Southern Observatory. 1997. Annual Report.
4. Pockley P 1998. Seventh heaven for Australia's optical astronomers. *Nature* **391**, 724. doi:10.1038/35678.
5. McGregor PJ, Conroy P, Bloxham G, van Harmelen J 1999. Near-infrared Integral Spectrograph (NIFS): an instrument proposed for Gemini. *Publications of the Astronomical Society of Australia* **16**, 273–287. doi:10.1071/AS99273.
6. Frame T, Faulkner D 2003. *Stromlo: An Australian Observatory*. Allen and Unwin, Sydney.
7. Lucas S 2002. ANU's $6.3 m contract out of this world. *Canberra Times*, 3 December, p. 1.
8. Burton M, Aitken DK, Allen DA, Ashley MCB, Cannon RD, Carter BD, DaCosta GS, Dopita MA, Duldig ML, Edwards PG, Gillingham PR, Hall PJ, Hyland AR, McGregor PJ, Mould JR, Norris RP, Sadler EM, Smith CH, Spyromilio J, Storey JWV 1994. The scientific potential for astronomy from the Antarctic Plateau. *Proceedings of the Astronomical Society of Australia* **11**, 127–150.
9. Overbye D 1991. *Lonely Hearts of the Cosmos*. Picador.
10. Mould J, Aaronson M, Huchra J 1980. A distance scale from the infrared magnitude/HI velocity-width relation: the Virgo cluster. *Astrophysical Journal* **238**, 458–470. doi:10.1086/158002.

11. Kirshner R 2002. *The Extravagant Universe*. Princeton University Press, Princeton.
12. Kennicutt RC, Freedman WL, Mould JR 1995. Measuring the Hubble Space Telescope. *Astronomical Journal* **110**(4), 1476–1491.
13. Aaronson M, Mould J 1986. The distance scale: present status and future prospects. *Astrophysical Journal* **303**, 1–9. doi:10.1086/164046.
14. Freedman WL, Madore BF, Mould JR, Hill R, Ferrarese L, Kennicutt RC Jr, Saha A, Stetson PB, Graham JA, Ford H, Hoessel JG, Huchra J, Hughes SM, Illingworth GD 1994. Distance to the Virgo cluster galaxy M100 from Hubble Space Telescope observations of Cepheids. *Nature* **371**, 757–762. doi:10.1038/371757a0.
15. Vandenberg DA, Bolte M, Stetson PB 1996. The age of the Galactic globular cluster system. *Annual Review of Astronomy and Astrophysics* **34**, 461–510. doi:10.1146/annurev.astro.34.1.461.
16. Freedman WL 2000. The expansion rate and size of the universe. In *The Scientific American Book on the Cosmos.* (Ed. DH Levy). Macmillan, London.
17. Sebo KM, Rawson D, Mould J, Madore BF, Putman ME, Graham JA, Freedman WL, Gibson BK, Germany LM 2002. The Cepheid period–luminosity relation in the Large Magellanic Cloud. *Astrophysical Journal* **Supplement 142**, 71–78.
18. Chaboyer B, Demarque P, Kernan PJ, Krauss LM 1998. The age of globular clusters in light of Hipparcos: resolving the age problem? *Astrophysical Journal* **494**, 96–110. doi:10.1086/305201.
19. Mould J 1998. The age of globular clusters. *Nature* **395**, A20–A22.
20. Freedman WL, Madore BF, Gibson BK, Ferrarese L, Kelson DD, Sakai S, Mould JR, Kennicutt RC, JrFord HC, Graham JA, Huchra JP, Hughes SMG, Illingworth GD, Macri LM, Stetson PB 2001. Final results from the Hubble Space Telescope Key Project to measure the Hubble Constant. *The Astrophysical Journal* **553**, 47–72. doi:10.1086/320638.
21. Bhathal R 2006. Interview with Brian Schmidt for the National Oral History Project on Significant Australian Astronomers and Physicists. National Library of Australia, Canberra.
22. Bhathal R 2012. Profile: Brian Schmidt. *Astronomy & Geophysics* **53**, 1.13–1.15.
23. Eastman RG, Kirshner RP 1989. Model atmospheres for SN1987A and the distance to the Large Magellanic Cloud. *Astrophysical Journal* **347**, 771–793. doi:10.1086/168168.
24. Schmidt B, Kirshner RP, Eastman RG, Phillips MM, Suntzeff NB, Hamuy M 1994. The distances to five Type II supernovae using the expanding photosphere method, and the value of H_0. *Astrophysical Journal* **432**, 42–48. doi:10.1086/174546.
25. Leibundgut B, Kirschner RP, Phillips MM, Wells LA, Suntzeff NB, Hamuy M, Schommer RA, Walker AR, Gonzalez L, Ugarte P, Williams RE, Williger G, Gomez M, Marzke R, Schmidt BP, Whitney B, Caldwell N, Peters J, Chaffee FH, Foltz CB, Rehner D, Siciliano L, Barnes TG, Cheng K-P, Hintzen PMN, Kim Y-C, Maza J, Parker JW, Porter AC, Schmidtke PC, Sonneborn G 1993. SN 1991bg – A new Type Ia supernova with a difference. *Astronomical Journal* **105**, 301–313. doi:10.1086/116427.
26. Phillips MM 1993. The absolute magnitudes of Type IA supernovae. *Astrophysical Journal* **413**, L105–L108. doi:10.1086/186970.

27. Hamuy M, Phillips MM, Schommer RA, Suntzeff NB, Maza J, Aviles R 1996. The absolute luminosities of the Calan/Tololo Type IA supernovae. *Astrophysical Journal* **112**, 2391–2397.

28. Filippenko AV 2001. Einstein's biggest blunder? High-redshift supernovae and the accelerating universe. *Publications of the Astronomical Society of the Pacific* **113**, 1441–1448. doi:10.1086/324512.

29. Garnavich PM,, Jha S, Challis P, Clocchiatti A, Diercks A, Filippenko AV, Gilliland RL, Hogan CJ, Kirshner RP, Leibundgut B, Phillips MM, Reiss C, Riess AG, Schmidt BP, Schommer RA, Smith RC, Spyromilio J, Stubbs C, Suntzeff NB, Tonry J, Carroll SM 1998. Supernova limits on the cosmic equation of state. *Astrophysical Journal* **509**, 74–79. doi:10.1086/306495.

30. Garnavich PM, Kirshner RP, Challis P, Tonry J, Gilliland RL, Smith RC, Clocchiatti A, Diercks A, Filippenko AV, Hamuy M, Hogan CJ, Leibundgut B, Phillips MM, Reiss D, Riess AG, Schmidt BP, Schommer RA, Spyromilio J, Stubbs C, Suntzeff NB, Wells L 1998. Constraints on cosmological models from the Hubble Space Telescope observations of High-Z supernovae. *Astrophysical Journal Letters* **493**, L53. doi:10.1086/311140.

31. Riess AG 1998. Observational evidence from supernovae for an accelerating universe and a cosmological constant. *Astronomical Journal* **116**, 1009–1038. doi:10.1086/300499.

32. Schmidt B, Suntzeff NB, Phillips MM, Schommer RA, Clocchiatti A, Kirshner RP, Garnavich P, Challis P, Leibundgut B, Spyromilio J, Riess AG, Filippenko AV, Hamuy M, Smith RC, Hogan C, Stubbs C, Diercks A, Reiss D, Gilliland R, Tonry J, Maza J, Dressler A, Walsh J, Ciardullo R 1998. The High-Z supernova search: measuring cosmic deceleration and global curvature of the universe using Type 1A supernovae. *Astrophysical Journal* **507**, 46–63. doi:10.1086/306308.

33. Tonry JL, Schmidt BP, Barris B, Candia P, Challis P, Clocchiatti A, Coil AL, Filippenko AV, Garnavich P, Hogan C, Holland ST, Jha S, Kirshner RP, Krisciunas K, Leibundgut B, Li W, Matheson T, Phillips MM, Riess AG, Schommer R, Smith RC, Sollerman J, Spyromilio J, Stubbs CW, Suntzeff NB 2003. Cosmological results from High-Z supernovae. *Astrophysical Journal* **594**, 1–24. doi:10.1086/376865.

34. Bhathal R 2008. Profile: Matthew Colless *Astronomy & Geophysics* **49**, 5.16–5.18.

35. Bhathal R 2005. Interview with Matthew Colless for the National Oral History Project on Significant Australian Astronomers and Physicists. National Library of Australia, Canberra.

36. Colless M, Ellis RS, Taylor K, Hook RN 1990. The LDSS deep redshift survey. *Monthly Notices of the Royal Astronomical Society* **244**, 408–423.

37. Colless M, Dunn AM 1996. Structure and dynamics of the Coma cluster. *Astrophysical Journal* **458**, 435–454. doi:10.1086/176827.

38. Ellis RS, Colless M, Broadhurst T, Heyl J, Glazebrook J 1996. Autofib Redshift Survey: I. Evolution of the galaxy luminosity function. *Monthly Notices of the Royal Astronomical Society* **280**, 235–251. doi:10.1093/mnras/280.1.235.

39. Geller MJ, Huchra JP 1989. Mapping the universe. *Science* **246**, 897–903. doi:10.1126/science.246.4932.897.

40. Shectman SA, Landy SD, Oemler A, Tucker DL, Lin H, Kirshner RP, Schechter P 1996. The Las Campas Redshift Survey. *Astrophysical Journal* **470**, 172–188. doi:10.1086/177858.
41. Colless M, Dalton GB, Maddox SJ, Sutherland WJ, Norberg P, Cole SM, Bland-Hawthorn J, Bridges TJ, Cannon RD, Collins CA, Couch WJ, Cross N, Deeley K, De Propris R, Driver SP, Efstathiou G, Ellis RS, Frenk CS, Glazebrook K, Jackson CA, Lahav O, Lewis IJ, Lumsden S, Madgwick DS, Peacock JA, Peterson BA, Price IA, Seaborne M, Taylor K 2001. The 2dF Galaxy Redshift Survey: spectra and redshifts. *Monthly Notices of the Royal Astronomical Society* **328**, 1039–1063. doi:10.1046/j.1365-8711.2001.04902.x.
42. Peacock JA, Cole S, Norberg P, Price I, Bland-Hawthorn J, Jackson C, Paterson BA, Colless M 2001. A measurement of the cosmological mass density from clustering in the 2dF Galaxy Redshift Survey. *Nature* **410**, 169–173. doi:10.1038/35065528.
43. Percival WJ,, Baugh CM, Bland-Hawthorn J, Couch W, Jackson C, Peterson BA, Colless M 2001. The 2dF Galaxy Redshift Survey: the power spectrum and the matter content of the universe. *Monthly Notices of the Royal Astronomical Society* **327**, 1297–1306. doi:10.1046/j.1365-8711.2001.04827.x.
44. Cole S, Percival W, Peacock JA, Norberg P, Baugh C, Frenk CS, Baldry IK, Bland-Hawthorn J, Bridges TJ, Cannon RD, Colless M, Collins CA, Couch WJ, Cross NJG, Dalton GB, Eke VR, De Propris R, Driver SP, Efstathiou GP, Ellis RS, Glazebrook K, Jackson CA, Jenkins AR, Lahav O, Lewis I, Lumsden SL, Maddox SJ, Madgwick D, Peterson BA, Sutherland WJ, Taylor K 2005. The 2dF Galaxy Redshift Survey: power spectrum analysis of the final data set and cosmological implications. *Monthly Notices of the Royal Astronomical Society* **362**, 505–534. doi:10.1111/j.1365-2966.2005.09318.x.
45. MacRobert A 2003. Turning a corner on the new cosmology. *Sky and Telescope* **May**, 16–17.
46. Norris JE, Ryan SG, Beers TC 2001. Extremely metal-poor stars. VIII. High-resolution, high signal-to-noise ratio analysis of five stars with (Fe/H) < -3.5. *Astrophysical Journal* **561**, 1034–1059. doi:10.1086/323429.
47. Christlieb N, Bessell MS 2002. A stellar relic from the early Milky Way. *Nature* **419**, 904–906. doi:10.1038/nature01142.
48. Freeman KC, Norris JE 1981. The chemical composition, structure and dynamics of globular clusters. *Annual Review of Astronomy and Astrophysics* **19**, 319–356. doi:10.1146/annurev.aa.19.090181.001535.
49. Cannon RD, Stobie RS 1973. Photometry of southern globular clusters. I Bright stars in ω Centauri. *Monthly Notices of the Royal Astronomical Society* **162**, 207–225.
50. Freeman KC, Rodgers AW 1975. The chemical inhomogeneity of Omega Centauri. *Astrophysical Journal Letters* **201**, L71–L74. doi:10.1086/181945.
51. Norris JE, Da Costa GS 1995. The giant branch of Omega Centauri. IV. Abundance patterns based on Echelle spectra of 40 red giants *Astrophysical Journal* **447**, 680.
52. Schilling G 2003. Cosmology's treasure map. *Sky and Telescope* **February**, 32–39.
53. *Science* **282**, 18 December, 1998. Cover illustration: John Kascht. Albert Einstein™ represented by the Roger Richman Agency, Beverly Hills.

2002–2007

11

Bushfires and a new beginning

Penny Sackett has a long record of research on galaxies, and was also very active in a large search for extrasolar planets, using microlensing techniques. She was very keen to see Australia involved in one of the giant telescope projects, and through her initiatives Australia is now part of the Giant Magellan Telescope (GMT) construction.[29]

On a personal note I probably take greatest pride in watching Mount Stromlo go from strength to strength after the fires. It taught me so much about the human spirit, and what's really important in life. What was really important were not the telescopes or the buildings, but the character of the people, and their spirit, resolve, and their will.[1]

Penny Sackett was the first woman scientist to become the Director of a major astronomical institution in Australia. She took over the directorship of the Observatory in July 2002. She was no stranger to Australia. She had visited the country for sabbatical work at the Anglo-Australian Observatory (now the Australian Astronomical Observatory) and knew some of the key players in astronomy in Australia. Her reason for applying for the directorship was, 'I knew Australia would be a lovely place to live, and I knew that the astronomers were friendly as well as knowledgeable. And of course Ken Freeman, who's a national treasure, was one of the first Australian scientists that I knew. I am not sure if I would have had the bravery at that time to apply to be Director of such a prestigious institution if someone hadn't suggested it to me. But the more I thought about it, the more I thought it would be really a wonderful job because of the reputation of the people that worked there. Also because of the place it held in world astronomy – many people like to visit Mount Stromlo Observatory, you know, for a sabbatical, so there are always good people going through – and because of the match of the observing facilities to my own work. I was once again up for a challenge, and so for all of these reasons I decided to throw my hat in the ring'.[1]

She arrived with a mission. She had inherited a very strong research institution from the previous Director, Jeremy Mould: 'The faculty that I inherited was just stunning. You know, Bessell, Freeman, Peterson, Norris, Dopita and

Figure 11.1: Penny Sackett, Director of the Mount Stromlo Observatory from 2002 to 2007. Photo: Mount Stromlo Archives.

Woman astronomers in Australia

Figure 11.2: Yarrum Parpur Tarneen, arguably Australia's first woman astronomer. Photo: JD Collection.

The honour of being the first woman astronomer in Australia belongs to Yarrum Parpur Tarneen, the daughter of the chief of the Morpor tribe in Victoria. She was a social-cultural astronomer and taught not only her own people but also explorers, scholars and government officials about the Australian night sky from an Aboriginal perspective.[2] Women have played and continue to play an important role in astronomy in Australia. In the 1940s Ruby Payne Scott became the first woman radio astronomer and, along with Joseph Pawsey, Australia's pioneer radio astronomer, made several significant contributions to the development of radio astronomy. Today, several Australian women astronomers are leaders and work at the international frontiers of astronomy.

Schmidt. Just incredible. The scientists were absolutely first-rate, splendid teachers almost to a person, and so really it was a matter of trying to reinvigorate that with some younger blood as well and watch the tradition continue.'

Bushfires destroy the Observatory

About six months into Sackett's directorship a terrible catastrophe occurred. It began on a hot summer's day on 18 January 2003. Fires had been burning in nearby rural areas for some time and extensive preparations had been made to protect life and property. On that afternoon more than one fire front converged on the mountain, leaving a trail of devastation that had never been witnessed before. According to Sackett, 'When the damage was assessed it was clear that we had lost all of our research facilities – that is, all of our research telescopes, all of our library facilities, and our workshop where we had built instruments for our own telescopes and telescopes for other organisations.' The $4 million Near-infrared Integral Field Spectrograph (NIFS) which was undergoing final testing before being shipped to the Gemini North Telescope in Hawaii had turned a blackened melted mass. A lot of thought had been given to moving it, and a decision to put it in the inner part of the workshop was carefully considered but not acted upon. Project Scientist Peter McGregor said it was heartbreaking.[3] He feared the worst for the instrument when he heard that the fires were approaching. But at the last moment, the instrument was left behind.

Figure 11.3: The NIFS was a $5 million instrument for one of the largest telescopes on Earth, the Gemini North telescope in Hawaii, and was ready for shipment when destroyed in the 2003 bushfires. It was quickly rebuilt according to the design drawings by the team at Auspace Pty Ltd. Photo: Mount Stromlo Archives.

Figure 11.4: The 1.9 m telescope and its complement of instrumentation was judged a total loss after the bushfires. It has been left as a 'managed ruin' and is not open to the public. (a) The external view of the dome and building. (b) The internal view. The optics were melted and the telescope structure was deformed by the heat. All auxiliary instrumentation and facilities were also destroyed. Photos: Mount Stromlo Archives.

Figure 11.5: Gases built up as the interior of the several domes heated and a spark or ember caused ignition, melting the aluminium domes and destroying the telescopes. (a) Aerial view of the historic 66 cm Yale–Columbia refractor burning during the bushfires. The visitor centre to the right remained largely unaffected, as were the astronomers' offices across from the Commonwealth Solar Observatory building. (b) A view of the Yale–Columbia post-fires, graphically showing the melted aluminium dome. Photos: Mount Stromlo Archives.

It was a trying time for Sackett. A lesser person would probably have given up. She had several issues to solve in the rebuilding process. 'First and foremost, there wasn't any money. The insurers had a different view of what the policy was than the university did. That meant very, very long, protracted battles that still were not over at the time that I left as Director. We're talking tens of millions of dollars of damage. The university put up some of its own funds promptly, and the Federal Government also gave $7.3 million to help us rebuild the workshops. So we were very grateful for that. That was probably the biggest issue, at least in the beginning.'[1] She also had to tackle the heritage issues, since the whole mountaintop was heritage-listed. The Commonwealth Solar Observatory building had been designed by the architect (J.S. Murdoch) who had designed old Parliament House. There were 'really excellent compromises', she said.

Figure 11.6: The 1.27 m telescope post-bushfire. This telescope began life in 1868 as the Great Melbourne Telescope. It was transferred to Mount Stromlo when the Melbourne Observatory was closed after the Second World War and has been refurbished several times. (a) The aluminium dome melted during the fires and left the telescope exposed. (b) The primary mirror, 1.27 m in diameter, was melted and cracked by the intense heat. Photos: Mount Stromlo Archives.

Figure 11.7: The heritage Commonwealth Solar Observatory building, in recent years serving as quarters for the library, admin staff and meeting rooms, was completely gutted. The library was reduced to ashes. The only facility spared was the precision gravimeter housed in the basement under the library, in what originally was the heliospectrograph room. This view from above shows the Farnham telescope dome to the left. Photo: Mount Stromlo Archives.

Figure 11.8: The historic 22 cm Oddie Telescope and building were destroyed by the bushfires. The Oddie was the first telescope on Mount Stromlo and went into operation for site testing in September 1911. The building was the first constructed by the Commonwealth government in the then new Federal Capital Territory. Photo: Mount Stromlo Archives.

The other problem was staff morale: 'As time went on, what became difficult was managing staff morale, because directly after the fire most people had, you know, this sense of fight in them, that "No, we're not going to let this get us down. No, this isn't the end of Mount Stromlo Observatory." But when the rebuilding began to take much longer than people had anticipated, and when the insurance agencies in particular were recalcitrant, it became a little bit harder to keep staff morale up.'[1]

Nevertheless, Sackett drew up a strategic plan for 2003–07 in consultation with the staff, that described where the Observatory should be heading. This included 'things like developing new leadership roles in the next generation of telescopes, increasing the number of graduate students that came to Mount Stromlo, increasing the efficiency of the telescopes at Siding Spring and supporting those that were most productive, expanding the breadth of the knowledge of the faculty, and maintaining and growing our instrumental and astro-engineering capacity'.

Finding an astronomical niche

Sackett had arrived at Mount Stromlo with a track record of astronomical research and highly cited publications. After graduating with a PhD in theoretical physics in 1984 from the University of Pittsburgh, it took her some time to find a niche. She completed postgraduate studies at a time when there were very few women physicists in the US and it was difficult for women physicists to get jobs. Linda Schiebinger, a historian of science at Pennsylvania University, noted that even as late as 1996 in the US the unemployment rate for female PhD physicists remained twice that of male peers (3.8% compared with 1.9%) after controlling for job experiences.[4] According to Sackett, 'It was difficult because there weren't very many women in physics. I think too that the networking opportunities simply weren't there. In general women were not mentored in the same way that men were mentored with respect to finding a job.' There also seemed to be some prejudice against women in a highly male-dominated discipline. The male locker-room mentality was still prevalent in US academic circles. This was also the case in Australian academic circles in the 1980s. Sackett gives a classic example of this: 'I remember particularly once a famous scientist coming to visit when I was in graduate school, and the male graduate students were asked to go out to dinner with him, but I was asked to babysit one of the faculty's children so that he could go. I didn't do the babysitting, but then I didn't go to dinner either.'[1]

She worked at several jobs: as a science journalist for *Science News*, as a teacher at Amherst College (a privately funded college attended by many of America's privileged minority), as a researcher at the Biological Sciences Department at the University of Pittsburgh and as a Program Officer at the National Science Foundation. Her breakthrough came when she won a small grant to study at the Kapetyn Astronomical Institute in the Netherlands, a major centre of astronomical research. Some of the well-known astronomers when she was there were Stuart Pottasch, who was known for his work on planetary nebulae, and Renzo Sancisi and Tjeerd van Albada, who were internationally recognised for their work on the distribution of dark matter in galaxies. The short stint provided her with the transition from being a theoretical physicist to becoming an observational astronomer. Two years (1992–94) at the Institute for Advanced Study at Princeton helped her to begin tying together several elements that she was interested in, namely the structure and dynamics of galaxies which were related to dark matter: 'And one way to study dark matter is through microlensing. So all of these connections were sort of forged while I was at the Institute for Advanced Study in Princeton.'[1] Princeton, she said, was 'extraordinarily important to me'. It provided a springboard for the work she was going to do at the Kapteyn Institute, where she worked for the next seven years, focusing exclusively on microlensing. It was during her time at Princeton that she realised 'through the work of Bohdan Paczynski and Andy Gould that microlensing could be a way to look for planets, and not only the so-called MACHOs (Massive Compact Halo Objects) that [were thought to] make up dark matter'.[1] This was the direction she began to increasingly explore at the Kapetyn Institute.

Extrasolar planets around a Sun-like star were first discovered by Swiss astronomers Michel Mayor and Didier Queloz in 1995. Mayor and Queloz were hailed as the Galileos of the 20th century and they opened up an area of research which has become very active and competitive. The method they used is called the radial velocity or the Doppler technique. The more direct method is called the transit technique. In this technique, astronomers look for planets that partially eclipse the star that they are orbiting. This causes a temporary but periodic dip in the brightness of the parent star. However, Sackett employed the gravitational microlensing technique to search for extrasolar planets. It is based on the fact that gravity affects light, an idea that originated with Einstein. The microlensing method is a statistical technique that produces data about the planet population as a whole, while the Doppler and transit techniques are good methods for studying individual planets, especially the most massive ones.

According to Sackett, 'Gravity not only affects objects with mass, but it affects light. It can actually cause light to bend. And so when a massive object

Giordano Bruno

Michel Mayor and Didier Queloz confirmed what Giordano Bruno had written in the 16th century in *On the Infinite Universe and Worlds* – that there were other planets circling around other Suns in our universe.[5] This was in conflict with the teachings of the Bible. For this and other 'heretical thoughts' Bruno got in trouble with the Pope and was burnt at the stake on 17 February 1600 at the Piazza Campo dei Fiori in Rome, where his statue now stands. Even the erection of the statue in 1889 met protests from the Vatican. In 2000, on the 400th anniversary of Bruno's burning, the Vatican offered regret for his death, but no apology. Cardinal Paul Poupard told a Jesuit-sponsored symposium on Bruno that Bruno's thinking was 'incompatible with Christian thought'. It is interesting to note that adjacent to the statue stands the 'Fahrenheit 451' bookstore, the name of which refers to Ray Bradbury's 1953 novel on freedom of thought (http://beliefnet.com/story/11801.html).

sits between a luminous object that's emitting light and a detector (an astronomer with a telescope) it can change the path of that light in such a way that more light reaches the observer than otherwise would, simply because there's some massive object sitting in-between. And that massive object is called a microlens, and you can think of it most often as just a normal star somewhere in the galaxy. It might be a star we can see, or it might be one we can't. If that star has a planet orbiting it, then the gravitational field will not just be the gravitational field of the star but the gravitational field of the star plus the planet. It's not so important that the field is stronger, because after all the planet is just a tiny fraction of the total mass of the star, but the distribution of that gravitational field is different. And that means that as a bright object such as a background star sends its light to the telescope, and it's moving as it does so because everything in the galaxy moves, it probes different parts of this gravitational field, which looks different in different places because of the presence of the planet. We can do what's called "inverting" that problem by taking the light curve that we observe with the telescope and deducing what the gravitational field must have been, and in particular whether it was due to a planet orbiting a star. The remarkable thing about microlensing is that you don't even have to see the star that the planet is orbiting in order to detect planets. You're using the natural effect of the lensing that Einstein talked about to create a telescope that is more powerful than the ones you actually have available to you here on earth.'[1]

In the same year (1995) that Michel Mayor and Didier Queloz discovered the first planet going around a normal star, Sackett formed the PLANET collaboration which consisted of astronomers from various parts of the world that could provide continuous time coverage. One of her motivations was the 'possibility of being the first to detect an extrasolar planet', although that honour was to go to Mayor and Didier. Her second motivation related to her previous work on dark matter in galaxies, in particular with respect to the physics of gravitational lensing and potential use of the technique to discover and characterise extra-solar planets. The chance to find out whether they understood the microlensing technique in their search for extrasolar planets came from their observations of the binary microlens MACHO-1997-BLG-41. According to Sackett, when they analysed their data they found that they were seeing 'not a planet and a star but a double star system'.[6] By itself, this was not extraordinary. What was extraordinary was the fact that they could 'deduce this just from the microlensing light curve alone'.[1] This gave them confidence in their ability to understand and use the technique.

Her observations of MACHO-98-SMC-1 in the Small Magellanic Cloud provided an interesting insight into MACHO events and dark matter. At the time they were doing this research, one of the big questions that microlensing was expected to answer was whether dark matter was made up of small compact objects, such as dead stars, or Jupiters or some unknown objects that could cause a microlensing event. Many astronomers at that time were looking at stars in the Magellanic Clouds and seeing whether the light from them exhibited a microlensing curve as if there was a dark unseen microlens between the observer and the Clouds. Whenever astronomers received warning from microlensing teams that a microlensing event was in progress, they could turn their telescopes to it. In the case of MACHO-98-SMC-1, the PLANET team obtained an absolutely stunning light curve: 'This was good enough for us to detect small things in this light curve that weren't – at least at that time – typically observed. And those included being able to tell when one part of the background star passed a particular point in the gravitational pattern, compared to when a different part of the star crossed.'[1] This allowed them to measure, among other things, the proper motion of the microlensing event. They found 'that alas, in this particular event, it wasn't some mysterious dark matter that was causing the microlensing, it was in fact a typical star in the Magellanic Clouds'.[7]

A lot of hard work goes into researching a problem. Quite often one goes down wrong paths and dead ends. However, sometimes the work is rewarded, like Archimedes, with a Eureka or 'Aha' moment. This happened to Sackett: 'it relates back to this problem of dark matter – it had been suggested by some that indeed the dark matter in our Galaxy must be in the form of these MACHOs (Massive Compact Halo Objects) that cause microlensing, and that in fact a colour magnitude diagram would actually show the population of these objects, some of which were stars. Now, what I mean by a colour magnitude diagram is that you plot the brightness of the star on one axis and the colour on the other, and if the objects are all the same age and same distance, then there'll be a pattern that develops on this plot that's quite characteristic. But if some of the stars are a little bit closer, and therefore, say, not in the Magellanic Clouds but between us and the Magellanic Clouds, then they would show up at a slightly different place on the colour magnitude diagram. And it was proposed that something like this was actually seen.'[1] There were several reasons why this suggestion bothered Sackett but, as she said, 'It was such a simple idea. It seemed to be there for all to see, this extra little addition in the colour magnitude diagram.' She was working at the Kapteyn Institute at the time, and to reach the second floor where she worked she had to walk by the rack of new journals. On the cover of one was new data from the Hipparcos satellite, showing the colour magnitude diagram not of the Large Magellanic Cloud but of the nearby stellar neighbourhood. In that, she saw the very same additional feature. And that, she said, 'was my aha moment, because I realised, this has nothing to do with the Large Magellanic Cloud. It has nothing to do with intervening objects that are in-between us and the Cloud. Nothing to do with dark matter. It actually has to do with stellar evolution. There is a period in stars' lives when they occupy this particular part of the colour magnitude diagram, and how many stars are there depend on how old the different populations are. And so with a postdoc we took many colour magnitude diagrams that he already had for the Large Magellanic Cloud. We analysed them this way, we showed that this extra feature was exactly where you would expect it to be, and it was occupied with precisely the number of stars you would expect it to be given the LMC's star formation history ... That aha moment was seeing in a completely different context the same signal, and therefore immediately knowing that this explanation was different than what had been originally proposed.'[8]

Extrasolar planets were not Sackett's only interests. Over the last two decades, several studies had been conducted to understand the evolution of

Figure 11.9: Polar ring galaxy NGC 4650A. This galaxy has material in unusual orbits around its centre, allowing astronomers to estimate the amount of unseen mass (dark matter) in the system. The outer ring of gas and stars is rotating too fast for the galaxy to have only the mass that is visible. Photo: J. Gallagher (UW-M) *et al.* and the Hubble Heritage Team (AURA/STScI/NASA).

galaxies. For example, there has been a realisation that interactions between galaxies may play an important role, and observations have been made of shells and rings around elliptical galaxies. Sackett and her colleagues carried out new observations and produced a photographic atlas of polar ring galaxies.[9] Some astronomers were interested in the history of these galaxies and wanted to know what happened to these odd-looking objects. Her interest in these fascinating galaxies arose from her desire to study the shape of dark matter haloes and how dark matter was distributed. Astronomers deduced the presence of dark matter in galaxies largely through rotation curves. They measured the speed of individual stars in the galaxy as a function of their distance from the centre, and this showed how much force they were experiencing and therefore how much mass the galaxy has that is acting upon that particular star or set of stars. This is how astronomers became confident that there must be more mass in the galaxy than we can see, and there must be dark matter. With Linda Sparke, Sackett studied the polar ring galaxy NGC 4650A when she was at the Kapetyn Institute for the first time, on a small grant from the National Science Foundation. NGC 4650A is a strikingly beautiful object with a central body of older stars in the middle, somewhat flattened into an elliptical shape. This is surrounded by a gigantic ring of gas and much newer stars that are orbiting in a completely perpendicular plane to the way the stars were orbiting in the central object.

They studied this galaxy because they could measure the rotation curve not only in the plane of the central body where there were stars, but also in a plane perpendicular to it where there were stars and gas they could analyse. According to Sackett, by studying these 'you could ask the question: Do these two curves look the same because the distribution of matter is spherical? If this was the case it would not matter in which direction you looked because the gravitational field would be spherical. Or does the polar ring

Figure 11.10: NGC 5907, also called the Knife Edge Galaxy, is a spiral dial galaxy seen edge-on. Deep images reveal sweeping tails of starlight that must be the remnants of a small galaxy torn apart as it passed by NGC 5907. The motions of these tails allow astronomers to estimate the amount and distribution of unseen matter in the galaxy. Photo: Jay Gabany, Blackbird Observatory, New Mexico.

have a different rotation curve to that of the central body, which might indicate that the dark matter preferentially lies in one plane or the other.'[10]

From a more detailed analysis of the motions of the stars in this galaxy, 'We found, after taking into account elliptical orbits and all sorts of other things, that the rotation curve of the stars and gas in the polar ring is different than that in the central region. That immediately tells you that the total gravitational mass is not spherically distributed.' They knew that some of the matter is not spherically distributed because the galaxy is flattened and has a ring, so they needed to subtract the effects of the mass they could see,then ask: of the mass that is left over, how is it distributed? 'And the answer', Sackett said, 'was as near as we could tell within the precision of our measurements that the dark matter was also flattened – not as thin as a stellar disk, but sort of as flat as a rugby ball or something like that.' This work was important for astronomers who were trying to understand the composition of dark matter, because certain proposals for dark matter were calling for a very flattened disk while others were calling for an absolutely spherical distribution. Sackett and Sparke gave them 'something in-between'.[11]

In Sackett's study of the spiral galaxy NGC 5907 she found the galaxy had a faint luminous halo. What surprised her and her colleagues was that the distribution of that light was more extended than had been previously recognised. It did not fall off as sharply as typically expected in stellar haloes. 'We asked the obvious question – Is the light distributed the way we think dark matter is distributed? Because if it was, maybe there was a connection between this light we were seeing and dark matter.' *Nature*, the British scientific journal, decided to make it the cover story[12] and it landed them in a controversy. Critics began saying that Sackett and her colleagues were claiming that they had 'found what dark matter was', despite the fact, she said, 'that they were as cautious as they could possibly be in the write up of their abstract'. Nevertheless, one benefit of the controversy was that it prompted other astronomers 'to look for faint light around other galaxies, and it was found in many other galaxies as well', Sackett said. 'When people looked deeper, what they discovered was that it wasn't smoothly distributed but was often on streamers that were at great distance from the primary galaxy. And some of the new thinking (and no doubt this will be modified in time as well) is that this faint distributed light is in the outskirts of galaxies where the dark matter is because it's caused by satellite galaxies that have dropped into the potential well of the primary galaxy and slowly over time have had their stars stripped away into beautiful long tails that surround the primary galaxy.'

Developing instruments for studying the universe

One of the major challenges after the fires was to rebuild the workshops, which had always been an integral part of Mount Stromlo. Many of the instruments that had been used to make significant astronomical discoveries were based on instruments designed and constructed in the workshops, which have excellent technical staff. A modern Advanced Instrumentation Technology Centre (AITC) was built in stages. Sackett was responsible for the building of Stage 1, which was opened in 2006. According to Sackett, 'The goal of the Centre was to replace the activities that took place in the workshops at Mount Stromlo that were destroyed in the fire.' She built the AITC for the future. It had a large integration and assembly hall attached to the electrical and mechanical design workshops and laboratories. The integration hall and the laboratories were designed to cater for the next

Figure 11.11: The AITC was built in two phases. The first was completed in 2006 and was dedicated by Minister Nick Minchin in 2007. The laboratories are large enough to handle production and testing of instruments and subsystems for the Giant Magellan Telescope. Photo: Mount Stromlo Archives.

Figure 11.12: Sackett and CEO Roger Franzen of Auspace with a completed NIFS. Auspace was a spin-off company of Mount Stromlo in 1983, and became Australia's premier space technology company. It was well placed, both in terms of capability and interest, to take on the NIFS instrument. Photo: Mount Stromlo Archives.

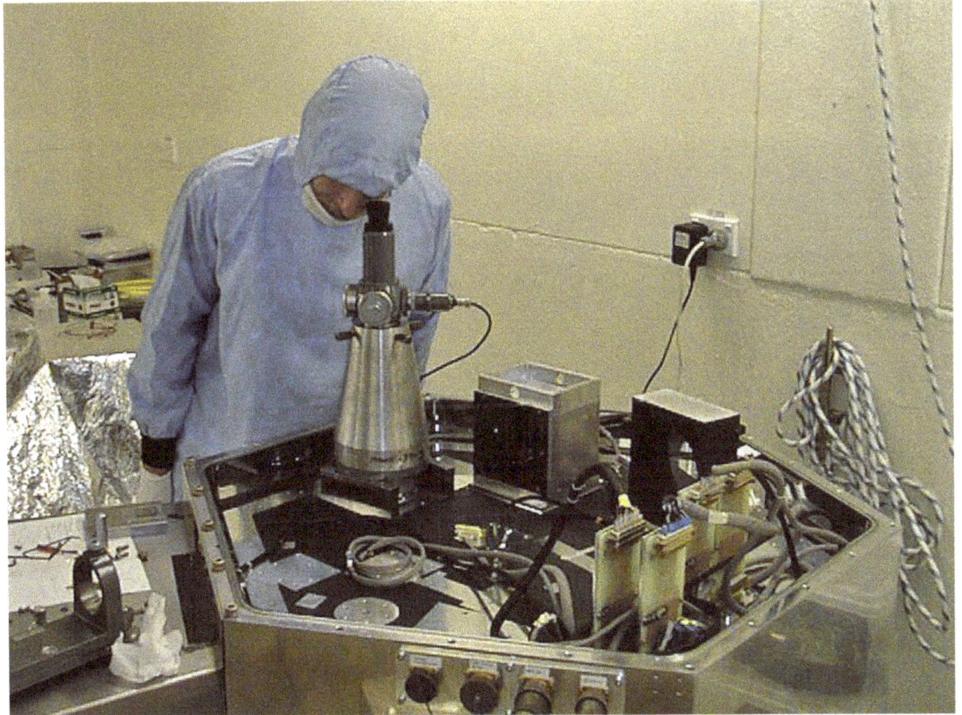

Figure 11.13: Astronomer Peter McGregor was the intellectual father of a series of integral field spectrographs developed at Mount Stromlo. Here he is aligning the optics of the NIFS. The detector and optics must be cooled to cryogenic temperatures, so it is essential to ensure that both optical and mechanical alignment at room temperature remain perfect after the instrument is made cold. Photo: Mount Stromlo Archives.

generation of instruments expected to be installed on Extremely Large Telescopes currently being planned by international consortia.[13]

The Observatory was contracted to build the NIFS for the Gemini North telescope, but it was completely destroyed by the fires just before it was to

Figure 11.14: (a) The Stromlo and Auspace teams travelled to the Mauna Kea Observatory to install the NIFS instrument under the 8.1 m primary mirror of the Gemini North telescope. (b) In operation, NIFS is used with a laser beam that characterises the blurring effects of the Earth's atmosphere, and an adaptive optics system to correct the blurring. Photos: Gemini Observatory/AURA.

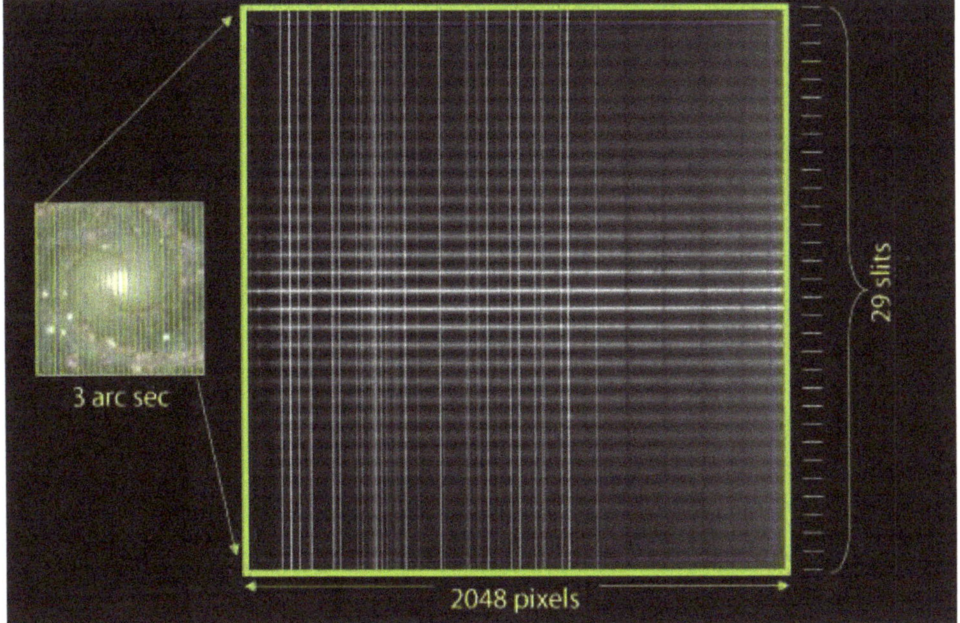

Figure 11.15: NIFS comprises an optical system that in effect slices its field of view and produces a spectrum of each point in the field. This is a typical image of a galaxy with the slices shown, together with the resulting constellation of spectra imaged onto the detector. Analysis of such spectra allows astronomers to learn about the temperatures, densities, compositions and motions of the gas and stars in the object under study. Photo: Mount Stromlo Archives.

leave the workshops for Hawaii. NIFS 2 was built by the Canberra-based aerospace company, Auspace Ltd, in collaboration with Mount Stromlo designers and engineers. The completed instrument was shipped to Hawaii in August 2005 and installed on the Gemini North 8.1 m telescope in October. Jan van Harmelen, the NIFS Project Manager and Peter McGregor, the Project Scientist, spent a month at the Gemini North Observatory supervising the tests and the initial observations. First observations on the Gemini North telescope showed unprecedented spatial and spectral detail in otherwise dust-obscured regions, allowing young stellar wind speeds and black hole masses to be measured.[13] The project was a great success and a credit to the technical and scientific staff at Mount Stromlo Observatory.

The Observatory also secured a contract to build the Gemini South Adaptive Optics Imager (GSAOI) for the Gemini South 8 m telescope in Chile. This is a near-infrared camera which is being used with Gemini's flagship Multi-Conjugate Adaptive Optics (MCAO) system on the Gemini South telescope. The MCAO system corrects the blur due to the Earth's atmosphere over a wider field than was previously possible. GSAOI records diffraction-limited images of astronomical objects with 0.02 arcsecond sampling over its full 85 × 85 arcsecond field of view. The GSAOI was designed and built by a team of engineers led by Peter McGregor. The team included John Hart, Dejan Stevanovic, Gabe Bloxham, Damien Jones, Jan van Harmelen, Jason Griesbach, Murray Dawson, Peter Young and Mark Jarnyk. The camera was shipped to

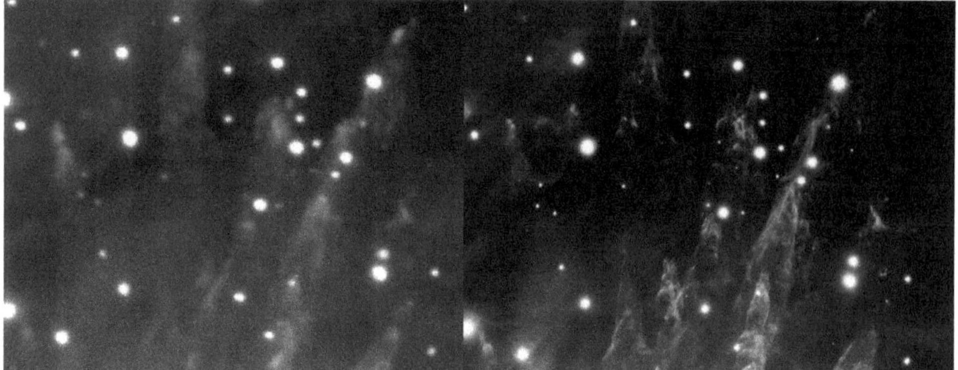

Figure 11.16: Comparison of images with and without adaptive optics. On the left is an image taken with the 3.9 m AAT at Siding Spring under excellent seeing conditions (minimal atmospheric blurring), compared with one taken by GSAOI in Chile. Photo: AAO/Gemini Observatory/AURA.

the Gemini South telescope in Chile in October 2006.[13,14] It lets astronomers take images in infrared light as sharp as can be achieved the Hubble Space Telescope in visible light.

Another instrument approved for construction in December 2005 was the Wide Field Spectrograph (WiFeS), which provides 950 simultaneous spectra throughout the optical frequency range and has a data-gathering capability of 10–100 times the rate of conventional spectrographs. It has been installed on the 2.3 m telescope at Siding Spring. Mike Dopita was the Project Scientist, Gabe Bloxham was the Optics Design Engineer and John Hart was the Mechanical Design Engineer.[13]

In 2005 Brian Schmidt's highly innovative telescope – the SkyMapper – was approved for fabrication. It is a 1.35 m diameter telescope located at Siding Spring Observatory, one of a new breed of dedicated survey telescopes that will conduct the Stromlo Southern Sky Survey which will produce the first deep digital map of the entire southern sky. The design was produced by Conroy, Granlund and Oates; it features a 268 million pixel CCD mosaic. In 2006 the advanced electronics of the camera achieved the ultra-high data rates that will be required for the massive 150 Terabyte Stromlo Southern Sky Survey. The survey will be the first fully digital and publicly accessible survey of the southern sky.[14]

After the fires the astronomers had a long and hard think about what they wanted in terms of telescopes for the future. It would have been easy to recreate what they had lost, but Sackett had rather bold ideas about future telescopes. According to Sackett, 'We thought that one of the legacies we would like to leave is that the future generation would have a facility that was to them what the AAT (Anglo-Australian Telescope) was in the 1970s and 1980s, namely, one of the biggest telescopes in the world that would be capable of doing things that otherwise weren't possible, but also a telescope that was reasonably versatile, that could have a variety of instruments and do different sorts of astronomy.'[1] Her biggest coup was to get Mount Stromlo involved in planning the world's next-generation very large optical telescopes, the Giant Magellan Telescope (GMT). The building to enclose the telescope will

Figure 11.17: Artist's impression of the GMT with its seven large mirrors, each as large as the primary mirrors of the Gemini telescopes on Hawaii and in Chile. Photo: Giant Magellan Telescope – GMT Corporation.

be as high as a 23-storey skyscraper and the instrument will have 10 times the resolution of images from the Hubble Space Telescope. It will have six segments, each 8.4 m in diameter, surrounding a seventh mirror of the same size. The total light-gathering power will be nearly seven times that of the Gemini telescopes. It will be a time machine which will allow astronomers to probe the darkest reaches of the universe and ask questions such as: When did the first stars, galaxies and black holes form? As Wendy Freedman from the Carnegie Institution for Science said, 'We are in the midst of a revolution in our understanding of the origin and evolution of the universe. The GMT will allow us to witness directly what we can only conjecture today.'

Sackett's achievement in getting Mount Stromlo Observatory involved with the GMT was the best and furthest-reaching decision she made for the future of the Observatory and Australian astronomy. She signed the Memorandum of Understanding in Texas on behalf of the Australian National University. By signing the document, Mount Stromlo joined an elite group of teaching and research institutions in the US (Carnegie Institution of Washington, Harvard University, Massachusetts Institute of Technology, the University of Arizona, the University of Michigan, the Smithsonian Institution, the University of Texas at Austin and the Texas A&M University) involved in the detailed design of the telescope.

The science objectives, Sackett, said, 'will be many, because it is meant to be a multi-purpose telescope. But in particular, the image quality of this telescope is something that is of prime concern, so that using a method called adaptive optics, which can undo the blurring caused by movement in the atmosphere, this telescope should be able to not only take pictures that are as crisp as those of the Hubble Space Telescope but in fact much crisper. That means seeing more detail. It means being able to pull out faint, very distant objects from the background or from brighter foreground objects. That means finding more distant stars in the early universe. It means being able to actually see those planets orbiting other stars, detecting the light from the planets themselves, and thereby being able to analyse the atmosphere, for example, of those planets, which I think will be tremendously exciting.'[1]

As this book goes to print, the telescope is set to pass its final design review. The site in Chile has been levelled and is ready for construction. Two of the seven giant mirrors have been fabricated. They posed a technical challenge as they are 'off-axis'. It is much harder to manufacture and polish an off-axis mirror. Furthermore, because of the required image quality, it has to be done at a much higher precision than had been done before. The mirror must not only be polished to have a highly reflective finish, its shape must match the desired optical prescription with a precision of 25 nanometres. According to Sackett, 'This was probably the most severe technical risk. It was the thing that kept astronomers awake at night. That technical risk has been largely retired with the successful completion of an off-axis mirror that meets the specifications.'[1] The hope is that construction may begin as early as 2014, assuming adequate funding becomes available.

Research frontiers

Despite the fires the Observatory remained very active in research, publishing over 500 papers in refereed international journals during the Sackett years. The research effort spread over several speciality areas: solar and extrasoloar planetary systems, stars and stellar populations, the Milky Way and galaxies, dark energy and dark matter, gamma-ray bursts, interstellar medium and galactic feedback, and adaptive optics. Because of space limitations, only some of the research is highlighted below.

In 2005 Brian Schmidt was awarded a Federation Fellowship by the Australian Research Council and in 2006 he won the Shaw Prize with Saul Perlmutter (University of California) and Adam Riess (Space Telescope Institute) for their ground-breaking discovery that the universe is accelerating. Michael Dopita was the other Federation Fellow at the Observatory and was awarded one of the inaugural Fellowships in 2001.

2002 was the first full year in which astronomers from the Research School for Astronomy and Astrophysics were able to compete for ARC funding. They were very successful, winning over $2.7 million for various projects proposed by Briggs, Da Costa, Driver, McGregor, Norris, Schmidt, Wood and Bicknell.[15] Over 50% of the School's funding was derived from the increasing success of its astro-engineering design and fabrication teams.

Sackett's arrival at the Observatory in 2002 brought the new field of extrasolar planets into the traditional astronomy programs which were concerned with stellar astrophysics, galaxies, globular clusters, metal deficient stars and supernovae. Together with the Research School of Earth Sciences,

Sackett established a new Planetary Science Institute in 2002. According to Sackett, the Institute was 'designed to cultivate a research program of planetary science combining the detailed studies possible in our Solar System with the recent wealth of information from the young field of extrasolar planets'.[1]

Figure 11.18: Ken Freeman is well-known worldwide for his work on the dynamics of the stars that make up galaxies. His career has centred around discovering observational trends and regularities that provide clues to the formation and evolution of galaxies. Photo: Mount Stromlo Archives.

One of the highlights in the year she arrived at Mount Stromlo was the discovery by Price, Schmidt and collaborators of evidence that a supernova coincided with a cosmological gamma-ray burst (GRB 011121). The evidence came from Hubble Space Telescope observations of the burst GRB 011121, whose light curve showed the tell-tale signature of an underlying supernova. Gamma-ray bursts have been detected by spacecraft since the 1960s and were the subject of investigation by the international REACT collaboration of which Price and Schmidt were members.[16,17]

Ken Freeman and Joss Bland-Hawthorn wrote a major review paper for the *Annual Reviews of Astronomy and Astrophysics 2002* which gave a comprehensive account of how our Galaxy formed.[18] A study of the motions and chemical composition of stars in the Galaxy provides a probe to investigate the conditions in our Galaxy long ago. It offers a view of the local universe as it was at high redshifts (distant past). This approach, called 'near-field cosmology', has blossomed into a new area of research activity worldwide.

Mike Bessell, an international expert in photometry, gave a comprehensive review of standard photometric systems[19] in the *Annual Review of Astronomy and Astrophysics 2005*. He is also involved in the study of metal-poor stars, with John Norris and others. Norris and Bessell, working with Christlieb and others on metal-poor candidates in the Hamburg/ESO Survey, discovered an object with an iron abundance significantly less than 1/10 000 that of the Sun. This was only the third such object known.[14]

Norris examined the oldest stars in the Milky Way Galaxy, which give clues about the types of objects that produced the first heavy elements. With his collaborators Aoki, Ando, Beers and Ryan, he obtained data for two such stars, in which metals, for example iron, are reduced in abundance by a factor of 6000 compared with that observed in the Sun. Their data for the extremely metal-poor star CS 22949–037 confirmed their earlier report that it exhibits extreme nitrogen enhancement, up by a factor 200 relative to what is seen for iron. The star is also known to have large relative overabundances, of order 10, of carbon and magnesium, relative to iron. The team also discovered the carbon-rich extremely metal-poor star CS 29498–043. High-resolution, high signal-to-noise spectra of the star, which is very similar to CS 22949–037, suggest the existence of a class of extremely metal-poor stars with large excesses of carbon, nitrogen, magnesium and silicon. The unusual abundances of these two stars is still the subject of inquiry.[20,21]

Anna Frebel, working with Norris, Bessell, Christlieb, Beers and others, used the ESO Very Large Telescope to obtain high-resolution, high-signal-to-noise spectra of HE 1523–0901, which turned out to have only 1/1000 the iron abundance of the Sun. However, the object was tremendously enhanced in the neutron capture elements relative to iron. Because of this they were able to determine accurately the abundances of the radioactive elements thorium and uranium. Using the radioactive decay of these elements as cosmic chronometers, Frebel was able to determine the age of HE 1523–0901 to be 13.1 Giga years, with an accuracy of 2 Giga years.[14,22]

Helmut Jerjen, together with Cabanac, Valls-Gabaud, Jaunsen and Lidman, discovered a partial Einstein Ring in the southern constellation Fornax. It was only the fourth ring of its kind ever observed. It is remarkably bright and is almost complete. It gave a rare glimpse to the time when galaxies were still in their infancy, in fact when the universe was ~12% of its present age.[23]

An international team of astronomers, including those from Mount Stromlo Observatory, completed a survey of the nearby universe. Their results, published in the Millennium Galaxy Catalogue,[24,25] are a cosmic stocktake of over 10 000 giant galaxies, each containing 10 million to 1000 billion stars. The survey also provided an answer to the question: Where did the normal matter that was produced 13.7 billion years ago in the Big Bang, go? It revealed that ~20% is in stars, 0.1% lies in dust expelled from massive stars (from which solid structures like the Earth and human beings are made) and ~0.01% is in super-massive black holes. According to the survey leader, Simon Driver from the University of St Andrew, now at the International Centre for Radio Astronomy Research in Perth,'the remaining material is almost completely in gaseous form lying both within and between the galaxies, and forming a reservoir from which future generations of stars may form'.[14]

Ten years after she founded the PLANET collaboration, Sackett and colleagues discovered a cool planet of 5.5 Earth masses, through gravitational lensing.[26] They had hoped for such a discovery within five rather than 10 years, but it was exciting since up to that time no planet that was close to Earth-like had been found. Most of the planets that had been discovered 'were 100 to 300 times the Earth's mass', Sackett said. With the advent of the Kepler mission many more potential Earth-size planets have been found. Since the last Kepler catalogue was released in 2012 the number of candidates discovered in the Kepler data now totals 2740 potential planets orbiting 2036 stars. The most dramatic increases are in the number of Earth-size and super Earth-size candidates discovered, which grew by 43% and 21% respectively. At the American Astronomical Society meeting in Long Beach in January 2013, Chris Burke from the SETI Institute announced that one in six stars has an Earth-size planet.

There is great variety among the over 200 known planets that orbit other stars, ranging from the recently discovered icy super-Earths (planets that weigh up to 15 times more than the Earth) to low-mass gas giants. Scientists are able to explain how planets form around Sun-like stars but were surprised by the discovery of planets around the relatively smaller and cooler red dwarf stars. Grant Kennedy and his collaborators Scott Kenyon (Smithsonian Astrophysics Observatory) and Benjamin Bromley (University of Utah) used some of the best available planetary models to show that many of the hefty planets may be huge icy 'snowballs' formed as the proto-planetary disks of young M dwarf stars cool at an early age.[27]

In its final stages of evolution, a star like the Sun will become an asymptotic giant branch (AGB) star. One of the characteristics at this stage is the strong outflow of matter in a 'super wind'. It is theoretically difficult to predict how much mass will be lost from the star and how much nuclear-processed material will be ejected into space to form the next generation of stars and planets. Wood, Groenewegen (Leuven), Sloan (Cornell) and a team of collaborators used the Spitzer Space Telescope to find an accurate empirical measure of the strengths of the winds emanating from AGB stars. They measured the mass rate loss in 60 AGB stars in the Magellanic Clouds and devised an empirical law which provides a relationship between mass loss and rotation period.[28]

Observations of the well-known group of galaxies called Stephan's Quintet, with a super-sensitive spectrograph on the Spitzer Space Telescope by a team of astronomers including Michael Dopita, revealed one of the biggest shockwaves ever seen. They discovered that one of the galaxies – NGC 7318b – is falling towards the others at nearly 1000 km per second, generating shockwaves bigger than the Milky Way as it falls into the cluster. Their observations provide a new diagnostic tool for studying conditions in merging and colliding galaxies, which were more prevalent in the early universe. The new results may indicate that some of the emission from the most luminous infrared galaxies seen in the very distant universe may actually be created not by stars, but by vast shocks in the gas between colliding galaxies. It has been suggested that it is likely that about 2 billion years from now our Galaxy will collide with the Andromeda Galaxy and create galactic-scale shocks of its own.[13]

Vision for the future

Despite the initial setback to her directorship, Sackett achieved most of what she had set out to accomplish in her five-year term. She had reinvigorated the institution and continued the growth of its tradition of excellence. She steered the Observatory towards a new beginning: her decision to embark on the Great Magellan Telescope was her most audacious and visionary decision. She will be remembered for her vision for Mount Stromlo Observatory and Australian astronomy in the era of extremely large telescopes that will bring unimaginable discoveries

In 2007 Sackett completed her term as the Director of Mount Stromlo Observatory and took up the position of Chief Scientist for Australia, the first woman appointed to that role.

Figure 11.19: Stephan's Quintet, a group of galaxies starting to collide. This image shows (in green) the location of a gigantic shockwave caused by a collision. Unexpectedly, the shock is visible in the light of hydrogen molecules. Photo: NASA.

References

1. Bhathal R 2013. Interview with Penny Sackett. 15 January 2013. National Oral History Project on Significant Australian Astronomers. National Library of Australia, Canberra.
2. Bhathal R 2012. *Aboriginal Astronomy.* ARI, Sydney.
3. Macdonald E 2003. Telescopes may not return to Stromlo. *Canberra Times,* 21 January 2003.
4. Schiebinger L 1999. *Has Feminism changed Science?* Harvard University Press, Cambridge, MA.
5. Boulting W 1972. *Giordano Bruno: His Life, Thoughts and Martyrdom.* Books for Libraries Press, New York.

6. Albrow M, Beaulieu J-P, Birch P, Caldwell JAR, Kane S, Martin R, Menzies J, Naber RM, Pel J-W, Pollard K, Sackett PD, Sahu KC, Vreeswijk P, Williams A, Zwaan MA 1998.. The 1995 Pilot Campaign of PLANET: searching for microlensing anomalies through precise, rapid, round-the-clock monitoring *Astrophysical Journal* **509**, 687–702. doi:10.1086/306513.
7. Afonso C, Alard C, Albert JN, Andersen J, Ansari R, Aubourg É, Bareyre P, Bauer F, Beaulieu JP, Bouquet A, Char S, Charlot X, Couchot F, Coutures C, Derue F, Ferlet R, Glicenstein JF, Goldman B, Gould A, Graff D, Gros M, Haissinski J, Hamilton JC, Hardin D, de Kat J, Kim A, Lasserre T, Lesquoy É, Loup C, Magneville C, Marquette JB, Maurice É, Milsztajn A, Moniez M, Palanque-Delabrouille N, Perdereau O, Prévot L, Regnault N, Rich J, Spiro M, Vidal-Madjar A, Vigroux L, Zylberajch S, Alcock C, Allsman RA, Alves D, Axelrod TS, Becker AC, Cook KH, Drake AJ, Freeman KC, Griest K, King LJ, Lehner MJ, Marshall SL, Minniti D, Peterson BA, Pratt MR, Quinn PJ, Rodgers AW, Stetson PB, Stubbs CW, Sutherland W, Tomaney A, Vandehei T, Rhie SH, Bennett DP, Fragile PC, Johnson BR, Quinn J, Udalski A, Kubiak M, Szymański M, Pietrzyński G, Woźniak P, Zebruń K, Albrow MD, Caldwell JAR, DePoy DL, Dominik M, Gaudi BS, Greenhill J, Hill K, Kane S, Martin R, Menzies J, Naber RM, Pogge RW, Pollard KR, Sackett PD, Sahu KC, Vermaak P, Watson R, Williams A 2000.. Combined analysis of the binary lens caustic-crossing event MACHO 98-SMC-1. *Astrophysical Journal* **532**, 340–352. doi:10.1086/308561.
8. Beaulieu JP, Sackett PD 1998. Red clump morphology as evidence against a new intervening stellar population as the primary source of microlensing toward the Large Magellanic Cloud. *Astronomical Journal* **116**, 209. doi:10.1086/300400.
9. Whitmore BC, Lucas RA, McElroy DB, Steiman-Cameron TY, Sackett PD, Olling RP 1990.. New observations and a photographic atlas of polar ring galaxies. *Astronomical Journal* **100**, 1489–1522.
10. Sackett PD, Sparke LS 1990. The dark halo of the polar-ring galaxy NGC 4650A. *Astrophysical Journal* **361**, 408–418. doi:10.1086/169206.
11. Sackett PD, Rix HW, Jarvis BJ, Freeman KC 1994.. The flattened dark halo of polar ring galaxy NGC 4650A: a conspiracy of shapes? *Astrophysical Journal* **436**, 629–641. doi:10.1086/174938.
12. Sackett PD, Morrison HL, Harding P, Boroson TA 1994. A faint luminous halo that may trace the dark matter around the spiral galaxy NGC 5907. *Nature* **370**, 441–443. doi:10.1038/370441a0.
13. Research School for Astronomy and Astrophysics 2005. Annual Report.
14. Research School for Astronomy and Astrophysics 2006. Annual Report.
15. Research School for Astronomy and Astrophysics 2002. Annual Report.
16. Price P, Berger E, Reichart D, Kulkarni SR, Yost SA, Subrahmanyan R, Wark RM, Wieringa MH, Frail DA, Bailey J, Boyle B, Corbett E, Gunn K, Ryder SD, Seymour N, Koviak K, McCarthy P, Phillips M, Axelrod TS, Bloom JS, Djorgovski SG, Fox DW, Galama TJ, Harrison FA, Hurley K, Sari R, Schmidt BP, Brown MJI, Cline T, Frontera F, Guidorzi C, Montanari E 2002.. GRB 011121: a massive star progenitor. *Astrophysical Journal* **572**, L51–L55. doi:10.1086/341552.
17. Bloom J, Kulkarni S, Price P, Reichart D, Galama TJ, Schmidt BP, Frail DA, Berger E, McCarthy PJ, Chevalier RA, Wheeler JC, Halpern JP, Fox DW, Djorgovski SG, Harrison

FA, Sari R, Axelrod TS, Kimble RA, Holtzman J, Hurley K, Frontera F, Piro L, Costa E 2002.. Detection of a supernova signature associated with GRB 011121. *Astrophysical Journal* **572**, L45–L49. doi:10.1086/341551.

18. Freeman K, Bland-Hawthorn J 2002. The new galaxy: signatures of its formation. *Annual Reviews of Astronomy and Astrophysics* **40**, 487–537.
19. Bessell M 2005. Standard photometric systems. *Annual Review of Astronomy and Astrophysics* **43**, 293–336. doi:10.1146/annurev.astro.41.082801.100251.
20. Norris J, Ryan S, Beers T, Aoki W, Ando H 2002. Extremely metal-poor stars. IX. CS 22949–037 and the role of hypernovae. *Astrophysical Journal* **569**, L107–L110. doi:10.1086/340665.
21. Aoki W, Norris J, Ryan S, Beers TC, Ando H 2002. Subaru/HDS study of the extremely metal-poor star CS 29498–043: Abundance analysis details and comparison with other carbon-rich objects. *Publications of the Astronomical Society of Japan* **54**, 933–949.
22. Frebel A, Christlieb N, Norris JE, Beers TC, Bessell MS, Rhee J, Fechner C, Marsteller B, Rossi S, Thom C, Wisotzki L, Reimers D 2006. Bright metal-poor stars from the Hamburg/ESO Survey. I. Selection and follow-up observations from 329 fields. *Astrophysical Journal* **372**, 630.
23. Cabanac RA, Valls-Gabaud D, Jaunsen AO, Lidman C, Jerjen H 2005. Discovery of a high-redshift Einstein ring. *Astronomy and Astrophysics* **436**, L21–L25. doi:10.1051/0004-6361:200500115.
24. Driver SP, Allen PD, Graham AW, Cameron E, Liske J, Ellis SC, Cross NJG, de Propris R, Phillipps S, Couch WJ 2006. The millennium galaxy catalogue: morphological classification and bimodality in the color-concentration plane. *Monthly Notices of the Royal Astronomical Society* **368**, 414. doi:10.1111/j.1365-2966.2006.10126.x.
25. Liske J, Driver SP, Allen PD, Cross NJG, de Propris R 2006. The millennium galaxy catalogue: a census of local compact galaxies. *Monthly Notices of the Royal Astronomical Society* **369**, 1547. doi:10.1111/j.1365-2966.2006.10411.x.
26. Beaulieu J-P, Bennett DP, Fouqué P, Williams A, Dominik M, Jorgensen UG, Kubas D, Cassan A, Coutures C, Greenhill J, *et al.*2006. Discovery of a cool planet 5.5 Earth masses through gravitational lensing. *Nature* **439**, 437–440. doi:10.1038/nature04441.
27. Kennedy GM, Kenyon SJ, Bromley BC 2006. Planet formation around low-mass stars: the moving snow line and super-Earths. *Astrophysical Journal* **650**, L139. doi:10.1086/508882.
28. Zijlstra AA, Matsuura M, Wood PR, Sloan GC, Lagadec E, van Loon JT, Groenewegen MAT, Feast MW, Menzies JW, Whitelock PA, Blommaert JADL, Cioni M-RL, Habing HJ, Hony S, Loup C, Waters LBFM 2006. A Spitzer mid-infrared spectral survey of mass-losing carbon stars in the large Magellanic Cloud. *Monthly Notices of the Royal Astronomical Society* **370**, 1961. doi:10.1111/j.1365-2966.2006.10623.x.
29. Freeman K 2012. Email to Ragbir Bhathal. 11 December.

2007–2013

12

Stromlo Fellows and a Nobel Prize

The ultimate accolade in science is the Nobel Prize. It marks off its recipients from all the other scientists in that particular year.[30]

It feels good. I'm not sure what old Albert would think but he'd be scratching his head [Brian Schmidt, on winning the Nobel Prize].[31]

With his long experience in technology management, Harvey Butcher revitalised the engineering group at Mount Stromlo, and vigorously pursued the opportunities for instrumentation for the GMT (Giant Magellan Telescope). He leaves the Observatory in excellent condition, with high morale and a large complement of outstanding young researchers.[32]

September 2007 was the second coming of Harvey Butcher to Mount Stromlo Observatory. But this time he came as Director of the institution at which he had as a young man cut his astronomical teeth. He had come in the first instance to Australia on the suggestion of Allan Sandage in the 1970s to do his PhD. He was attracted to Mount Stromlo by 'two scientific reasons. One was the southern sky. I'd never seen the southern sky and the Magellanic Clouds. The second was that I knew that the Coudé spectrograph at the Stromlo 74-inch [1.9 m] telescope was built by Theodore Dunham.' Dunham had earlier built the spectrograph for the 2.54 m at Mount Wilson, so Butcher reckoned that the later edition at Mount Stromlo would likely be even better. Another, non-scientific reason for coming to Australia arose when 'I discovered that there were no lecture courses at Mount Stromlo! I was completely fed up with sitting in lecture courses and taking exams. The idea of having four years to do nothing but a research project was unbelievably attractive. When I arrived I found students were treated almost as staff members as regards access to telescopes and other facilities. In hindsight it was one of the best decisions I ever made.'[1]

For his PhD he carried out a study of nucleosynthesis in the Galaxy under the supervision of Mike Bessell, then an up-and-coming young astronomer who was making a name for himself in the study of stellar astrophysics, and of Alex Rodgers who became Director of the Observatory in 1986. It was a

Figure 12.1: Harvey Butcher was the ninth Director at Mount Stromlo, serving from September 2007 to January 2013. Photo: Mount Stromlo Archives.

topic Butcher had chosen himself, based on the theory that stars convert light elements into heavier ones via nuclear reactions, which had been worked out by Burbidge, Burbidge, Fowler and Hoyle in the 1950s. 'That was the bible that showed the different elements come from different nuclear processes in different stars. And the question when I came on the scene', he said, was, 'Is the result the same throughout the whole history of the Galaxy, or is there evidence of secular, relative abundance evolution for elements produced by nuclear reactions under very different conditions?'[1] Butcher wanted to measure differential chemical abundances in dwarf stars of r- and s-process elements, which are produced in different stars and over widely different time-scales.

However, he soon found that the available gratings in the 1.9 m Coudé were not suitable for the work. With advice and help from Bessell and Rodgers, he put together in the Coudé one of the first high-resolution echelle spectrographs in astronomy.[2,3,4] 'And what I found was, over a very large range of ages and over mean abundance levels differing by a factor of 30, basically there was very little or no measurable difference in the relative abundances. I thought that was a problem for the concept of stellar nucleosynthesis being a vigorous, ongoing phenomenon in the Galaxy.'

This approach to doing research, of developing new instrumental capabilities to make new observations possible, characterised the rest of Butcher's professional career.

On a visit to Mount Stromlo in 1973, Peter Strittmatter from the University of Arizona in Tucson offered Butcher a job following his thesis defence. Butcher left Mount Stromlo in 1974 to work at the Steward Observatory as a Bart Bok Fellow. Two years later he moved across the street to the Kitt Peak National Observatory, where he worked for seven years.

By 1983, Butcher had moved up the organisation at Kitt Peak to a position where he was influential in determining policy, but even so decided it was time to try something new. By chance, the Dutch became interested in his expertise as an astronomical instrumentalist. 'I was asked to consider a job in the Netherlands specifically to help out in a collaboration with the UK to build a new observatory on La Palma in the Canary Islands. So my instrumentation background was what motivated them to offer me a professorship at the Groningen University.'[1]

Butcher lived in Netherlands for over 25 years and took up Dutch citizenship. Dutch forthrightness was initially a shock, but after a while 'you begin to really appreciate that, because you know what people are thinking, there are generally no hidden agendas'.[1] He was also impressed with the way the

The Butcher-Oemler effect

Figure 12.2: Augustus Oemler Jr worked with Butcher in the mid 1970s, using early digital panoramic detectors to study the colours of galaxies in distant galaxy clusters. Photo: The Observatories, Carnegie Institution for Science.

While in Tucson Butcher became friends with Gus Oemler and Roger Lynds at the Kitt Peak National Observatory, eventually joining them on the Kitt Peak staff. Lynds had a particular interest in the new panoramic digital detectors and was kind enough to involve Butcher in helping to test and implement them on the telescope. Oemler interested him in trying to observe the evolution of galaxies over cosmic time. They decided to try to use the new digital detectors to look at rich galaxy clusters, which were ideal targets for the relatively small fields of view of these early vidicon and CCD devices. Butcher noted, 'It is hard to appreciate today that in the 1970s, received wisdom was that galaxies formed early and essentially didn't evolve visibly over recent cosmic time.'[1] But Butcher thought S0 galaxies (which are disk systems without any current star formation) might just be very old spiral galaxies in which the gas had all been converted into stars. Oemler felt that might be the case, but that probably in clusters their gas gets stripped away by the ambient cluster medium. To test the two hypotheses for the origin of S0 galaxies, they used the new detectors to observe what was happening in clusters seen 3–5 billion years ago. 'And lo and behold', said Butcher, 'we found lots of blue galaxies in clusters at modest redshifts, which shouldn't have been there according to all the then current ideas. Some of the galaxies had changed dramatically in recent cosmological times.'[5,6]

Senior astronomers were scathing about the claim that many cluster members were forming stars rapidly and the result became dubbed the Butcher-Oemler effect. According to Butcher, 'If you had an effect named after you that tended to mean that nobody believed it and it wasn't going to turn out to be correct in the long run. So it was not a positive thing to have a Butcher Oemler effect at that time. It was an unpleasant period in my life.'[1]

In the early 1980s, Barry Newell at Mount Stromlo teamed up with PhD scholar Warrick Couch to study a dozen high redshift galaxies, the photometry being derived from deep photographic plates mostly from the Anglo-Australian Telescope at Siding Spring Observatory. According to Couch, this work 'made the very important step of independently confirming the Butcher-Oemler effect and showing it to be wide-spread and hence generally a universal property of rich centrally-concentrated clusters at redshifts beyond 0.2.'[7] Today, undergraduate textbooks carry a description of the Butcher-Oemler effect, which is examinable.

Butcher was at Kitt Peak in Tucson for about seven years. While there, he worked to perfect CCD detector systems and spearheaded their use for imaging and multi-aperture spectroscopy for observing very faint high redshift galaxies. He was also project scientist for several new observing instruments, including an early spectrograph for obtaining spatially resolved spectra at resolutions approaching the diffraction limit.

Dutch do things. 'The Dutch think on the long term. So whether they're organising the country or their science, they don't think of the next two years, they think of the next 10 to 20 years. I found that very attractive.'[1]

While in Groningen Butcher continued his work with Oemler, but branched out to several other areas requiring new instrumentation. He developed an early stellar seismometer to test stellar models for the star Alpha Centauri.[8,9] He showed that a radioactive chronometer could also be devised for the whole Galaxy,[10] and participated in the specification of instrumentation for the European Southern Observatory's (ESO) Very Large Telescope (VLT).

By 1991, he was again looking for a change, but with his family settled in the Netherlands the options were limited. As chance would have it, he was offered the directorship of Netherlands Foundation for Research in Astronomy (ASTRON), a government-financed institution in Dwingeloo that specialised in radio astronomy but was starting to develop visible light instrumentation as well. Dutch astronomers were divided at the time as to whether future investments should focus on optical astronomy using facilities at La Palma and ESO, or on radio astronomy, which would at least allow new facilities to be located in the country. The compromise reached was for

Astronomy in the Netherlands and ESO

Groningen has long had a strong reputation in astronomy, together with the university in Leiden. In the early part of the 20th century J.C. Kapteyn at Groningen was a strong advocate for travelling to the very best overseas observatories to take the highest-quality data, which would then be analysed at home in Groningen. This is a common enough approach today but was unusual for the time. It was only after the Second World War, when Dutch astronomers moved into radio astronomy, that they could observe the skies seriously from the Netherlands itself. The Westerbork Synthesis Radio Telescope was the first common-user radio telescope and went into operation from 1970. For optical astronomy, the Dutch were a founding member of the European Southern Observatory in the 1960s. In the 1980s the Dutch decided also to invest with the British and Spanish in the Isaac Newton Group of telescopes on La Palma, which occasioned the search for an astronomer with Butcher's background.

In the mid 1980s, Butcher became involved at ESO with developing the scientific specifications for what would become the VLT. The design of an efficient, high-resolution, stellar spectrograph was a major challenge, and led to the development in Groningen of an innovative prototype instrument, later called FRINGHE, a heterodyned holographic spectrometer.[11] 'The idea was to image the Fourier Transform of the spectrum onto a two-dimensional CCD detector, thereby to gain both throughput (Jacquinot) and multiplex (Fellgett) advantages. You could build quite a small instrument for a very large telescope this way. So it was a way of making a cheap, quite high-resolution spectrometer for the VLT.'[1] They tested the concept and it worked very well. The only problem was that the available detectors were relatively small, so the wavelength coverage that could be achieved in a single integration was also limited. In the end, ESO chose a conventional and much more expensive solution, UVES, which provided a wider wavelength coverage. According to Butcher, 'the [FRINGHE] instrument was cute. I really enjoyed developing it.'[1]

modest investments in both. Butcher spent the next 16 years managing ASTRON, and did both radio astronomy research and instrumentation for radio astronomy and infrared astronomy. The organisation ran the Westerbork radio telescope and helped build the Isaac Newton Group of optical telescopes on La Palma and the James Clerk Maxwell sub-mm telescope in Hawaii. It built instruments not only for each of those facilities but also for the ESO's VLT and the James Webb Space Telescope.

SKA collaboration

The idea of the Square Kilometre Array (SKA) radio telescope became a matter for serious discussion in 1993. It was realised that observations of the neutral hydrogen content of galaxies in the early universe would provide essential information about galaxy evolution and the development of large-scale structure in the cosmos. But a radio telescope larger than any built before would be required, one having a signal collecting area of a million square metres – a square kilometre array.

By 1996, at the General Assembly of the International Union of Radio Science in Lille, there began a collaboration with the Division of Radio Physics at CSIRO to develop the concept. Ron Ekers led the effort on the Australian side, while Butcher ensured that Dutch R&D at ASTRON would focus effectively on the necessary technologies and scientific development. These would include the development of aperture and focal plane phased array detection, new correlator approaches, and a long-term enhancement program for the Westerbork radio telescope that would deliver new, relevant science in the meantime. Butcher commented, 'The successes at ASTRON

Figure 12.3: LOFAR antennas in the Netherlands. Butcher led development efforts in the Netherlands as a technological and scientific precursor for the SKA radio telescope. To make sharp images of the radio sky, antennas must be placed over a large area. LOFAR antennas span an area 2000 km in diameter, while the SKA is planned to cover 6000 km. Photo: ASTRON.

Figure 12.4: LOFAR remote station antennas. For both LOFAR and SKA, the antennas are grouped in small areas, with their signals combined locally (note the small hut at centre-left) then sent over glass fibre cables to a central supercomputer. For LOFAR, IBM provided a special version of its BlueGene/P machine as central processor. Photo: ASTRON.

during my directorship are proof that when directors support their best people, really good things happen.'[1]

One of the major outcomes of the ASTRON program was the Low Frequency Array (LOFAR). Butcher was responsible for helping define and secure the funding for the $130 million LOFAR pathfinder to SKA. He viewed it as an essential technological step towards the SKA as well as a way to do innovative astronomy at radio frequencies just above the ionospheric cut-off.

LOFAR is one of the world's largest radio telescopes having sensitivity in the frequency domain 15–240 MHz, where the ionosphere causes major imaging difficulties. According to Butcher, the problem would bedevil the SKA at higher frequencies as well. In addition, there is a serious cost problem with the SKA: 'the problem with conventional radio telescopes has been that much of the cost is in the steel of the giant dishes that move. The cost of steel is not going down with time, so such mechanical systems are and will stay very expensive'.[1] Jan Noordam and Jaap Bregman pointed the way to the solution and LOFAR was based on their advice, with the costs shifted much more towards electronics and software. 'That means because of Moore's law it'll get cheaper with time rather than more expensive. And so the idea was to build a telescope with no moving parts, in which the pointing and focusing is done in software, and with the antennas very low to the ground to mitigate the RFI (Radio Frequency Interference). The resulting instrument becomes much less expensive than people have been able to do with conventional designs.'[1] Indeed, the cost and the use of new technology in a pathfinder for the SKA was a major motivation to build the telescope, although it is also making high-quality observations at these frequencies possible for the first time.[12]

An initial idea to site LOFAR in the outback of Western Australia, near a proposed SKA site, had to be abandoned. Constraints on the use of the Dutch funding, the only funding available at the time, made investment outside the Netherlands difficult if not impossible. Most of the antennas are located in the Netherlands, although now there are others scattered across Europe. 'It is really big, essentially two thousand kilometres in size', said Butcher.

The telescope has an ambitious science program and its design was driven by four fundamental applications: (i) the epoch of re-ionization, (ii) extragalactic surveys and their exploitation to study the formation and evolution of clusters, galaxies and black holes, (iii) transient sources and their association with high-energy objects such as gamma-ray bursts, and (iv) cosmic ray showers and their exploitation to study the origin of ultra-high energy cosmic rays.[13]

In 2006, after almost 16 years at ASTRON, Butcher was planning to retire in the Netherlands. However, one fine day in mid 2006 he had a phone call from Ken Freeman, who said that 'the directorship at Mount Stromlo would soon be vacant, and encouraged him to apply for the position'. He applied for the job and towards the end of that year got a phone call from Australian National University Vice-Chancellor, Ian Chubb, who 'was sceptical I was really interested, in so many words accusing me of planning to use an offer from ANU as leverage for some other job I might be interested in. I said I had nothing else in my plans and if I thought I could help out I would most likely accept an offer. Shortly thereafter I was invited to visit for an interview, at which I said I had concluded that with my background I could probably help the School out, that I would be interested only in a single 5 year term.'[14]

On arrival, Butcher confronted about a million dollar deficit in the School's and Observatory's annual operating budget. This had arisen because of the severe budget cuts in the previous five or six years. The situation was a combination of general budget cuts to ANU, at about the 25% level, which had been passed on to the School, and a decision early that decade that the ANU would join the Australian Research Council grants scheme, resulted in the loss of more operating budget, which could not be recouped through the ARC grant process.

He also found that several excellent academic staff had left since the 2003 bushfires and technical staff had also been let go. Attracting talented students and postdocs was going to be very difficult. The Observatory was suffering an image problem both in Australia and overseas. Added to all this, there were problems with the way things were administered. The local administration was decoupled from the School and took its orders from ANU's head office rather than from the astronomers. At Siding Spring, Butcher found the campus was 'in a parlous state generally'.[14]

However, there was a silver lining. Two internationally important projects begun during the Sackett years were under construction: Michael Dopita's Wide Field Spectrograph (WiFeS) and Brian Schmidt's SkyMapper. Both were highly innovative projects. 'Placing the WiFeS on the 2.3 m', Butcher said, 'would give the telescope a new lease of life for about 6 to 8 years until international competition caught up with it.'[14]

The SkyMapper project was to be used for producing a map of the southern sky with a telescope, using detectors and filters that would revolutionise the field of galactic archaeology. The NIFS and GSAOI projects led by Peter McGregor were doing very well on the 8 m Gemini telescopes. Butcher found that 'Not only were senior staff scientifically productive, but they were clearly very capable of leading ground-breaking new international projects at the highest level.' He felt that 'in concert with scientific discovery, the instrumentation program would be the way forward to resurrect the Observatory's international reputation'.[14]

Being a strategic thinker, Butcher rolled up his sleeves and worked out a plan to get the Observatory back on its feet. He talked extensively to staff, both academic and technical/operational, and by December 2007 had drawn up a plan that built on the foundation laid by Penny Sackett. The priorities for his directorship were, 'First, to recruit new, young researchers of the highest international standing by using the ARC Fellowship programs and the School's endowment funds. Second, to complete the construction, commissioning and scientific exploitation of the new SkyMapper (Southern Sky Survey) and Wide Field Spectrograph (WiFeS) instruments. And third, the realization of the [full financial] participation in a major international project to develop a next generation very large telescope, assumed to be the Giant Magellan Telescope.'[14]

Figure 12.5: Lisa Kewley did her doctoral research at Mount Stromlo, receiving her PhD in 2002. She subsequently built up an international reputation through her research on the evolution of galaxies using the largest telescopes in the world. She returned to Mount Stromlo as an ANU Professor in 2011. Photo: Karen Teramura, University of Hawaii.

One of the first things he did was to devise plans to recruit young talented astronomers under a new Stromlo Fellowship program. The first of the Fellows was Chiaki Kobayashi, an expert in the numerical simulation of galaxy evolution. Butcher also enticed two top mid-career astronomers to come back to Mount Stromlo. One was Martin Asplund, who had taken leave of absence in 2007 to assume the directorship of the Max-Planck Institute for Astrophysics in Garching in Germany. The other was Lisa Kewley, a former Stromlo graduate who had made a name for herself in the US. Both returned in 2011.

He also began a program to attract good students to take up honours and PhD studies at Mount Stromlo, and noted that foreign students tended to prefer to stay on Mount Stromlo close to their workplace, rather than be distracted by Canberra's social and cultural life. He got a loan to build accommodation for them. The building was called the Faulkner Court 'after Don Faulkner, a well-loved former academic' on the staff. Faulkner Court has been a great success with a full occupancy rate.

SkyMapper and WiFeS

SkyMapper, a 1.35 m very wide field telescope with a 268 Megapixel digital camera, is the brainchild of Brian Schmidt. It photographs an area of sky 29 times the area of the full moon every minute. Its scientific goal is to map the southern sky, covering it 36 times with six special filters that give astronomers information about the ages, masses and temperatures of stars. It is designed to provide a deep census of the sky that will also allow rare and variable objects to be discovered. The telescope is completely automated, sensing when the sky is clear, taking data as programmed and sending the observations over a fibre optic cable to the supercomputer at ANU in Canberra, where it is calibrated, reduced and sent to the astronomers for analysis. The instrument aims to provide essential information on the history of the Milky Way galaxy, on distant exploding stars, and on rare solar system objects.

Figure 12.6: Brian Schmidt's Sky Mapper project garnered widespread and enthusiastic support as a bold initiative to explore deep space with innovative technologies. Photo: *The Weekend Australian Magazine*.

Figure 12.7: SkyMapper is located at Siding Spring Observatory, a very dark sky site in the central west of New South Wales. Photo: Jamie Gilbert, AAO.

Figure 12.8: The team at Mount Stromlo that built the SkyMapper telescope and data reduction facilities. (1) Mike Petrovic; (2) Timothy Preston; (3) Patrick Tisserand; (4) Patrick Oates; (5) Colin Vest; (6) Gabe Bloxham; (7) Mark Waterson; (8) Brian Schmidt; (9) Andrew Granlund; (10) Andre Degans; (11) Marian Szczepkowski; (12) Tony Martin-Jones; (13) Stefan Keller; (14) Bernard Keys; (15) Errol Kowald; (16) Annino Vaccarella; (17) Peter Conroy. Photo: Mount Stromlo Archives.

SkyMapper is an elegant and even beautiful instrument of its kind, and also pushes technology in several areas. During its design it needed a systems engineering approach much like that required for space instrumentation, but the resources were not available. According to Butcher, 'Shortly after my arrival I predicted a long trail of problems that would have to be solved one by one and that would likely delay the planned Southern Sky Survey for quite some time.'[14] He was worried that the data processing software would take longer than expected to finish and test, a phenomenon common in such projects. He was particularly concerned about the choice of

Figure 12.9: Dopita explaining his WiFeS to staff and students. The instrument, operating at the 2.3 m Advanced Technology Telescope at Siding Spring, allows astronomers to obtain spectra at every position across its field of view. It has proved invaluable for determining compositions, temperatures, densities and velocities in a wide variety of nebulae and galaxies. Photo: Mount Stromlo Archives.

Figure 12.10: WiFeS mounted on the 2.3 m Advanced Technology Telescope at Siding Spring. Its ability to obtain many spectra simultaneously over the entire optical spectral range makes the instrument a favourite among astronomers and allows this modestly sized telescope to be very competitive with much larger telescopes. In addition, the telescope and instrument can be operated remotely from Mount Stromlo, or indeed from anywhere that has a good internet connection. Photo: Mount Stromlo Archives.

the closed-cycle Helium refrigeration unit as it could excite resonance modes in the telescope structure. 'This risk had not been evaluated in the factory. In the end, this risk became reality and was not completely resolved even as I departed as Director five years later.' He concluded that 'we should not be skimping on systems engineering and a seriously professional team of engineers would have to be recruited'.[14]

The WiFeS is a project initiated by Dopita and designed and built in the Advanced Instrumentation and Technology Centre (AITC) at Mount Stromlo, to operate on the 2.3 m telescope. It is an integral field spectrograph and draws on the design concepts of the NIFS spectrograph built for the Gemini North 8 m telescope in Hawaii by Peter McGregor and his team, and on the experience of the Double Beam Spectrograph built by Rodgers in the late 1980s.[15] According to Dopita, 'The WiFeS instrument is designed as a facility instrument. Consequently, the science mission of the WiFeS spectrograph is very broad and encompasses observations of single stars, clusters of stars, extended nebulosities, galaxies in the nearby and distant universe and gamma-ray bursts.'[15] Exciting projects that astronomers will be able carry out with the instrument include solving the chemical abundance controversy for HII regions, the ages and metallicities of LMC globular clusters, the nature of the filaments of Cen A and Galaxy mergers and metallicity. 'With the advent of WiFeS, the 2.3 m telescope has once more been transformed into a fully-competitive major research facility', Dopita said.[15] 'It has', Butcher agreed, 'become the workhorse instrument on the 2.3 m'.

Giant Magellan Telescope

Penny Sackett, the previous Director, had signed the ANU to the Giant Magellan Telescope (GMT) project in 2006, as mentioned in Chapter 11. It is a billion-dollar project currently in the final stages of design. A Go–No-Go decision to proceed with the construction is expected early in 2014, and full operations are planned from ~2020. The effort is a collaboration of 10 partner institutions: seven in the US (Harvard University, Carnegie Institution, University of Chicago, University of Arizona, University of Texas at Austin, Texas A&M and the Smithsonian Astrophysical Observatory), all of which have built and operated large telescopes previously, one in Korea (the Korea

Figure 12.11: A computer rendition of the Giant Magellan Telescope. Seven of the largest mirrors currently possible to fabricate are on a single mounting structure and together they capture up to 100 times more light than the Hubble Space Telescope can. The light is reflected to seven small mirrors at the upper end of the structure, then to scientific instruments located behind the large mirrors. The GMT is designed to be able to detect and study galaxies and black holes in the earliest epochs of the universe, and to investigate the properties of planets circling other stars in the Milky Way Galaxy. Photo: Giant Magellan Telescope – GMT Corporation.

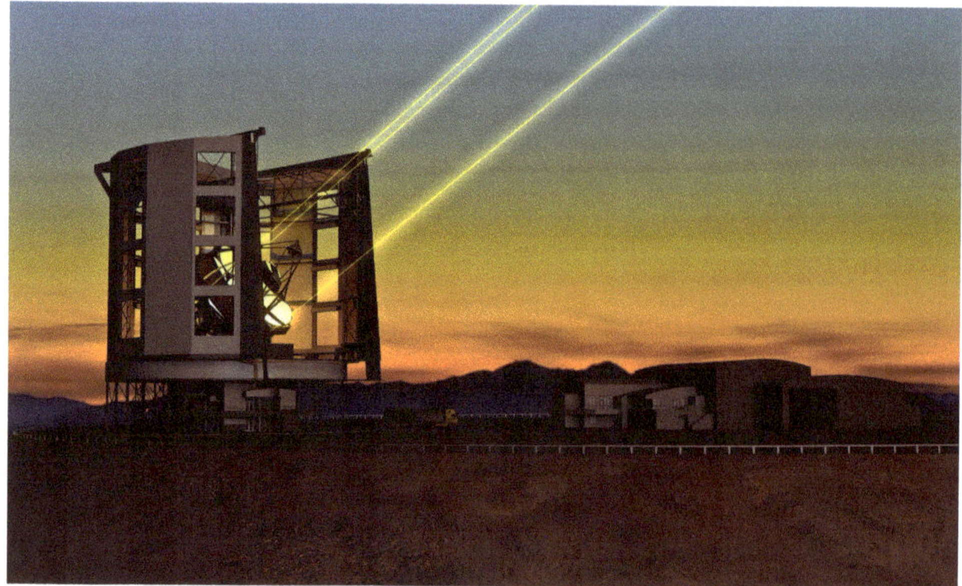

Figure 12.12: The GMT will be equipped with a laser system and adaptive optics. The lasers help quantify the blurring of images caused by the Earth's atmosphere. The adaptive optics can then correct the blurring, allowing the GMT to make images of the sky 10 times sharper than the Hubble Space Telescope. Photo: Giant Magellan Telescope – GMT Corporation.

Astronomy & Space Institute, a government-run centre) and two in Australia (ANU and Australia Astronomy Ltd). However, at the time the ANU could contribute only a limited amount to show its commitment to the project. Unexpectedly, the global financial crisis (GFC) in 2008 was a blessing in disguise for Mount Stromlo astronomers. In response to the GFC the Australian federal government set up the Educational Investment Fund (EIF) as part of an economic stimulus package to not only tackle the long-standing chronic need of Australian universities for additional and upgraded infrastructure but also to improve education generally.

Butcher saw a tremendous opportunity and grabbed it. According to him, 'I proposed [full] participation in the GMT to the Vice-Chancellor, Ian Chubb and somewhat to my surprise he enthusiastically agreed to have ANU propose the project.'[14] In fact, Chubb chaired the Mount Stromlo delegation to the formal interview with the evaluation panel. 'Our proposal was for $88.4 million, to include $65 million as our contribution to the international project (i.e. to be sent to the project office in Pasadena and yield a 5% access for ANU researchers plus 5% access through Astronomy Australia Ltd. for the whole university community, including ANU)'.[14]

It is surprising that there was considerable opposition from astronomers at other universities even though there was no cost to them. Some even lobbied against the proposal. The EIF proposal was a great success, however, and a contract was signed with the government on 10 February 2010. According to Butcher, 'This I view as probably my most important and enduring achievement as Director'.[14] Butcher will be remembered in years to come for his bold and successful proposal for the full participation in the GMT. It was one of the most far-reaching proposals for the Australian astronomical community, and a valuable complement to investments at CSIRO in the SKA.

Butcher used $21.4 million from the grant to complete the AITC and to do R&D so as to compete for GMT instrumentation and other engineering projects. He appointed Arnold Swart, a systems engineer and project manager from South Africa, to implement the EIF–GMT project. Unfortunately after a year in the job Swart resigned; he had been struggling against obstacles placed in his path by bureaucrats from ANU administration. He was replaced by Roger Franzen, a well-known Australian space technology engineer who for a decade had been the CEO of Auspace, the spin-off company from Mount Stromlo during the Mathewson years in the 1980s.

The completion of the AITC with funding from the EIF–GMT grant allowed Mount Stromlo not only to appoint new engineering staff but also to compete for one of only two or three first-light scientific instruments. Peter McGregor led a team, with systems engineer Simon Parcell, to propose and win an instrument, GMTIFS, that essentially is an evolution and combination of the NIFS and GSAOI instruments previously built for the Gemini observatories. According to Butcher, 'The international evaluation panel organized by the GMT project office judged it the best prepared and most convincing of the 6 or 7 instruments proposed'.[14]

The aim of GMTIFS is to allow medium-resolution spectra to be taken of galaxies in the early universe, of the surrounds of black holes and of exoplanet systems. It will map such objects in the near infrared at the highest spatial resolution possible with the GMT (diffraction limited or as close as

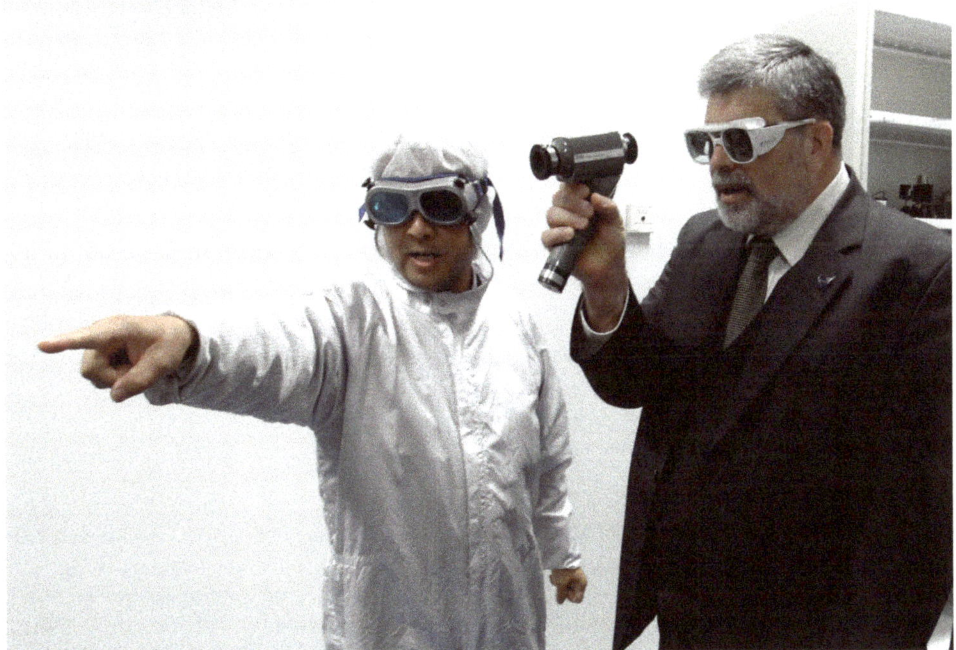

Figure 12.13: Minister for Innovation, Industry, Science and Research, Kim Carr, being shown the AO laser laboratory by Yue Gao during the dedication of the completed AITC on 18 April 2011. Photo: *Sydney Morning Herald*.

achievable – in any case some 10 times sharper images than the Hubble can take). GMTIFS relies on a functioning adaptive optics (AO) system on the telescope. Because of its reliance on an AO system, Butcher said, 'I therefore early on took the decision to have our program also become involved in AO technologies, and to compete for contracts to help design and build the AO system on the GMT. In that way I felt we could best guarantee that GMTIFS would end up scientifically productive at the earliest possible stage. We recruited AO specialists from overseas – Rod Conan from Canada, Francois Rigaut and Celine D'Orgeville from France via Chile as well as system engineers, project managers and optics specialists – to complement our existing in-house expertise.'[14]

Since Mount Stromlo did not have an established reputation in AO, 'we needed actually to build a working system to show our team could really perform'. Interestingly, the EOS-Space Systems company was involved in satellite laser ranging measurements from Stromlo, for example to maintain knowledge of the orbits of GPS satellites to adequate accuracy. They had developed an ability to detect and monitor the orbits of space debris to unprecedented accuracy, which they hoped to turn into a profit-making business (saving commercial and military satellites from collisions with space debris). It was concluded that their effectiveness could be dramatically

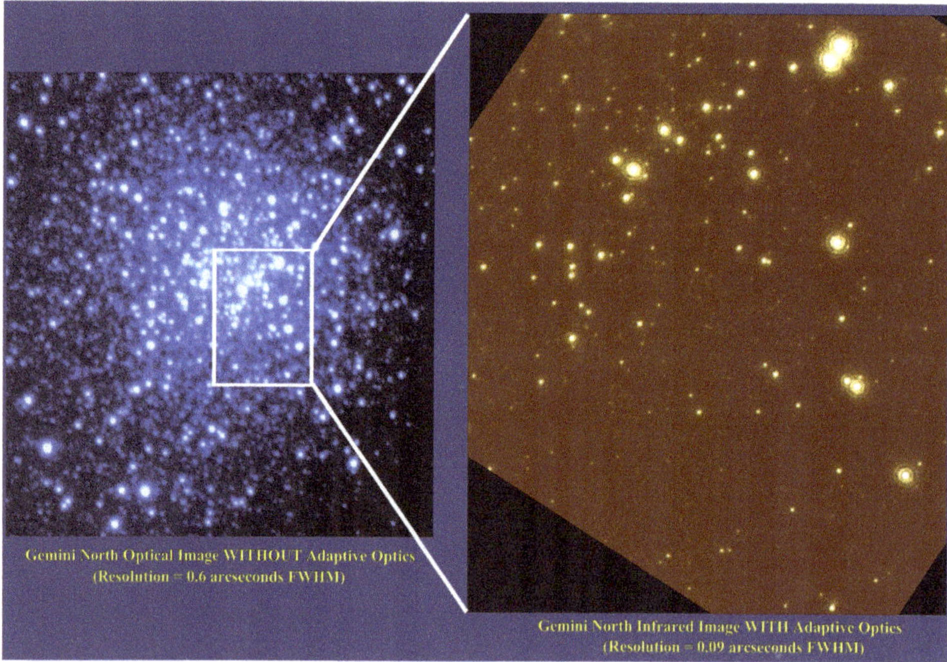

Figure 12.14: AO at work on the Gemini North telescope in Hawaii. At the left is an image of a cluster of stars, as obtained with the AO systems switched off (but under normal to good atmospheric blurring). At the right is a section of the same field of view, with the AO systems turned on. The rings around the brighter stars are an indication that the system is correcting exceedingly well for the blurring effects of the atmosphere above Hawaii. Photo: NOAO.

improved if they implemented an AO system on their laser ranging telescope. According to Butcher, 'We entered into a MoU (Memorandum of Understanding) on 10 Nov 2009, jointly to invest and work together to develop an AO demonstrator system on the 1.8 m EOS-SS telescope at Mt Stromlo. From our side, this would also demonstrate our capability to deliver on a potential contract for parts of the AO system on the GMT. As part of the ANU contribution we agreed to host the EOS-SS R&D team at the AITC. Their team would bring laser physics expertise as well as a telescope and funding to the project.'[14] However, the ANU administration balked and would not convert the MoU into a formal contract. It took a great deal of Butcher's and Franzen's time to convince the ANU lawyers that 'the university would not be taken to town with the deal'. The contract was finally signed on 29 August 2011, after 22 months of delay.

As soon as the EIF-GMT was granted, Butcher did a stocktake and critical analysis of the financial affairs of the Research School of Astronomy and Astrophysics. He realised that in the long term 'it would be unrealistic to expect the School to finance even part of the technical program from the School's operating budget'. He had to find a solution to this problem, which would become acute following the expiry of the EIF-GMT grant in 2014. The solution came from the new Australian space initiative announced by the federal government.

Space research

On 12 November 2008, the government released its report, *Lost in Space? Setting a New Direction for Australia's Space Science and Industry Sector*. The report recommended a return to investment in a national space effort, in particular to train a new generation of engineers with knowledge of space technologies and satellite operations.

'Realizing that the engineering protocols, tools and test facilities would be very similar to what we would need for GMT instrumentation', Butcher said, 'I thought this could be the hook to acquire additional funding. I subsequently spent a great deal of my time following the EIF award networking into the space community, greatly helped by Roger Franzen of course. And we agreed with government to use some of the EIF funding to provide a national engineering capability at Stromlo not only for GMT and astronomy but also for space research.'[14]

In its 2009–10 Budget, the government announced the establishment of a Space Policy Unit which would oversee the $48.6 million Australian Space Science Program's four years of activity. However, it specifically excluded funding for astronomy since the government felt that astronomers had received enough support, for the SKA and GMT projects. Nevertheless, Butcher cleverly used his contacts to get the Research School to participate in and acquire funding from five projects: Antarctic Broad-Band led by Aerospace Concepts Pty Ltd, Australian Plasma Thruster led by ANU Physics, Space Debris Detection and Monitoring Systems led by EOS-Space Systems Pty Ltd, Greenhouse Gas Monitor led by VIPAC Engineering & Scientists Pty Ltd, and GRACE Follow-on led by ANU Earth Sciences and Physics. These collaborations would estab-

Figure 12.15: EOS Space Systems built and operates this Satellite Laser Ranging observatory at Mount Stromlo. Its primary purpose has been precision orbit determinations for geodesy, such as for satellite navigation systems using GPS satellites. Photo: EOS Space Systems.

lish Mount Stromlo and the AITC as an important player in the space research field, and hopefully lead to larger projects in the future. To further this goal, Naomi Mathers was recruited at the AITC as an industry liaison officer with the task of building the space research side of the Research School's program into a major source of funding and research. She has a degree in space engineering from the Royal Melbourne Institute of Technology.

Siding Spring Observatory

Annual reductions to the Observatory's operations budget over a decade had led to a chronic maintenance problem for the facilities at Siding Spring. As Sackett had pointed out, the Research School's facilities were at a tipping-point and serious choices would have to be made. Butcher's view was that the ANU telescopes would have a useful research life of perhaps another five to seven years, or at least until the SkyMapper Southern Sky Survey could be completed. At that point, unless budget increases were forthcoming, ANU would be forced to abandon the site. He therefore undertook four initiatives.

First, he convinced ANU to transfer responsibility for site maintenance to its Division of Facilities and Services, which had formal responsibility for the basic maintenance of the other university buildings and grounds. The transfer left the Research School responsible only for its own operational telescopes – the 2.3 m, SkyMapper and the Uppsala Schmidt.

Second, he arranged for the original team of engineers led by Herman Wehner, who had worked with Don Mathewson to design and build the 2.3 m telescope, to return from retirement and refurbish the instrument. In particular, its drive system electronics, originally developed by Gary Hovey, had run out of spare parts and could no longer be kept in regular, reliable operation. With a very minimal budget but a great deal of dedication and intimate knowledge of the system, the team brought the telescope system back to a state at which it might remain operational for another five to seven years.

Third, the conviction that learning to observe hands-on with small telescopes is important to the education of young astronomers led Butcher to attract several other research institutions to Siding Spring. To the Anglo-Australian Observatory, University of New South Wales instruments and the Las Cumbres Faulkes telescope would be added facilities from Korean, Polish and US institutions. Siding Spring would become an international observatory, similar to if smaller in scale than the Mauna Kea Observatory on Hawaii or the Roque de los Muchachos Observatory on La Palma. Each new facility would contribute to communal site maintenance costs but otherwise enjoy a rent-free site; in exchange, each would grant ANU astronomers a level of

Figure 12.16: The Great Comet of 2007. Siding Spring is an important site at which international groups carry out observations. For example, NASA and the University of Arizona have supported a highly successful hunt for potentially dangerous asteroids using the Uppsala Schmidt telescope. In the course of that work, nearly 100 comets have also been discovered, including C/2006 P1, the Great Comet of 2007. At its brightest this spectacular comet could be seen with the naked eye during the day. It was discovered on 7 August 2006 by ANU astronomer Rob McNaught. It began as a as a faint blur, moving slowly across the sky, but by the end of January 2007 it had developed into the most brilliant comet for decades. Photo: © Robert McNaught.

observing access. Even when the Research School would have to abandon its own facilities, its students would be able to carry out projects at Siding Spring.

Finally, the Anglo-Australian Agreement between the British and Australian governments would come to an end on 30 June 2010, providing an opportunity to reorganise activities at Siding Spring. The 3.9 m telescope and headquarters in Sydney would henceforth be called the Australian Astronomical Observatory (still with the initials AAO and same logo!) and become a division of the Department of Innovation, Industry, Science and Research of the Australian federal government. Its new purpose would be to operate observing facilities for Australian astronomers and support Australian access to international observatories. Matthew Colless, then the AAO Director, said, 'We now build instruments not only for our own telescope, but also we've built them for the Very Large Telescope at ESO in Chile, we've built one for Subaru and are proposing to build a very ambitious spectrograph for the Gemini Observatory. But, as well as that, we've actually contributed in a number of areas of fundamental technology, and the most recent example of that is an invention here by Joss Bland-Hawthorn, which is optic fibres that actually suppress the air-glow emission of the night sky.' Butcher proposed that it would be logical, given the AAO mission and its transition to an all-Australian organisation, for the government to consider taking over the operation and maintenance of all national facilities at Siding Spring. After all, the AAO would likely remain far and away the largest facility and have the most staff on the site, hence would have the most to lose when the ANU could no longer manage the site at an appropriate level. Butcher even proposed that rather than duplicate engineering and office facilities, the new AAO could move to Canberra and co-locate with the Research School at Mount Stromlo. Unfortunately, the senior echelons at ANU would not consider relinquishing management control of the Siding Spring campus and the astronomical community clearly feared that ANU might come to dominate the AAO program, so neither proposal gained support.

The future of the Siding Spring Observatory was advanced during the Butcher years, but was not fully resolved.

Research at the highest level

During Butcher's directorship the main thrust of the scientific research programs was in galactic archaeology (content and evolution of galaxies), observational cosmology (content and evolution of the whole cosmos) and the astrophysics of stars, planets and black holes. The group of academic researchers at the Observatory could be compared with the best anywhere in the world and continued to reap the recognition they deserved. Although a large number of papers were published during the Butcher years we are, because of space limitations, able to highlight here only some of the exciting and innovative research and major awards received by staff during this period.

Brian Schmidt topped the list of researchers at Mount Stromlo by being awarded the 2011 Nobel Prize for Physics, jointly with US astronomers Adam Riess (Johns Hopkins University) and Saul Perlmutter (University of California, Berkeley) for their ground-breaking research on the accelerating universe (see Chapter 10).

Figure 12.17: Schmidt receiving the Nobel Prize for Physics 2011 from King Carl XVI Gustav of Sweden, 10 December 2011. It was granted for the discovery in the 1990s that the universe is not only expanding but that the expansion is accelerating, a completely unexpected and unexplained result. Photo: Copyright © The Nobel Foundation.

When Schmidt received a call one evening from Stockholm, telling him that he had won the 2011 Nobel Prize for Physics with Adam Riess and Saul Perlmutter, he thought at first it was a prank call. When he realised it was real, 'He felt weak in his legs and somewhat amazed'. It was almost 96 years since an Australian had won a Nobel Prize in Physics – the first was won by the father and son team, William and Lawrence Bragg, for their work on X-ray crystallography.

Shortly after the 8.30 p.m. call from Stockholm, ABC *Lateline* called Schmidt for an interview and dispatched a film crew to his property near Canberra, where he grows pinot noir grapes as a hobby. The phone then went ballistic and Schmidt did not stop answering calls until ~1.30 a.m. The next morning the story hit the front pages of the *The Australian* and the *Sydney Morning Herald,* with stories by science writers Leigh Dayton and Deborah Smith. The Observatory shot into international fame and every Australian suddenly became aware of Schmidt and Mount Stromlo. Schmidt gave talks and media interviews all over Australia and became an immediate media celebrity both in Australia and overseas.

Schmidt also won the 2007 $0.5 million Gruber Prize for Cosmology with members of the High-Z SN Research Team, and was elected in 2008 as a Fellow of the Australian Academy of Science, a Fellow of the US National Academy of Science and Foreign Member of the Spanish Royal Academy of Sciences. He was also elected a Fellow of the Royal Society of London. In the 2013 Australia Day Honours, he was made a Companion of the Order of Australia.

Figure 12.18: Stromlo astronomer Ken Freeman received the Prime Minister's Prize for Science in 2012 from Julia Gillard. The Prize is Australia's pre-eminent recognition of excellence in science. Photo: AGS.

His main instrumental project during the Butcher years was the SkyMapper to survey the southern sky. According to Schmidt, it is a '1.3 m telescope built by the Australian National University as a replacement for the Great Melbourne Telescope lost in the fires that destroyed Mt Stromlo Observatory in 2003'.[16]

Schmidt continued his studies of Type Ia supernova, realizing that it is important that the use of SNe Ia standard candles be explored in as much detail as possible. The basis for the study is that these objects should be able to be used as standard candles if they really can be calibrated adequately.

Ken Freeman is probably the most prolific author of high-quality papers at Mount Stromlo, and indeed of the whole astronomical community in Australia. He has been a trend-setter in several areas of astronomy and astrophysics. With Joss Bland-Hawthorn of the University of Sydney, he is celebrated as the father of the field of galactic archaeology and is recognised as an early proponent of the theory that dark matter must form a large fraction of the mass of galaxies. He is a Fellow of the Australian Academy of Science and the Royal Society of London and in 2012 was awarded the prestigious Prime Minister's Prize for Science for his entire oeuvre. In January 2013 the Australian Academy of Science awarded him the Matthew Flinders Medal. Also in January 2013 he was awarded the prestigious Henry Norris Russell

Lecture by the American Astronomical Society in recognition of a lifetime of seminal contributions to astronomy, including work on the structure and dynamics of galaxies including the Milky Way Galaxy. Freeman said, 'I feel very honoured to receive this prize for lifetime achievement from the American Astronomical Society. Many of my old friends, mentors and colleagues are on the list of past recipients of this prize, and it is a great pleasure to be listed with them.'

Figure 12.19: John Norris was elected a Fellow of the Australian Academy of Science in 2012. His research on the oldest stars having the least amounts of heavy chemical elements has demonstrated the complexity of the chemical evolution of the Milky Way galaxy, and forced changes in the current theory of stellar evolution. Photo: Australian Academy of Science.

Freeman and Van der Kruit from the University of Groningen in the Netherlands published the culmination of 40 years of the study of galaxy disks,[18] in a paper that will be the standard work in the field for a generation. The study of the chemical evolution of the Milky Way Galaxy is sometimes called 'near-field cosmology' (as opposed to 'far field cosmology', the study of galaxies very far away). Near-field cosmology is increasingly being called galactic archaeology. Having spearheaded the concept of galactic archaeology, Freeman was able to convince the astronomical community to support a new multi-million dollar spectrometer, HERMES, for the Anglo-Australian Telescope, which is specially designed to measure exceedingly efficiently the abundance of key elements in stars across the Galaxy.

John Norris and Mike Bessell have long and distinguished careers at Mount Stromlo and are well-known in international astronomical circles. Both have made significant contributions in international astronomy. John Norris was made a Fellow of the Australian Academy of Science in 2012. In awarding the Fellowship, the President of the Australian Academy of Science said, 'John Norris' research has forced major revision of several basic concepts in astronomy, including changing the concept of how the Galaxy formed. John's discovery of the most metal-deficient stars has illuminated the complexity of chemical evolution in the early Universe with his studies of the lithium abundance in the oldest stars putting a strong upper limit on the baryonic content of the Universe. In addition, John's work on globular cluster chemistry has opened up major new areas of research and initiated significant changes in the current theory of stellar evolution.'

Norris and Bessell, together with David Yong and Martin Asplund, carried out an extensive study of the most metal-poor stars.[19] It is currently believed that only hydrogen and helium and a very small amount of lithium were generated by the Big Bang. All the heavier elements were supposed to have been built in nuclear reactions in the centres of stars, either slowly during their lifetimes or rapidly as they exploded at the ends of their lives. Each succeeding generation of stars created new elements, some of which were returned to the interstellar gas via supernovae, novae or stellar winds, thereby continually enriching that gas. Furthermore, low-mass stars like the Sun, once formed, live a very long time, as long or longer than the Galaxy, and still exhibit in their outer layers the chemical composition of the gas out of which they were born. By observing the atmospheric compositions of long-lived stars of all ages we should get a picture of the chemical evolution of the Milky Way Galaxy. Martin Asplund's earlier seminal work showed how to account for the physics of stellar atmospheres, to get the most reliable abundances. Indeed, the strange abundance patterns found in the most metal-poor stars are clear indications they were formed out of gas that had a very different history from that of the Sun and most other, younger stars. Astronomers concentrate on carbon because it is the first abundant heavy element to be built up in stars – and it shows interesting abundance variations in these stars. The results represent a major study in the area and will be quoted for many years.

The research carried out by Kaviraj, Mike Dopita and colleagues[20] on how dominant old stars in today's galaxies form followed in a direct line the work done by Eggen, Lynden-Bell and Sandage.[21] It used early data from the new camera on HST – WFC3 (Hubble Space Telescope – Wide Field Camera 3). Dopita, a Fellow of the Australian Academy of Science, was on the science team for this camera and is profiting from early access. The observations look at massive spheroidal galaxies in the early universe, and demonstrate that processes other than major mergers (e.g. violent disk instability driven by cold streams and/or minor mergers) likely play a dominant role in building massive spheroidal galaxies, and creating a significant fraction of the old stellar populations that dominate today's universe. The direct relevance to the Milky Way halo is unclear, but likely so for the inner if not the outer halo (which at least in part is a result of many mergers of small galaxies with our own).

Stefan Kellar, Dougal Mackey and Gary Da Costa[22] were involved in the study of the formation of globular clusters which are large balls of stars. In the Milky Way Galaxy they are all old and, as far as we can tell from their colour-magnitude diagrams, are the result of a single burst of star formation in the early Galaxy. The only exceptions are several clusters that seem to be mergers of two clusters, and the frequent observation that there are light element abundance differences among the member stars. While the Galaxy has no young or intermediate age globular clusters, the Large Magellanic Cloud (LMC) does. Their study with very high-quality data of LMC intermediate age clusters showed that they exhibit evidence for episodes of star formation lasting 50–300 million years. Such an extended period of star formation would not be detectable in clusters in our Galaxy. The authors hypothesised that all globular clusters, including those in the Galaxy, have had multiple episodes of star formation, and make a clear prediction, namely that the LMC clusters will also show light element abundance variations similar to those in the galactic globulars. If this hypothesis pans out, it will lead to a new way of understanding not only the evolution of globular clusters in the early Galaxy but also the evolution of light element abundances in stellar systems.

Chiaki Kobayashi and Amanda Karakas'[23] study of the evolution of isotope ratios was a major theoretical contribution to the Mount Stromlo research theme of galactic archaeology. It involves chemical 'tagging' to isotopic ratios – that is, from elemental abundances to isotopic abundances. The study uses the most advanced galactic evolution models and incorporates the most recent ideas on the contributions of asymptotic giant branch stars (which dredge up newly formed elements and disperse them in a stellar wind). The study concludes that such isotopic data can help determine that production time-scales for the chemical elements in the various stellar components of the Galaxy – thin disk, thick disk, bulge, halo – differ and can point to the origins of the different elements (from supernovae, AGB stars etc.). It predicts that it should be possible to locate metal-rich stars (more heavy elements than in the solar system, produced by supernovae) that were formed in the very early universe. A challenge to the observers!

Mike Dopita and his team at Mount Stromlo – Lawrence Krauss, Ralph Sutherland, Chiaki Kobayashi and Charles Lineweaver[24] – investigated the re-ionization epoch of the universe. The expanding gas from the Big Bang creation event was very hot and fully ionized; it was opaque to electromagnetic radiation. About 300 000 or 400 000 years after the Big Bang, it had expanded and cooled enough that it became neutral and transparent (allowing the thermal radiation to 'escape' and become visible to us today as the 3° background radiation). But the gas between the galaxies today is fully ionized, so sometime between 400 000 years after the Big Bang and now, the medium was re-ionized. The first luminous objects were probably responsible for the re-ionization, that occurred about a thousand million years after the Big Bang. But we do not know what those first objects were. Theoreticians make model calculations which show that ultraviolet light from stars, even the very massive ones expected to be the first stars, is probably not strong enough to do the re-ionization. Dopita and his team came up with the idea that as gravitational instabilities began to form the first potential wells of dark matter in the universe (wells which would later be the sites for galaxies to form), the neutral gas falling by chance into those wells would collide with itself and form very strong shockwaves, which in turn would radiate enough ionizing X-ray radiation to ultimately re-ionize essentially all the gas between the potential wells. They combined their separate skills in advanced computational astrophysics and in the theory of structure formation in the early universe to show that this mechanism should have contributed to the re-ionization process, but probably was not strong enough in and of itself to cause the whole re-ionization.

Helmut Jerjen worked on the problem of current cosmological models, in which dark matter in the early universe clumps and the clumps then aggregate to provide the gravitational potential wells in which galaxies form. The models predict large numbers of satellite galaxies to our own Galaxy, but such large numbers are not observed. Other discrepancies with the predictions are also becoming evident. Jerjen's work involved an investigation of the distribution and motions of our satellite galaxies, to try and make sense of their origins. His research with Kroupa and colleagues[25] examined all the available evidence, including several new and as yet uncertain observations, and concludes that the standard cosmological model, popularly termed Cold Dark Matter, is unable to explain the observations. The authors spoke the unspeakable – that, as far as they can see, the Modified Newtonian Dynamics

(MOND) of Milgrom, Sanders and others better explains the observations than does conventional gravity theories (Newtonian or General Relativity)! In MOND, in the very weak field limit gravity departs from the standard Newtonian behaviour. While there is no known physics involved and the consequences for all of astrophysics have not yet been fully explored, MOND does seem to explain naturally many of the problems of the current standard model. Their research gave a good survey of the situation and made no claim that MOND is required to explain the observations, but it did suggest that more study should go into working through the consequences of MOND.

Several other papers were written in the area of astrophysics during the Butcher years. Here we discuss two of them. With Dopita and Sutherland, David Nicholls, a mature PhD student who had retired early from the public service to take up his love of astronomy, decided to derive chemical abundances of gaseous regions in our and other systems. He was made aware of discrepancies with solar and other abundance determinations. He discovered that he could explain the discrepancies if one of the basic assumptions of all previous work was not quite correct, namely that the energy distribution of the thermal electrons is not Maxwellian, but has a tail to higher energies.[26] He found that such tails exist in laboratory plasmas and in solar wind and other plasmas. This was a neat result that has profound implications for using the interstellar gas for galactic archaeology.

Observations of the Solar System show that there has been significant chemical fractionation among the planets, with the inner objects composed of rocky material and the outer ones of gas and ices. One could imagine that the atmospheres of stars with planets might show evidence of such fractionation compared to stars with no planets. Martin Asplund and international colleagues for the first time managed to determine chemical abundances in stars with sufficient accuracy to show such an effect. They showed that compared to an average star the Sun exhibits abundance anomalies as a function of the condensation temperature of the element.[27] This is a spectacular result that promises a technique to identify stars most likely to have terrestrial planets around them.

Ever since Mount Stromlo Observatory was established, new instrumentation for the telescopes has been built to expand the frontiers of astronomy. The new instrumentation has allowed the Observatory not only to keep up with international competition but also to make major discoveries. Like the astronomers, the instrumentalists have published the results of the construction of their wonderful instruments in international journals and presented papers at international conferences.

Adaptive Optics is playing a significant role in astronomical instrumentation today and its technologies have been improving steadily. The GeMS system on the Gemini South 8 m telescope is the first AO system to incorporate Multi-Conjugate Adaptive Optics, enabling wide field imaging at the diffraction limit. Francois Rigaut and Celine D'Orgeville designed GeMS and Peter McGregor's team at the Research School of Astronomy and Astrophysics built the imaging camera GSAOI that is an integral part of that system. The first results from GeMS were presented by Rigaut, Peter McGregor and colleagues[28] in 2012 at a conference organised by the International Society for Optical Engineering.

Damjanov, McGregor and Rigaut discussed the optimisation of extragalactic fields for AO in a paper[29] in the *Publications of the Astronomical Society of the Pacific* in 2011. It is one of those papers that will likely be extremely useful to observers in the coming decades. CCD detectors are very sensitive but can be overloaded if a brightish star appears in the field of view. Hence to study the early universe, Hubble and other telescopes have chosen fields completely devoid of foreground stars. These fields have been studied *ad nauseam* in recent years. AO is now becoming available and promises to provide 10 times the spatial resolution of Hubble, but it requires a modestly bright star somewhere near the field of study to determine the lowest order corrections to the distorting effects of atmospheric turbulence. The paper provides future AO observers with optimised fields to study with the most powerful telescopes today and in the future (the GMT).

As these research results and accompanying prizes and awards attest, the Butcher years saw the Observatory go from strength to strength in international recognition, in particular with the award of a Nobel Prize to one of its astronomers. Butcher noted, 'in 2009 we were placed number 10 in the world in space science by the Thomson-Reuters citation ranking – ahead of Cambridge, UC Berkeley and Harvard'.[14] At the start of 2012 Mount Stromlo Observatory and the Research School were supporting an Australian Research Council Centre of Excellence, two ARC Laureate Fellows, three ARC Future Fellows, an ARC Research Fellow, a DECRA Early-Career Fellow, 17 ARC Discovery grants, an ARC-LIEF grant and a major EIF grant worth $84.4 million. The Observatory had two Fellows of the Royal Society of London and memberships of the national academies of the US, Netherlands and Spain, as well as the Australian Academy of Science.

According to Butcher, at least some of our recent success may be a result of his management philosophy: 'I have tried during my tenure to adopt a philosophy that the Director should insulate research and technical staff from the administrative activities of the University to the extent possible. I took on much of the burden myself together with our local admin staff. I especially observed that the general dysfunction at the College level within the ANU tended to promote clerical and reporting activities being shifted onto (highly paid) academic research staff. I discouraged participation in the myriad of committees on the grounds that they took a great deal of time while occurring in a culture of poor preparation, little actual authority, bureaucratic turf disputes and avoidance of real issues. I believe the general satisfaction with my Directorship among staff is at least partly due to this approach of shielding scholarship from administration.'

A grand vision

In January 2013, Butcher retired from the position of Director. He had helped fulfil Duffield's dream – 'It is my earnest desire that we should take our place among the great observatories of the world.' Butcher will also be remembered for achieving the full participation of Mount Stromlo Observatory in the international Giant Magellan Telescope, one of his most far-reaching decisions for Mount Stromlo and for the Australian astronomical community. He placed the Observatory in position to reap the benefits of discoveries in the era of very large telescopes by this and by future generations of astronomers. This will be remembered as his most important and enduring legacy as Director of Mount Stromlo Observatory.

References

1. Bhathal R 2011. Interview with Harvey Butcher. National Oral History Project on Significant Australian Astronomers. National Library of Australia, Canberra.
2. Butcher HR 1972. Studies of heavy-element synthesis. I. Separation of r- and s-process abundances. *Astrophysical Journal* **176**, 711–722. doi:10.1086/151673.
3. Butcher HR 1975. Studies of heavy-element synthesis. II. A survey of e-, r- and s-abundances. *Astrophysical Journal* **199**, 710–717. doi:10.1086/153742.
4. Butcher HR 1975. An Echelle grating for the 74 inch Coudé. *Publications of the Astronomical Society of Australia* **2**, 21–24.
5. Butcher HR, Oemler A 1978. The evolution of galaxies in clusters. I. ISIT photometry of Cl 0024+1654 and 3C 295. *Astrophysical Journal* **219**, 18–30. doi:10.1086/155751.
6. Butcher HR, Oemler A 1984. The evolution of galaxies in clusters. V. A study of populations since z ~ 0.5. *Astrophysical Journal* **285**, 426–438. doi:10.1086/162519.
7. Bhathal R 2006. Interview with W Couch. National Oral History Project on Significant Australian Astronomers. National Library of Australia, Canberra.
8. Butcher HR, Hicks A 1985. A Fabry-Perot based stellar seismometer. In *Proceedings NATO Advanced Research Workshop.* (Ed. D Gough). Cambridge University Press.
9. Pottasch EM, Butcher HR, van Hoesel FHJ 1992. Solar-like oscillations on Alpha Centauri A. *Astronomy and Astrophysics* **264**, 138–146.
10. Butcher HR 1987. Thorium in G-dwarf stars as a chronometer for the Galaxy. *Nature* **328**, 127–131. doi:10.1038/328127a0.
11. Douglas NG, Butcher HR, Melis WA 1990. Heterodyned, holographic spectroscopy: results with the FRINGHE spectrometer. *Astrophysics and Space Science* **171**(1–2), 307–318. doi:10.1007/BF00646870.
12. Butcher HR 2004. LOFAR: first of a new generation of radio telescopes. *Proceedings of the Society for Photo-Instrumentation Engineers* **M5489**, 537–544.
13. Rottgering HJA 2006. LOFAR: opening up a new window on the universe. In *Cosmology, Galaxy Formation and Astroparticle Physics on the Pathway to the SKA*. (Eds H-R Klockner, M Jarvis and S Rawlings). Oxford University Press, Oxford.
14. Butcher HR 2013. Notes for Ragbir Bhathal. 1 February 2013.
15. Dopita M, Rhee J, Farage C, McGregor P, Bloxham G, Green A, Roberts B, Nielson J, Wilson G, Young P 2010. The Wide Field Spectrograph (WiFeS): performance and data reduction. *Astrophysics and Space Science* **327**, 245–257. doi:10.1007/s10509-010-0335-9.
16. Schmidt B 2012. SkyMapper: surveying the southern sky. American Astronomical Society Meeting 20, no. 426.01.

17. Blondin S, Matheson T, Kirshner RP, Mandel KS, Berlind P, Calkins M, Challis P, Garnavich PM, Jha SW, Modjaz M, Riess AG, Schmidt BP 2012. The spectroscopic diversity of Type Ia supernovae. *Astronomical Journal* **143**, 126–158.

18. van der Kruit PC, Freeman KC 2011. Galaxy discs. *Annual Review of Astronomy and Astrophysics* **49**, 301–371. doi:10.1146/annurev-astro-083109-153241.

19. Norris JE, Yong D, Bessell MS, Christlieb N, Barklem PS, Asplund M, Murphy SJ, Beers TC, Frebel A, Ryan SG 2013. The most metal poor stars. I, II, III, IV. *Astrophysical Journal* **762**, 25/26/27/28.

20. Kaviraj S, Cohen S, Ellis RS, Peirani S, Windhorst RA, O'Connell RW, Silk J, Whitmore BC, Hathi NP, Ryan RE, JrDopita MA, Frogel JA, Dekel A 2013. Newborn spheroids at high redshift: when and how did the dominant, old stars in today's massive galaxies form? *Monthly Notices of the Royal Astronomical Society* **428**, 925–934. doi:10.1093/mnras/sts031.

21. Eggen OJ, Lynden-Bell D, Sandage AR 1962. Evidence from the motions of old stars that the galaxy collapsed. *Astrophysical Journal* **136**, 748–766. doi:10.1086/147433.

22. Keller SC, Mackey AD, Da Costa GS 2011. The extended main-sequence turnoff clusters of the Large Magellanic Cloud: missing link in globular cluster evolution. *Astrophysical Journal* **731**, 22–30. doi:10.1088/0004-637X/731/1/22.

23. Kobayashi C, Karakas AI, Umeda H 2011. The evolution of isotope ratios in the Milky Way Galaxy. *Monthly Notices of the Royal Astronomical Society* **414**, 3231–3250. doi:10.1111/j.1365-2966.2011.18621.x.

24. Dopita M, Krauss LM, Sutherland RS, Kobayashi C, Lineweaver CH 2011. Re-ionizing the universe without stars. *Astrophysics and Space Science* **335**, 345–352. doi:10.1007/s10509-011-0786-7.

25. Kroupa P, Famaey B, de Boer KS, Dabringhausen J, Pawlowski MS, Boily CM, Jerjen H, Forbes D, Hensler G, Metz M 2010. Local-Group tests of dark-matter concordance cosmology. Towards a new paradigm for structure formation *Astronomy and Astrophysics* **523**, A32.

26. Nicholls DC, Dopita MA, Sutherland RS 2012. Resolving the electron temperature discrepancies in H II regions and planetary nebulae: κ-distributed electrons. *Astrophysical Journal* **752**, 148–163. doi:10.1088/0004-637X/752/2/148.

27. Meléndez J, Asplund M, Gustafsson B, Yong D 2009. The peculiar solar composition and its possible relation to planet formation. *Astrophysical Journal* **704**, L66–L70. doi:10.1088/0004-637X/704/1/L66.

28. Rigaut F, Neichel B, Boccas M, d'Orgeville C, Arriagada G, Fesquet V, Diggs SJ, Marchant C, Gausach G, Rambold WN, Luhrs J, Walker S, Carrasco-Damele ER, Edwards ML, Pessev P, Galvez RL, Vucina TB, Araya C, Gutierrez A, Ebbers AW, Serio A, Moreno C, Urrutia C, Rogers R, Rojas R, Trujillo C, Miller B, Simons DA, Lopez A, Montes V, Diaz H, Daruich F, Colazo F, Bec M, Trancho G, Sheehan M, McGregor P, Young PJ, Doolan MC, van Harmelen J, Ellerbroek BL, Gratadour D, Garcia-Rissmann A 2012. GeMS: first on-sky results. Adaptive Optics Systems III. In *Proceedings of SPIE Astronomical Telescopes and Instrumentation*. Vol. 8447, pp. 844701–844701-15. International Society for Optics and Photonics.

29. Damjanov I, Abraham RG, Glazebrook K, McGregor P, Rigaut F, McCarthy PJ, Brinchmann J, Cuillandre J-C, Mellier Y, McCracken HJ, Hudelot P, Monet P 2011.

Extragalactic fields optimized for adaptive optics. *Publications of the Astronomical Society of the Pacific* **123**, 348–365. doi:10.1086/658931.

30. Merton R 1968. The Matthew effect in science *Science* **159**, 56–63.
31. Schmidt B 2011. Interview with Leigh Dayton, *The Australian*. 5 October 2011, p. 1.
32. Freeman K 2012. Email to Ragbir Bhathal, 11 December 2012.

13

Brian Schmidt's Nobel Lecture 2011: accelerating expansion of the universe through observations of distant supernovae

Introduction

Einstein postulated that nothing can travel faster than light. This created a problem, because the force of gravity seems to act instantly over great distances. Einstein resolved the dilemma by assuming that gravity causes space-time apparently to change shape and thereby affect the motions of objects. His General Theory of Relativity has been spectacularly successful in explaining the action of gravity in the nearby universe. Astronomers would dearly like to know how gravity works not only nearby but how it influenced the formation and evolution of the universe as a whole.

Now, space is transparent. This simple fact allows powerful telescopes to see and analyse stars and galaxies across the entire universe. Because it takes the light from distant objects a certain time to travel to us, we see them as they were at an earlier epoch. Today's astronomers are busy unravelling the history of the universe by studying objects at ever larger distances. Understanding the physics of gravity at the largest scales and over longest times lies at the core of this quest.

Galaxies and stars appear fainter and smaller with increasing distance, of course. But determining their precise distances is the hardest task faced by observational astronomers. To understand the action of gravity throughout cosmic history, a way had to be found to derive precise distances to remote galaxies. During his PhD studies at Harvard University, Stromlo astronomer Brian Schmidt learned about exploding stars called supernovae. When they explode, these stars become as bright as whole galaxies and can be seen across cosmic distances. At Mount Stromlo he led an international team of colleagues from Chile, Europe and the US to show that certain supernovae make good distance indicators and can be used to reveal the shape of space at the largest scales. Their conclusion was that a new force must be at work, causing the universal expansion of space actually to be accelerating. This force has become known popularly as Dark Energy, and its discovery was awarded the Nobel Prize for Physics in 2011.

Here we reprint Brian Schmidt's Nobel Lecture. It tells a fascinating story of discovery as well as provides a window on how fundamental science unfolds in practice.

Nobel Lecture: Accelerating expansion of the Universe through observations of distant supernovae*

Brian P. Schmidt

(published 13 August 2012)
DOI: 10.1103/RevModPhys.84.1151

This is not just a narrative of my own scientific journey, but also my view of the journey made by cosmology over the course of the 20th century that has lead to the discovery of the accelerating Universe. It is complete from the perspective of the activities and history that affected me, but I have not tried to make it an unbiased account of activities that occurred around the world.

20th Century Cosmological Models: In 1907 Einstein had what he called the "wonderful thought" that inertial acceleration and gravitational acceleration were equivalent. It took Einstein more than 8 years to bring this thought to its fruition, his theory of general Relativity (Norton and Norton, 1984) in November, 1915. Within a year, de Sitter had already investigated the cosmological implications of this new theory (de Sitter, 1917) which predicted spectral redshift of objects in the Universe dependent on distance. In 1917, Einstein published his Universe model (Einstein, 1917)—one that added an extra term—the cosmological constant—with which he attempted to balance gravitational attraction with the negative pressure associated with an energy density inherent to the vacuum. This addition, completely consistent with his theory, allowed him to create a static model consistent with the Universe as it was understood at that time. Finally, in 1922, Friedmann published his family of models for an isotropic and homogenous universe (Friedmann, 1922).

Observational cosmology really got started in 1917 when Vesto Slipher (to whose family I am indebted for helping fund my undergraduate education through a scholarship set up at the University of Arizona in his honor) observed about 25 nearby galaxies, spreading their light out using a prism, and recording the results onto film (Slipher, 1917). The results confounded him and the other astronomers of the day. Almost every object he observed had its light stretched to redder colors, indicating that essentially everything in the Universe was moving away from us. Slipher's findings created a conundrum for astronomers of the day: Why would our position as observer seemingly be repulsive to the rest of the Universe?

The contact between theory and observations at this time appears to have been mysteriously poor, even for the days before the internet. In 1927, Georges Lemaître, a Belgian monk who, as part of his MIT Ph.D. thesis, independently derived the Friedmann cosmological solutions to general relativity, predicted the expansion of the Universe as described now by Hubble's

* The 2011 Nobel Prize for Physics was shared by Saul Perlmutter, Adam G. Riess, and Brian P. Schmidt. These papers are the text of the address given in conjunction with the award.

law. He also noted that the age of the Universe was approximately the inverse of the Hubble constant, and suggested that Hubble's data and Slipher's data supported this conclusion (Lemâitre, 1927). His work, published in a Belgium journal, was not initially widely read, but it did not escape the attention of Einstein who saw the work at a conference in 1927, and commented to Lemâitre, *"Your calculations are correct, but your grasp of physics is abominable."* (Gaither and Cavazos-Gaither, 2008).

In 1928, Robertson, at Caltech (just down the road from Edwin Hubble's office at the Carnegie Observatories), predicted the Hubble law, and claimed to see it when he compared Slipher's redshift versus Hubble's galaxy brightness measurements, but this observation was not substantiated (Robertson, 1928). Finally, in 1929, Hubble presented a paper in support of an expanding universe, with a clear plot of galaxy distance versus redshift—it is for this paper that Hubble is given credit for discovering the expanding Universe (Hubble, 1929). Assuming that the brightest stars he could see in a galaxy were all the same intrinsic brightness, Hubble found that the faster an object was moving away from Slipher's measurements, the fainter its brightest stars were. That is, the more distant the galaxy, the faster its speed of recession. It is from this relationship that Hubble inferred that the Universe was expanding.

With the expansion of the Universe as an anchor, theory converged on a standard model of the Universe, which was still in place in 1998, at the time of our discovery of the accelerating Universe. This standard model was based on the theory of general relativity, and two assumptions: one, that the Universe is homogenous and isotropic on large scales; and two, it is composed of normal matter—matter whose density falls directly in proportion to the volume of space which it occupies. Within this framework, it was possible to devise observational tests of the overall theory, as well as provide values for the fundamental constants within this model—the current expansion rate (Hubble's constant), and the average density of matter in the Universe. For this model, it was also possible to directly relate the density of the Universe to the rate of cosmic deceleration: the more material, the faster the deceleration; and the geometry of space: above a critical density, the Universe has a finite (closed) geometry, below this critical density, a hyperbolic (open) geometry.

In more mathematical terms: If the universe is isotropic and homogenous on large scales, the geometric relationship of space and time are described by the Robertson-Walker metric,

$$ds^2 = dt^2 - a^2(t)\left[\frac{dr^2}{1 - kr^2} + r^2 d\theta^2\right] \tag{1}$$

In this expression, which is independent of the theory of gravitation, the line element distance (s) between two objects depends on coordinates r and θ, and time separation, t. The Universe is assumed to have a simple topology such that if it has negative, zero, or positive curvature, k takes the value {–1, 0, 1}, respectively. These universes are said, in order, to be open, flat, or closed. The Robertson-Walker metric also requires the dynamic evolution of the Universe to be given through the evolution of the scale factor $a(t)$, which gives the radius of curvature of the Universe—or more simply put, tracks the relative size of a piece of space over time. This dynamic equation of the Universe is derived from general relativity, and was first given by Friedmann in the equation which we now name after him:

$$H^2 \equiv \left(\frac{\dot{a}}{a}\right)^2 = \frac{8\pi G\rho}{3} - \frac{k}{a^2} \tag{2}$$

The expansion rate of the Universe (H), called the Hubble parameter (or the Hubble constant, H_0, at the present epoch), evolves according to the content of the Universe. Through the 20th century, the content of the Universe was assumed to be dominated by a single component of matter with density, ρ_i, compared to a critical density, ρ_{crit}. The ratio of the average density of matter compared to the critical density is called the density parameter, Ω_M and is defined as

$$\Omega_i = \frac{\rho_i}{\rho_{crit}} \equiv \frac{\rho_i}{\left(3H_0^2/8\pi G\right)} \tag{3}$$

The critical density is the value where the gravitational effect of material in the Universe causes space to become geometrically flat [$k = 0$ in Eq. (1)]. Below this density, the Universe has an open, hyperbolic geometry ($k = -1$); above, a closed, spherical geometry ($k = +1$).

As experimentalists, what we need are observables with which to test and constrain the theory. Several such tests were developed and described in detail in 1961 by Allan Sandage (Sandage, 1961) and are often described as the classical tests of cosmology. These tests include measuring the brightness of an object as a function of its redshift. The redshift of an object, z, indicates the amount an object's light has been stretched by the expansion of the Universe and is related to the scale factor such that

$$1 + z = \frac{\lambda_{obs} - \lambda_{emit}}{\lambda_{emit}} = \frac{a(z - 0)}{a(z)} \tag{4}$$

The redshift is measured from the observed wavelength of light, λ_{obs}, and the wavelength at which it was emitted, λ_{emit}.

The luminosity distance, D_L, is defined from the inverse square law of an object of luminosity, L, and observed flux, f,

$$D_L \equiv \sqrt{\frac{L}{4\pi f}} \tag{5}$$

This was traditionally solved as a Taylor expansion to Eqs. (1) and (2),

$$D_L \approx \frac{c}{H_0}\left[z + z^2\frac{(1 - q_0)}{2}\right] \tag{6}$$

where c is the speed of light, H_0 the current cosmic expansion rate has units of velocity over distance, and the deceleration parameter, q_0, is defined as

$$q_0 \equiv -\frac{\ddot{a}_0}{\dot{a}_0^2} \qquad a_0 = \frac{\Omega_M}{2} \tag{7}$$

The equivalence of Ω_M and q_0 is provided through solutions of the Friedmann equation assuming a universe consisting solely of normal matter. The Taylor expansion is accurate to a few percent over the region of interest of the day ($z < 0.5$), but was perfected by Mattig (1958), who found a closed solution,

$$D_L = \frac{c}{H_0 q_0^2}\left[q_0 z + (q_0 - 1)\left(\sqrt{1 + 2q_0 z} - 1\right)\right] \tag{8}$$

These equations provide one of the classic tests of cosmology—the luminosity distance versus redshift relationship. For an object of known luminosity, a single measurement at a moderate redshift [not so low that gravitationally induced motions, typically $z \sim 0.002$, are important—and not so high that the second order term in Eq. (6) is important], of its redshift and brightness, will yield an estimate of H_0. By measuring a standard candle's (an object of fixed luminosity) brightness as a function of redshift, one can fit the curvature in the line, and solve for q_0.

In principle, from Eq. (6), measuring H_0 does not appear to be difficult. An accurately measured distance and redshift to a single object at a redshift between $0.02 < z < 0.1$ is all that it takes, with their ratio providing the answer. But making accurate absolute measurements of distance in astronomy is challenging—the only geometric distances that were typically available were parallax measurements (measurement of the wobbles in the positions of nearby stars due to the Earth's motion around the Sun) of a handful of nearby stars. From these few objects, through a bootstrapping process of comparing the brightnesses of similar objects in progressive steps, known as the extragalactic distance ladder, researchers came to conclusions which varied by more than a factor of 2, a discordance which persisted until the beginning of the new millennium.

Measuring q_0 required making accurate measurements of the relative distances [absolute not required since the Hubble constant can be normalized out of Eqs. (6) and (8)]. Attempts made in the 1950s (Humason, Mayall, and Sandage, 1956), based on the brightest objects in the sky, giant galaxies in the center of clusters, provided a range of answers. Ultimately, Tinsley (1972) showed that these galaxies should change dramatically in brightness as we look back in time, making them problematic cosmological probes. Progress in measuring q_0 required a precise standard candle bright enough to be seen to $z > 0.3$, where curvature in the luminosity distance redshift relationship could be accurately measured.

Supernovae and my Career: The beginning of my astronomical career started in 1985 when I arrived as a bright-eyed freshman at the University of Arizona studying physics and astronomy. In my first astronomy class I felt daunted by all of the astronomy majors, many of whom seemed to me to have encyclopedic knowledge of everything from white dwarf stars to quasars. I understood physics, but I knew nothing of all of these things, so I looked around for something to do at Steward Observatory to increase my knowledge, and started working for John McGraw on his CCD transit instrument (CTI, Fig. 1).

This instrument was 15 years ahead of its time, and made the first large digital maps of the sky. Employing charged coupled devices (CCD), Cerro

Tololo Inter-American Observatory (CTIO) did not track the sky, instead it let the night sky pass overhead, and followed the motion caused by the Earth's rotation electronically, using a technique known as drift scanning. The Sloan Digital Sky Survey applied this technique with its highly successful survey starting in 2000. In 1985, CCDs were still very young, and the data rates that this telescope achieved in the mid 1980s were staggering. This data rate pushed the software and computational hard-ware capabilities of the day to the detriment of the telescope's overall scientific impact. As with all undergraduates, my progress was slow, but by the end of my 3rd year I had a real job within the group, to try to come up with ways to discover exploding stars known as supernovae in this data set. With a newly minted classification, type Ia supernovae were reputed to be good standard candles, and the CTI instrument had the opportunity to obtain the first digital light curves of a set of objects at redshifts greater than $z > 0.01$ where they could be tested as standard candles. The task was hard because the data set was enormous and, for computational reasons, we only had the ability to search catalogs of objects. Supernovae, though, usually occur in galaxies, and when making catalogs it is difficult to discern new objects in the complex structure of a galaxy. By the time I finished my undergraduate degree, I had managed to discover a possible object. Unfortunately, it was in data that was more than a year old, and was therefore never confirmed.

Supernovae: Supernovae (SN), the highly luminous and physically transformational explosions of stars show great variety, which has lead to a complex taxonomy. They have historically been divided into two types based on their spectra. Type I supernovae show no hydrogen spectroscopic lines, whereas type II supernovae have hydrogen. Over time, these two classes have been further divided into subclasses. The type I class is made up of the silicon rich type Ia, the helium rich type Ib, and the objects which have neither silicon nor helium in abundance, type Ic. The type II class is divided into II-P, which have a $\approx$ 100 day "plateau" in their light curves, II-L which have a "linear" decline in their light curves, and II-n which have narrow lines in their spectrum (Filippenko, 1997).

Massive Star Supernovae: Massive stars typically undergo core collapse as the last amount of silicon is burned to iron in their cores. As pressure support is removed by the loss of heat previously supplied by nuclear reactions, their interiors collapse to neutron stars, and a shock wave is set up by neutrino deposited energy outside of the neutron star region. A massive star that has a substantial, intact hydrogen envelope produces a SN II-P. Other variants are caused by different stages of mass loss. SN Ib represent a massive star which has lost its hydrogen envelope, and SN Ic are objects which have, in addition, lost their helium envelope.

Thermonuclear Detonations: These explosions are the result of the rapid burning of a white dwarf star. The entire star is burned, mainly to ^{56}Ni, but also to intermediate mass elements such as sulfur and silicon. The actual mechanism has long been assumed to occur when a white dwarf star accretes mass from a companion, and approaches $1.38M_{\odot}$. In 1931, Chandrasekhar showed at this point a white dwarf's self-gravity will exceed the pressure support supplied by its electron degenerate gas (Chandrasekhar, 1931). As the star approaches this critical juncture, the high pressure and density in the star's core initiates carbon burning near its center, which eventually leads to the entire star being consumed by a rapidly expanding thermonuclear burning front. We now suspect that it may be possible to ignite such an explosion in a variety of ways. These include sub-Chandrasekhar explosions initiated by a surface helium detonation which compresses the star's center to its nuclear flash point, and super-Chandrasekhar explosions involving the merger of two white dwarfs via gravitational radiation.

Graduate School at Harvard: Late in 1988, I applied to a number of universities with the hope of receiving a scholarship to work on my Ph.D.—I was not particularly optimistic as I had heard horror stories from others about how competitive the process was. To my surprise, on my 22nd Birthday (24 Feb 1989), I received a call from Bob Kirshner at Harvard University, telling me of my acceptance to Harvard's Ph.D. Astronomy program. It was the best birthday gift of my life. This call was followed up with several more offers in the coming hours and days, and I had a hard choice of deciding where to study. Either I could stay in the west of the United States where I was comfortable, or move to the east, which was tantamount to a foreign country to me. After visits to several campuses, Harvard had risen to the top of my list, a decision which I finalized when Bob Kirshner visited Tucson to give the first Aaronson Memorial Lecture by asking if I could work with him on my Ph.D.

When I arrived at Harvard to work with Bob Kirshner, I decided to focus on studying supernovae rather than discovering them. The idea of measuring the Hubble constant appealed to me, and so we took the tack of building on my supervisor's thesis, to calibrate the luminosity of type II supernovae and use them to measure the extragalactic distance scale (Kirshner and Kwan, 1975). SN 1987A, the nearest observed supernova to the Earth in almost 400 years had created a frenzy of activity in the subject, and Bob's finishing Ph.D. student, Ron Eastman, had developed a sophisticated computer code to model how radiation emerged from this supernova. My thesis involved applying Ron's theory to several supernovae at sufficient distances so that we could reliably estimate the Hubble constant. Type II-P supernovae are well suited to this purpose because they have simple hydrogen based atmospheres, whose emergent flux is close to a blackbody. In addition, their expansion is unaffected by gravity, enabling us to infer their radius by making measurements over time using absorption lines in their spectra to indicate the velocity of the material from which the supernova's flux is emerging. Put together, the emergent flux calculations and expansion rate allow the distance to a supernova to be determined on a purely physical basis. We named the method the expanding photosphere method (EPM).

In addition to observations, this method has as an essential ingredient, atmospheric models to calculate the correction to the blackbody assumption. Ideally, these calculations would be handcrafted for each SN, but the calculations took weeks to run, and instead, we used an approximation where we found that the blackbody correction depended almost entirely on the SN's temperature, and not on other factors. For my thesis, I used this technique to measure the distances to 14 SN II at redshifts between $0.005 > z > 0.05$, and found a 95% range for the value of the Hubble constant to be $61 < H_0 <85$

FIG. 1 (color). CTI Telescope at Kitt Peak Arizona.

km=s=Mpc (Schmidt *et al.*, 1994). This result was completely independent of the cosmic distance ladder—the bootstrapping of distances from our solar system to the nearest galaxies, but was in almost perfect accord to galaxies who distances were determined using Cepheid variable stars as part of the Hubble Key Project. The accepted value today is $67 < H_0 < 75$ km=s=Mpc. Work on using type II SN to measure distances continues, and while some of the approximations made during my thesis have been challenged, the fundamental technique remains in place.

After my thesis, the next step was obviously to use these objects to measure the deceleration parameter, q_0, but SN II and the expanding photosphere method have three significant drawbacks in measuring the global properties of the Universe. The first is that SN II are difficult to observe beyond a $z > 0.3$ with current instrumentation—they are too faint. The second is that they require significant observations to obtain each distance—multiepoch high quality spectra with simultaneous photometric observations, making them observationally prohibitively expensive for measuring q_0. The final difficulty is that the EPM distance precision, while not poor at about 15%, means many objects need to be observed to make a sufficiently precise measurement of q_0 to be interesting. The principal advantage of EPM, that objects were calibrated in an absolute sense, while essential for H_0 measurements, was

FIG. 2. Bob Kirshner examing my thesis results at Harvard in 1993.

irrelevant in q_0 measurements. Fortunately, during my Ph.D., I was exposed to the rapidly emerging work directed at measuring distances to type Ia supernovae. More importantly, I had got to know and work with the world experts on these objects, and these relationships were ultimately the basis of forming the High-Z SN Search Team.

The Foundations of the High-Z Team: When I arrived at Harvard in 1989, I arrived with Bob's newest postdoc, Swiss national, Bruno Leibundgut. Bruno, rather than studying SN 1987A and its sibling type II supernovae like most of the world was doing at the time, had concentrated on understanding just how standard of candles type Ia supernovae were. Type Ia supernovae, and their antecedents type I supernovae, had developed a reputation from less than ideal data for being essentially identical, making them potentially very good cosmological probes.

For his thesis, Bruno spent many a night on telescopes in Chile, taking photographic images to discover objects in a project lead by his supervisor Gustav Tammann, and collaborator Allan Sandage. While this project successfully discovered supernovae, the search was unable to deliver a data set useful for testing the veracity of SN Ia as standard candles. So Bruno used the entirety of data collected over the previous 5 years, and by other groups previously, to develop a standard template of the average SN Ia light curve which could be used as a reference to test the homogeneity of the SN Ia family. The results were extremely encouraging—all of the SN Ia seemed to fit a single template (Leibundgut, 1988). Now at Harvard, Bruno was able to use Harvard facilities, the new 1.2 m telescope equipped with a CCD to monitor the light curves of nearby SN Ia as they were discovered, and the huge Multiple Mirror Telescope to obtain their spectra. Our first observing trip together, soon after we both arrived to Harvard, resulted in what I believe are the only ill feelings ever between Bruno and myself. We had trouble understanding each other's enthusiasm for thinking we knew the right way to observe. The fact that Bruno was the postdoc and I the student did not occur to me at the time as being a key factor in the discussion. Within a few months, though, we grew to know and respect each other—and to this day, if Bruno challenges anything I say or do, I listen first, and ask questions later.

Bruno's first scientific big break at Harvard came with SN 1990N, an object that was discovered in the summer of 1990, just as Bob and I were off to Europe for a summer school on Supernovae at Les Houches in the French Alps. This object was discovered extremely soon after explosion, and its spectrum showed some funny features that persisted and were different to other SN Ia. But SN 1990N's light curve was well matched by Bruno's template (Leibundgut *et al.*, 1991).

In Les Houches I realized just how lucky I was to be an astronomer. A gorgeous village at the base of Mount Blanc, the summer school immersed me for 5 weeks in a group of students from around the world, tutored by the greats of the field. I consider it to be the greatest 5 weeks of my life. There I met a young Chilean, Mario Hamuy, who was working at Cerro Tololo Inter-American Observatory as a research assistant for CTIO staff astronomer Nick Suntzeff. I was familiar with Mario by reputation, for the photometric data he and Nick had amassed on SN 1987A in the Large Magellanic Cloud, which I was using to measure this supernova's distance as part of my thesis.

Mario told us of a new project, the Calan/Tololo survey, which would use the Curtis Schmidt telescope at CTIO to discover objects at redshifts more distant than the objects we were all studying. By discovering SN at $0.02 < z < 0.1$, the Calan/Tololo survey aimed to test rigorously SN Ia as standard candles, using the redshift as an accurate proxy for relative distance. The members of this group, Mario Hamuy, Nick Suntzeff, and Mark Phillips, at CTIO, and Jose Maza at the University of Chile, were starting their program that year. In addition to Nick and Mario's work on SN 1987A, Jose Maza had lead a highly successful SN search from Calan in the 1980s, while Mark had made an impact in the field by observing SN 1986G in the nearby Centaurus A galaxy. SN 1986G was one of the first objects to be observed with a CCD, and showed a light curve that was ultimately accepted as being unusual compared to the Leibundgut template.

Partially as a result of the Les Houches school, and mainly due to subsequent work that Bob Kirshner was doing with Mark Phillips and Nick Suntzeff on SN 1987A, a 5 week trip for me to visit Cerro Tololo was planned for the end of 1991. There I would use data from the Calan/Tololo survey on type II SN for my thesis—and learn the techniques that were being used at CTIO to accurately measure the light curves of SN Ia using CCDs, and apply them to my SN II. To ensure that the cultural shock was not too great, my trip was sequenced with the arrival to CTIO of another of Bob Kirshner's graduate students, Chris Smith, who was about to start his first postdoc at the observatory.

I arrived in Santiago from the long flight from Miami, and was taken to the bus station for a 6 hour bus trip to La Serena. There I met Pete Challis from the Space Telescope Science Institute who was also on his way to CTIO, but in his case for a long observing run. In the 6 hours to La Serena, Pete and I covered a lot of ground, and we soon established that Pete had been at Michigan as an undergraduate with my Ph.D. supervisor, Bob Kirshner, and was interested in changing jobs. I told Pete that Bob was looking for someone to help him manage his Hubble Space Telescope observations, and in that way Pete and Bob were reconnected, and they continue to work together to this day.

When we arrived at La Serena, I was met by Mario Hamuy and Mark Phillips, who were, in addition to picking me up, putting a wooden box full of photographic plates from the Curtis Schmidt Telescope onto the bus for its return journey to Santiago. While I slept, the photographic plates made their journey south to the University of Chile where Jose Maza and his team would search them the following day for supernovae.

The Calan/Tololo survey used this technique to efficiently discover more than 50 objects from 1990–1993. The Calan/Tololo survey had regularly scheduled CTIO-4 m and CTIO-1.5 m time to obtain spectra as they were sufficiently regular at discovering objects that they could plan in advance on their discoveries. For photometry, because they only needed a small amount of time, they borrowed time from cooperative astronomers observing on CTIO telescopes who enjoyed the excitement of observing an astronomical object that changed over the course of a few nights.

A few days after arrival, I asked Mario how his work on SN Ia was going, and he said he was depressed. He showed me his first couple of objects, and one of them, SN 1990af, looked pretty normal with respect to its spectrum, but compared to Bruno's template, it clearly rose and fell more quickly, More significantly, SN 1990af was significantly fainter than the other objects in their sample, despite being at the same redshift. He felt that the Calan/Tololo program to use SN Ia to measure H_0, and eventually q_0, had run into a snag—the objects they were planning to use to measure distances were not living up to their reputation—they were not standard candles.

1991 was a transformational year for SN Ia. Early in the year, a nearby galaxy hosted SN 1991 T. In a paper lead by Mark Phillips that included both the Tololo and Harvard groups (Phillips *et al.*, 1992), as well as a paper lead by Alex Filippenko (Filippenko *et al.*, 1992), the object was shown to be highly unusual. Its spectrum had extra features early on but was largely missing the most recognized feature of the class, a strong silicon line at 6130 Å. In addition, the light curve rose and fell significantly more slowly than average and it seemed to be too bright given its host galaxy's distance. Between uncertainties in the amount of dust obscuring the object and the distance to its host galaxy, we could not be absolutely certain that this object was brighter than other SN Ia, although work done by Jason Spyromilio at the Anglo-Australian Observatory indicated that SN 1991 T produced more iron than is typical for normal SN Ia (Spyromilio *et al.*, 1992).

Later in the year, another object, SN 1991bg, occurred in a nearby elliptical galaxy. In papers lead by Leibundgut (Leibundgut *et al.*, 1993) (Harvard and Tololo groups) and Filippenko (Filippenko, 1992), this object was shown to have a different spectrum from the norm, and a light curve that faded much more quickly than average. In this case, the object was so much fainter than average, with no evidence of any obscuring dust, that the case was clear.

By 1993, based on the range of objects being studied in the nearby universe, and consistent with the picture that was emerging from the Calan/Tololo survey, Mark Phillips wrote his seminal paper which compared the rate that an object faded to its luminosity, finding that faster evolving objects were systematically fainter than their slower evolving siblings (Phillips, 1993). I remained a bit skeptical—while SN 1991bg was clearly different, the objects in Mark's 1993 paper were all in the nearby Universe, and the objects' distances uncertain. I felt that it was possible that the whole correlation

FIG. 3 (color). Brian Schmidt, Pete Challis, and Nick Suntzeff discussing the High-Z SN Search at Cerro Tololo.

might go away, if only SN 1991bg were thrown out. But this paper got the world thinking, and amongst those were Bob Kirshner, whose new Ph.D. student Adam Riess was looking for a project for his thesis. Bob focused Adam's attention at using the statistical expertise of Bill Press (who was one floor down at the Center for Astrophysics) to develop a technique to model SN Ia light curves and estimate their distances.

I was finishing my thesis on SN II-P during this time, but spent a lot of time talking to Adam about his project. The emerging picture of SN Ia was just so interesting, despite the need for me to write up, I could not stop from thinking about how to use SN Ia to measure distances. I submitted my thesis in August 1993, and stayed on at the Center for Astrophysics as a Harvard-Smithsonian Center for an Astrophysics Postdoctoral Fellow, where I had the benefit of a fellowship to do anything I wanted, but with the opportunity of being embedded in the expertise of Bob Kirshner's group.

Early in 1994, Mario Hamuy from the Calan/Tololo group visited. The Calan/Tololo group had expanded to include Bob Schommer, a CTIO astronomer who had experience in measuring the Hubble constant using the Tully-Fisher technique, and Chris Smith (another Kirshner student), whose all-around observational and analysis experience was being used to help analyze the SN light curves. Mario was armed with Calan/Tololo's first 13 SN Ia light curves and redshifts and to me what was an astonishing discovery. If they applied Mark Phillips' relationship to this independent set of objects, the scatter about the Hubble law dropped dramatically and demonstrated that tSN Ia provided distances with a precision better than 7% per object. This was much better than anything I thought could ever be achieved. The Calan/Tololo group allowed Adam to train up his new statistical method with these data—they were at sufficient distance that their relative distances could be inferred with high accuracy from their redshifts—thereby removing one of the principal problems in previous SN Ia distance work.

A month later, during one of the groups observing runs at the Multiple Mirror Telescope (MMT), Bob Kirshner, Adam Riess, and Pete Challis received a call from Saul Perlmutter of the Supernova Cosmology Project (SCP) to follow up a high redshift supernova candidate of theirs. The SCP had struggled to find distant SN Ia over the previous 5 years, but I was excited by the spectrum I saw the following morning from the MMT. Pete had already reduced the data and had eye balled it as a type Ia SN at a redshift of $z = 0.42$, something I confirmed during the day from the comfort of my CfA office. In the ensuing weeks, as we negotiated with Saul's team to publish the spectrum in the International Astronomical Union Circulars, we realized that this event was not alone—the SCP had discovered several such objects in the previous months.

These two events—the development of the ability to measure precise distances with SN Ia, and the capacity to discover these objects in the distant Universe, were the ingredients necessary to finally mount a successful campaign to measure the deceleration parameter. The Supernova Cosmology Project had been working towards this goal since 1988, but it became clear that they had significantly different views on how to approach the problem—especially with respect to measuring precise distances—than my supernova colleagues and I had.

The High-Z Team: Measuring the Deceleration Rate of the Universe: In mid-1994 I went to CTIO for an observing run for a project on clusters that ultimately did not pan out. While I stayed on at CTIO after observing, Nick Suntzeff and I hatched a plan to use the CTIO 4 m to mount our own campaign to measure q_0, given that the two essential ingredients were suddenly in place. Measuring q_0 had always been part of the plan of the Calan/Tololo survey, but opportunity knocked a few years earlier than the group had anticipated.

Type Ia supernovae are not common objects, they occur in a galaxy like the Milky Way a few times per millennium. Since SN Ia take approximately 20 days to rise from nothingness to maximum light, observing the same piece of sky twice with a one month separation (which equates to 20 rest-frame days at $z = 0.5$) will yield objects which are typically near maximum light, and therefore young enough to be useful for measuring precise distances. The CTIO 4 m telescope was equipped with a state-of-the-art 2048 × 2048 pixel CCD that covered the widest field of view of a 4 m telescope at the time. The weather at CTIO was also impeccable through the Chilean summer—so there would be virtually no chance of being weathered out in a supernova search. This was essential because the experiment required images taken a month apart to be compared, and additional preplanned telescope time afterwards to follow up the candidates. Bad weather at any of these times would prove fatal for the experiment, leaving no candidates and lots of telescope scheduled to observe objects that did not exist. This is a problem I knew that the SCP had faced many times.

Nick and I soon enlisted Mark Phillips, Mario Hamuy, Chris Smith, and Bob Schommer (CTIO), and Jose Maza (University of Chile) from the Calan/Tololo SN Search. We also brought on Bruno Leibundgut and Jason Spyromilio, who were now at the European Southern Observatory, as well as Bob Kirshner, Pete Challis, Peter Garnavich, and Adam Riess from Harvard. This provided the observational fire power for both discovering SN Ia, and for following up our discoveries.

Observing Proposal
Cerro Tololo Inter-American Observatory

Date: September 29, 1994	*Proposal number:*

TITLE: A Pilot Project to Search for Distant Type Ia Supernovae

PI: N. Suntzeff — Grad student? N — nsuntzeff@ctio.noao.edu
CTIO, Casilla 603, La Serena Chile — 56-51-225415

CoI: B. Schmidt — Grad student? N — brian@cfanewton.harvard.edu
CfA/MSSSO, 60 Garden St., Cambridge, MA 02138 — 617 495 7390

Other CoIs: C. Smith, R. Schommer, M. Phillips, M. Hamuy, R. Aviles (CTIO); J. Maza (UChile); A. Riess, R. Kirshner (Harvard); J. Spyromilio, B. Leibundgut (ESO)

Abstract of Scientific Justification:
We propose to initiate a search for Type Ia supernovae at redshifts to $z \sim 0.3 - 0.5$ in equatorial fields using the CTIO 4m telescope. This program is the next step in the Calán/Tololo SN survey, where we have found ~ 30 Type Ia supernovae out to $z \sim 0.1$. The proposed program is a pilot project to discover fainter SN Ia's using multiple-epoch CCD images from the 4m telescope. We will follow up these discoveries with CCD photometry and spectroscopy both at CTIO and at several observatories in both hemispheres. With the spectral classification and light curve shapes, we can use our calibrations of the absolute magnitudes of SN Ia's from the Calán/Tololo survey to place stringent limits (Figure 2) on q_0 in a reasonable time-frame. Based on the statistics of discovery from the Calán/Tololo SN survey, we can expect to find about 3 SNe Ia per month.

FIG. 4. The original High-Z SN search team proposal.

The proposal was due as my first child, Kieran, was being born, and Nick Suntzeff and Bob Schommer polished up our team's proposal, and submitted it on September 29, 1994 (Fig. 4).

I had successfully applied for a postdoctoral fellowship in Australia at the Mount Stromlo Observatory, and so in my last few months at the CfA at the end of 1994, I started writing a supernova discovery pipeline. Supernovae are not always easily identified as new stars on galaxies— most of the time they are buried in their hosts, and cannot just be identified without a more sophisticated technique. From colloquia, I knew that the SCP had developed some sort of image subtraction pipeline, and that was the technique they had used to successfully discover their distant objects.

As part of my thesis I had developed techniques of automatically aligning images, but the Earth's atmosphere blurs each image differently, making the shape of a star on each image, known as its point spread function, unique. I had met Drew Phillips at CTIO, and he had developed a technique for convolving images with a kernel to match two images' point spread functions, thereby enabling a clean image subtraction. I used this package as the basis of our pipeline, and set about developing a series of scripts to automatically subtract the massive amounts of data we would get in early 1995. These programs were meant to take the gigabytes of imaging data that we gathered in a night, align it with the previous epoch, and then match and scale the image point spread functions between the two epochs to make the two images as identical as possible. These two images are subtracted with the differenced image searched for new objects, which stand out against the static sources that have been largely removed in the differencing process.

During my last months at the CfA, Bob Kirshner's new postdoc, Peter Garnavich, arrived. Peter was busy principally working on SN 1987A and another nearby object, SN 1993J, during this time, but he was a new colleague with fresh ideas with whom I could discuss the High-Z program. We instantly became friends, and despite our short overlap at the CfA, Peter is a colleague I have always known I could trust through good times and bad. By the time I left for Australia, using some test data, I felt I had a discovery program that more or less worked.

When I arrived in Australia, I had a few weeks to get myself settled before our first observing run started in Chile. I had decided that I would stay put in Australia, rather than travel to Chile, since we were still in the middle of moving, with my wife starting her job and our 4 month old son proving not to be the great sleeper we had hoped for. As we started to implement the pipeline at CTIO it became clear we had a problem or two. The CTIO computing system, which I thought was a lot like my own in Australia, had substantial differences which prevented the software from running. To confound matters, the internet connection between Australia and Chile was about 1 character per second—making it almost impossible for me to do anything remotely. Working with a very patient Mario Hamuy, we slowly marched through the problems. I would email Mario snippets of code to be inserted in the subtraction program, with Mario reporting back how it worked.

Our first observations were taken on February 25th, 1995, and we had another night's data on March 6th. The processing of these data was an unmitigated disaster—nothing seemed to work, and I could not get the data to Australia to diagnose what was going wrong. We used a courier company to express tapes of data to Australia so I could work to fix problems, but that delivery was lost and never arrived. Now, working with the entirety of the CTIO collaboration, we slowly pieced the pipeline together, making it email tiny 16 × 16 pixel stamps of interesting things to me in Australia. These little mini images, combined with as vivid descriptions as could be mustered by telephone, were all that I had to figure out what was going wrong, or right. We had two nights on March 24th and 29th, and a proposal to write for a continuation of our program, due on the 30th of March. Around the 27th of March, suddenly the stamps that were being sent to me started producing objects that looked interesting. Several were asteroids—we could tell they

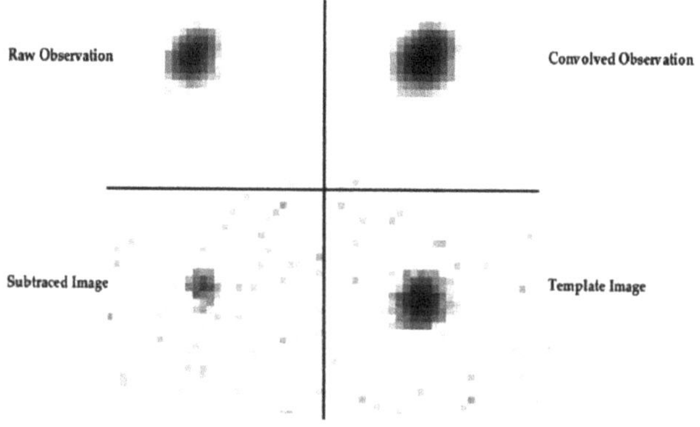

FIG. 5. Original stamps for candidate C14—complete with typos. This object was confirmed as SN 1995 K, which at $z = 0.479$, was the most distant SN Ia yet discovered in April, 1995. The observation taken on March 29th (upper left) was matched (upper right) to the observation taken in February (lower right), and subtracted (lower left).

were moving—but one was on the outskirts of a galaxy. This object was detected on March 6th, but was not visible on the data of March 24th (the data from this night were poor, so we could not confirm that it was not an asteroid). With these candidates, we submitted the continuation of our program, and set about searching the data from March 29th. Stamp after endless stamp arrived in Australia, and suddenly one, C14 as it was named, looked interesting. It was a new object, buried in a spiral galaxy—it did not move, and it appeared possibly fainter in our poor data from March 24th (Fig. 5). I excitedly called CTIO and the report back from looking at the whole image was positive. Yes, it looked like a supernova.

Using the CTIO-4 m spectrograph, Mark Phillips was able to obtain a spectrum of the galaxy—it was at a redshift of $z = 0.48$—making it potentially the most distant SN yet detected. But this spectrum showed no hint of the supernova, its light was overwhelmed by its host galaxy's. Bruno and Jason had follow-up time with the European Southern Observatory (ESO) New Technology Telescope (NTT) at La Silla on April 3rd. Through heroic effort (they observed the object all night) and data reduction—it took a week, the NTT spectrum showed the object was indeed a SN Ia. In writing the IAU circular, we needed to come up with a name for our team—for lack of anything better, we settled upon the High-Z SN Search team.

In the days that followed, Nick, Mark, and Bob Schommer convinced Allan Dressler at Carnegie to take a series of images of SN 1995 K with the DuPont telescope at Las Campanas. That data, combined with that taken at ESO and CTIO, provided what is still a very good light curve of a distant SN Ia. We presented the light curve and its place on the Hubble diagram September 1995 in our application for telescope time. While 1995 K showed $q_0 = -0.6$, the uncertainty was such that we required at least 10 objects to make a statistically significant measurement, and we did not give the actual value much thought (Fig. 6).

Supernova aficionados Alejandro Clocchiatti (Catolica University) and Alex Filippenko (Berkeley), along with non-supernova experts John Tonry

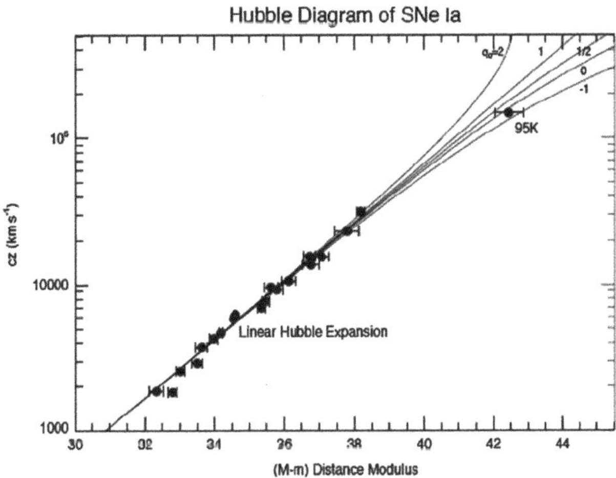

FIG. 6. SN 1995 K on the Hubble Diagram from our September 1995 telescope proposal.

(Hawaii), Chris Stubbs, and Craig Hogan (University of Washington), were all recruited to the team in 1995 bringing along specific skills and additional telescope time resources.

Alejandro Clocchiatti undertook his Ph.D. thesis at the University of Texas studying type Ib/c SN—likely contaminants in our experiment which we needed to control using his expertise. Alejandro was also resident in Chile where we could use his physical presence in helping executing observations, as well as providing us additional access to Chilean telescopes.

Alex Filippenko, a member of the community that studied supernovae, had approached me in 1995 to join the High-Z Team. We turned him down on the basis that we did not want to be seen as poaching a member from a competing team. By the end of 1995 it became clear that Alex's expertise and access to Keck were going to be essential for us to successfully undertake our experiment to measure on q_0. So when he asked again in 1996 to join our team, we immediately said yes.

John Tonry, in addition to providing access to telescope time through the University of Hawaii, is widely regarded as one of the most capable observational astronomers of our era. On my trips to Hawaii, John and I would discuss the current deficiencies in our experiment, and John would inevitably write new programs to assist with our discovery and analysis of SN Ia. In these bursts of programming, John developed our interactive search tool, our spectral analysis tool (SNID, which is still widely used by the community), and the core of our photometric analysis pipeline.

Chris Stubbs was one of the members of the MaCHO gravitational microlensing experiment which operated the Mount Stromlo 50 inch telescope, and he brought significant experience in analyzing large data sets, which our group sorely lacked.

Craig Hogan was an eminent theorist who had taught me cosmology at the University of Arizona before he moved to the University of Washington. I felt (and still feel) that it was important to have at least one theorist on any large observational program, and Craig was someone whose theoretical grounding was well matched to the needs of our team.

Given the dispersed nature of our team, we had to gather each year to discuss how the observational program was progressing, and how we were going to turn all of the our data into a definitive measurement of q_0. Our first meeting was in 1996 at Harvard. We had just been awarded Director's Discretionary time with the Hubble Space telescope, and we needed to plan on how to use this great resource effectively. We decided to expand our discovery platform to include the new wide field camera on the Canada France Hawaii Telescope, in Mauna Kea, with University of Hawaii astronomer and High-Z team member, John Tonry, providing access to this unique facility. Running SN searches on two telescopes, twice per year, made me and the team very busy people.

Each observing run was organized chaos. I would arrive a week early, with the latest version of the software. Since we did not have dedicated equipment, the whole pipeline would be rebuilt at the beginning of each run—and this never proceeded smoothly. Each facility had its own sets of operating systems that, while all UNIX, were sufficiently different that the code had to be individually compiled for each system. Because of the size of our data set, we needed to operate across multiple machines and disks—hardware that changed each run. This week inevitably ended with the entire team working 20 h days to ensure that we were able to promptly discover supernovae. This level of effort led to interesting coping strategies—Bob Schommer was famous for playing James Brown at high volume in the telescope control room. It also lead to the occasional mistake. One night in the CTIO-4 m control room as Alejandro Clocchiatti watched as I frenetically typed, Alejandro suddenly turned pale and said, "I do not think you wanted to do that." I had just accidentally deleted the night's data. While we pondered how to tell Nick (who was manning the telescope) the news, Nick suddenly screamed, "What happened to all the data?" I saw my career flash before my eyes, but we soon realized the data were stored (in a way that I had previously thought was inane) such that we were able to restore our files and continue on observing.

Spectroscopic follow up, principally using the Keck 10 m telescopes through time allocated to Alex Filippenko (through the University of California), and John Tonry (through the University of Hawaii), was scheduled just a few days after our search runs. Failure to quickly identify candidate supernovae meant our discoveries would be effectively useless. Despite the chaos, through 1995–1997, we did manage to discover, spectroscopically confirm, and photometrically follow 16 distant SN Ia—enough to make a statistically robust measurement of the deceleration parameter.

In early 1997, most of the team assembled in Seattle at the University of Washington, and we agreed that each paper would be led by a student or young postdoc from within the group. I would write the first paper where we laid out our program and presented our first object, SN 1995 K. Peter Garnavich was selected to write the next major paper, one that would include objects observed with the Hubble Space Telescope (HST), and would likely tell us our first statistically significant measurement of q_0. And finally, Adam Riess was selected to write the next paper that would refine the value of q_0 based on several years' data. The data grunts of the group (myself, Adam Riess, Pete Challis, Saurabh Jha, Alejandro Clocchiatti, David Reiss, and Al Diercks) stayed on in Seattle to work together for a week. Initially, the week was supposed to be a working bee where I would tutor the group on how to make photometric measurements of distant SN Ia, and we would as a group analyze our data set. While the week did not lead to an analysis of our data set, it instead became an intense workshop where we thought through most of the outstanding issues necessary to complete the experiment. It was one of the most memorable weeks of the High-Z team for me. While Hale-Bopp blazed invisibly above the continual Seattle drizzle, we clocked in 16 h days from the basement of the University of Washington Physics Department—taking a break to all see the movie "Swing Blade" at the request of Adam.

Over the course of the next few years, my life was dominated by SN discovery runs, photometric data reduction, and writing the paper on our SN program and SN 1995 K. The paper was largely complete in 1996, but the ever increasing data load made it challenging for me to finish. In addition, the complication that SN 1995 K was most consistent with negative acceleration made aspects of the analysis challenging. In addition, there were many possible systematic effects that could derail this experiment into giving an incorrect answer, and I was investigating these at this time, one by one.

Systematic Effects: In the nearby universe, we see SN Ia in a variety of environments, and about 10% have significant extinction. Since we can correct for extinction by observing the colors of SN Ia, we can remove any first order effects caused by the average extinction properties of SN Ia changing between $z = 0$ and $z = 0.5$. As part of his thesis, Adam Riess had developed techniques to correct for dust based on the colors of supernovae (Riess, Press, and Kirshner, 1996). This was essential work to accurately measure the relative distances to SN Ia, and is an essential ingredient in all supernova distance measuring techniques today.

Our supernova discoveries suffer from a variety of selection effects, both in our nearby and distant searches. The most significant effect is Malmquist bias—a selection effect which leads magnitude limited searches finding brighter than average objects near their brightness limit. This bias is caused by the larger volume in which brighter objects can be discovered compared to their fainter counterparts. Malmquist bias errors are proportional to the square of the intrinsic dispersion of the distance method, and because SN Ia are such accurate distance indicators, these errors are quite small—approximately 2%. In 1995, I developed Monte Carlo simulations to estimate these effects, and remove their effects from our data sets.

As SN are observed at larger and larger redshifts, their light is shifted to longer wavelengths. Since astronomical observations are normally made in fixed bandpasses on Earth, corrections need to be made to account for the differences caused by the spectrum of a SN Ia shifting within these bandpasses. The SCP had showed that these effects can be minimized if one does not stick with a single bandpass for nearby and distant objects, but by instead choosing the closest bandpass to the redshifted rest-frame bandpass (Kim, Goobar, and Perlmutter, 1996). The High-Z SN search took this one step further, designing new band-passes, specifically made to emulate the $z = 0$ bandpass at several redshifts.

SN Ia are seen to evolve in the nearby universe. The Calan/Tololo survey plotted the shape of the SN light curves against the type of host galaxy (Hamuy *et al.*, 1996). Early hosts (ones without recent star formation) consistently show light curves which evolve more quickly than those objects which occur in late-type hosts (objects with on-going star formation). This could be a terminal problem for using SN Ia to measure q_0 if it were not for the observation that once corrected for light curve shape, the corrected luminosity shows a much smaller correlation as a function of the characteristics of the host.

Cosmology Beyond Normal Matter: Since 1917, when Einstein first added the cosmological constant to his equations, this fudge factor had been trotted out on several occasions to explain observations of the Universe that did not conform to the standard model described earlier. The cosmological constant had developed a bad reputation as being incorrectly asserted as the solution to what were ultimately found to be bad observations.

In 1995, I had served as the referee of a paper by Goodbar and Perlmutter (Goodbar and Perlmutter, 1995) exploring if the meaningful limits on the value of the cosmological constant could be made by high redshift SN Ia measurements. In my referee report I expressed concern of the relevance of the paper—I felt that the paper failed to demonstrate that a meaningful limit could be made on the cosmological constant. If there was no cosmological constant, then the uncertainty in a SN Ia-based measurement would be sufficiently large as not to be interesting (see their Fig. 2). I had failed to grasp—so strong were my priors against a cosmological constant—that if there was a cosmological constant (see their Fig. 3), that a meaningful measurement could be made.

The cosmological constant was not new to me. Sean Carroll had written a review on the topic in 1992, while we shared an office during graduate school (Carroll, Press, and Turner, 1992). I remember that as he worked through hundreds of yellow post-it notes scrawled on his manuscript by his referee, Allan Sandage, I teased him about writing about something as ridiculous as the cosmological constant. This review ended up being extremely useful as I came to grips with how to interpret SN 1995 K, and the range of negative q_0 values it implied.

As part of my paper describing the High-Z SN search (Schmidt *et al.*, 1998), the team theorist, Craig Hogan, encouraged me to go beyond the notion of q_0. He was particularly interested in breaking the assumption that the Universe was made up of only normal matter, postulating that it could be composed of other things as well. In our paper, we adapted our measurement to the standards of particle astrophysics. That is, we adapted the Friedmann Eq. (2) to reflect all species of matter

$$H^2 \equiv \left(\frac{\dot{a}}{a}\right)^2 = \frac{8\pi G\rho_{\text{tot}}}{3} - \frac{k}{a^2} \tag{9}$$

describing each species of matter by their fraction of the critical density,

$$\Omega_i = \frac{\rho_i}{\rho_{\text{crit}}} \equiv \frac{\rho_i}{\left(3H_0^2/8\pi G\right)} \tag{10}$$

and this matter's equation of state,

$$w_i = \frac{P_i}{\rho_i c^2} \tag{11}$$

The equation of state for normal matter is $w = 0$, the cosmological constant, $w = -1$, and photons $w = 1/3$. This formulation made for a less trivial expression for the luminosity distance,

$$D_L H_0 = c(1+z)\Omega_k^{-1/2} S\left\{\Omega_k^{1/2}\int_0^z dz'\left[\Omega_k(1+z')^2 + \sum_i (1+z')^{3+3w_i}\right]^{-1/2}\right\} \tag{12}$$

where $S(x) = \sin(x)$, x, or $\sinh(x)$ for closed, flat, and open models, respectively, and Ω_k the curvature parameter, is defined as $\Omega_k = 1 - \Sigma_i \Omega_i$. With multiple forms of matter, Mattig's formulation for D_L [Eq. (8)], is no longer valid, but the q_0 expansion, Eq. (6) is still valid, except that q_0 is given by the expression

$$q_0 \equiv \frac{-\ddot{a}(t_0)a(t_0)}{\dot{a}^2(t_0)} = \frac{1}{2}\sum_i \Omega_i(1+3w_i) \tag{13}$$

The Discovery of Acceleration: By the middle of 1997, the High-Z team had HST observations of 4 objects, and 10 more distant objects to tackle our ultimate goal, measuring q_0. But there were some complications that needed to be sorted out dealing with statistics. In principle, measuring q_0 from several SN distances and redshifts is straightforward. The redshifts have negligible uncertainty, and the distance estimates had distances with uncertainties well described by a normal distribution. A classic X^2 method seemed entirely appropriate. Except our data were in a part of parameter space where Mattig's exact formula [Eq. (8)] was invalid, and at a redshift where the Taylor expansion solution (5) was not very accurate. On the other hand, Eq. (12) covered all possibilities, but there were regions of parameter space which were not allowed, like negative matter. In discussions at CTIO with members of the SCP in 1996, it became clear that we were both grappling with how to deal with these statistical issues— it was not that they had not been solved by science, it was just that we were in new territory for us, and we were struggling to figure out a solution. Adam Riess, who had become adept at statistics in his thesis, in discussions with Bill Press, came up with the solution of converting X^2 to a probability, applying priors to this probability space (e.g., no negative matter), and integrating over this space to find the probability distribution for the parameters of interest. It seems so passé now, but in 1996, none of us had ever seen this technique used before in astronomy. Computationally, this was not trivial, and Adam Riess, Peter Garnavich, and I all wrote our own versions of codes that did these calculations.

The HST data that Peter Garnavich was analyzing were of very high quality, and were consequently the easiest to reduce. By September he had finished his analysis—the data clearly showed $q_0 \neq 0.5$, a flat universe composed of normal matter was ruled out—but this seemed at odds with a paper put out by the SCP at this same time (Perlmutter *et al.*, 1997). Peter's draft created a range of reactions within the team—what were our controls on systematic errors, and how could we demonstrate the result was robust? This lead us into examining all sorts of possible systematic errors, and while we never quite reached agreement (Chris Stubbs, who had a particle physics background, was particularly critical of our ability to control all errors) it did mean the team had already grappled with this issue when things got substantially more interesting a few months later. My wife and I had just had our second child, and I have to admit to not doing a good job at getting the team to work together constructively around these issues.

In November of 1997, Adam Riess had finished his first pass at measuring his collection of supernovae—a feat that was achieved due to his unique ability to focus on this one thing with all of his might. He sent me a figure which a subject line of, "what do you think?" I looked at the figure and it showed that his group of SN Ia were, on average, definitively fainter than even a $q_0 = 0$ model. The Universe seemed to be accelerating. I remember thinking, "What has Adam done?" and thus opened up an intense exchange between the two of us, checking the result, and refining the analysis. At the same time, I was working to submit my paper, which I swore would be submitted in 1997—just managing to get it in before New Year's Eve. Finally, on the 8th of January 1998 (Australian Time), Adam and I agreed on all details of the

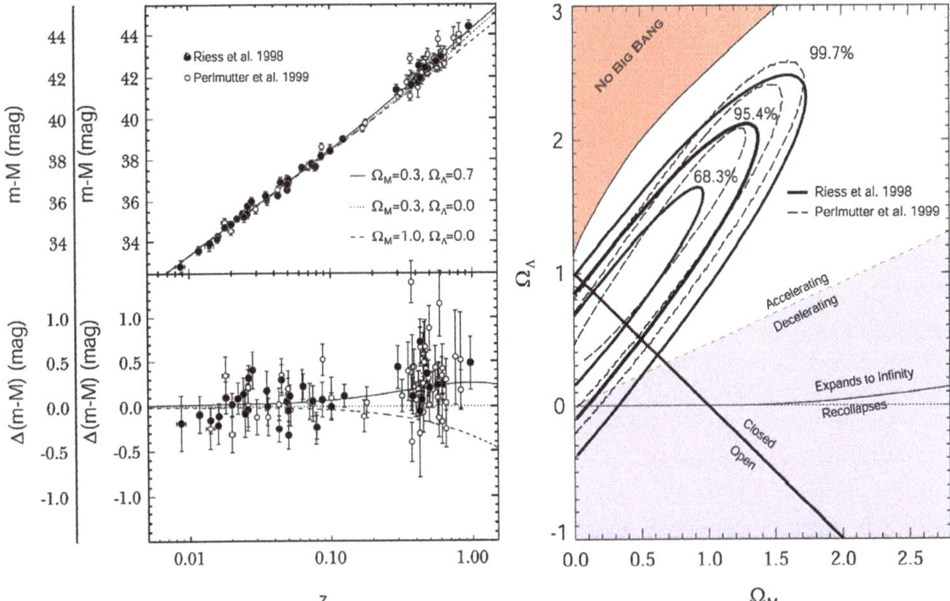

FIG. 7 (color). (a) (left): Top panel: Hubble diagrams of SN Ia showing the High-Z team and SCP data, with 3 sets of cosmological parameters. Bottom panel: Data from the top panel with the model of containing normal matter (30% of the critical density) subtracted. (b) (right): Probability contours for cosmological fits to the SCP and High-Z teams' data. The results from the two projects show remarkable consistency in their conclusion that the Universe has a significant matter component consistent with the equation of state of a cosmological constant.

calculation that showed that the Universe was accelerating, and I sent him an email with "Hello Lambda" as the subject line, and a figure of my calculations. Most of the High-Z team had not been shown the analysis at this point. Adam had shown his work to Alex Filippenko, and we told Peter Garnavich, who was presenting his paper, described above, at the American Astronomical Society Meeting, the next day.

The result was perplexing to me, the cosmological constant had a long history of being proposed to explain a set of observations which was later on shown to be fatally flawed. And then there were the results of the other team. The 1997 SCP paper was at such odds to what we are seeing, I felt no one would take us seriously with such a crazy result. What I had not seen was the SCP's new paper which appeared on the 17th of December on the astrophysics archive—only learning about it after the AAS press conference on January 8th (Perlmutter *et al.*, 1998). This paper indicated that their value of q_0 was much lower than they had previously presented.

On January 9th, I came into work to get a report of the AAS press conference from Peter Garnavich. In addition to presenting his HST data from his *Nature* paper showing the Universe was not decelerating quickly, Saul Perlmutter had given the audience a peak of his entire collection of 40 objects—and these objects *did* seem to be showing the same thing that we were seeing. Saul's objects were systematically fainter than could be explained in a universe composed only of normal matter. But Saul's team had not yet corrected for dust, a correction that was built into our analysis from the beginning. Adam had chosen this week to get married, and when he returned from a short honeymoon, we had a lot of explaining to do to the team, recounting all of the steps in our analysis. The team's reactions were mixed—some were excited, others were in disbelief, and still others felt that we had a long way to go to show the result to be robust to errors. While I shared in the skepticism, I also felt that it would be wrong not to publish a result just because we did not like it. I challenged the team to suggest tests that they felt needed to be made before we published. Over the remainder of January and February, under Adam's leadership, the team worked through all of the tests requested, such that by the end of February, the team had agreed to the contents of the paper, and we were ready to announce our result. Alex Filippenko presented our team's work at a meeting in California at the end of February, and it created a media sensation in the United States. Our paper was submitted a week later to the Astronomical Journal, "Observational Evidence for a Cosmological Constant and an Accelerating Universe." Over the next few months, in addition to continuing our punishing program of SN observations, Peter Garnavich did the first analysis to show that whatever was causing the acceleration, it seemed to have an equation of state, a lot like the cosmological constant.

While I felt that we had done all that was possible with our supernovae to understand our uncertainties, I could not help worrying that something unexpected would turn up, and nullify our results. In the language of a U.S. Secretary of Defense, we had controlled the known unknowns, but there were always the unknown unknowns—and this was a crazy result. I expected the community to be skeptical, and most probably scathing in the assessment of our results.

During this time, the SCP was working frenetically on their own paper—it soon emerged that the conclusions of the two independent experiments were virtually identical (Perlmutter *et al.*, 1999). Their experiment had more objects than ours, but less signal per object—in the end the overall significance of the two experiments was about the same. If combined, the two experiments achieved more than 4σ detection of acceleration (Fig. 7).

To my surprise, the accelerating Universe was received with a warmer reception than what I was expecting. The positive reception was due, I believe, partially to the fact that two highly competitive teams arrived independently to the same answer. But the discovery also provided a solution to some major failings of the prevailing cold dark matter model (CDM)—a model in which initial conditions were set by a period of inflation (Guth, 1981). This model predicted a geometrically flat universe with a distribution of initial fluctuations described as a nearly-scale-invariant Gaussian random field. CDM was in conflict with the distribution of galaxies on large scales, as were the prevailing combination of measurements of the Hubble constant, matter density, and age of the Universe. It was realized that the addition of a cosmological constant could fix all of these problems (Efstathiou, Sutherland, and Maddox, 1990; Krauss and Turner, 1995; and Ostriker and Steinhardt, 1995).

In 2000, the MAXIMA and Boomerang experiments made measurements of the cosmic microwave background which demonstrated that the Universe was flat to within 10%—i.e., $\Omega_k \sim 0$ in Eq. (10) (Hanany *et al.*, 2000 and Bernardis *et al.*, 2000). This measurement was essentially impossible to reconcile with our supernova distances unless the Universe was full of something like a cosmological constant. It was at that moment in 2000 that I finally felt secure that our findings would stand the test of time.

Concluding Remarks: In the 13 years since the discovery, the accelerating cosmos has received intense scrutiny throughout physics. On the observational side, increasingly large samples of type Ia supernovae have improved the precision of the measurements of acceleration to the point where they are now systematically, rather than statistically limited (Wood-Vasey, 2007; Hicken *et al.*, 2009; Kessler *et al.*, 2009; and Guy *et al.*, 2010).

Measurements of the cosmic microwave background have established an increasingly precise measurement of the angular size distance to a redshift of approximately $z \sim 1090$, as well as the physical conditions of the Universe from just after the big bang through to the time of recombination (Komatsu *et al.*, 2011). The scale of Baryon acoustic Oscillations, whose size are understood through modeling of the cosmic microwave background, have been traced over time through their imprint into the population of galaxies. Astronomy can now connect the scale of the Universe from $z \sim 1080$ to $z = 0.2$ (Percival *et al.*, 2010), $z = 0.35$ (Eisenstein *et al.*, 2005), and $z = 0.6$ (Blake, 2011). Together, the measurements listed above, and most others, remain consistent with a universe where the acceleration is caused by Einstein's cosmological constant ($\Omega_\Lambda \sim 0.73$, $w = -1$), the Universe is geometrically flat, and the remainder of the matter is dominated by pressureless ($w = 0$) matter (Sullivan, 2011), split between baryons ($\Omega_B \sim 0.045$) and cold dark matter ($\Omega_{CDM} \sim 0.225$). This basic model is often described as the flat -CDM model.

An enormous body of theoretical work has been undertaken in response to the discovery of the accelerating Universe. Unfortunately, no obvious breakthrough in our understanding has yet occurred—cosmic acceleration remains the same mystery that it was in 1998. The future will see bigger and better experiments that will increasingly test consistency of our Universe with the flat -CDM model. If a difference were to emerge, thereby disproving a cosmological constant as the source of acceleration, it would provide theorists with a new observational signature of the source of the acceleration. Short of seeing an observational difference emerge, we will need to wait for a theoretical revelation that can explain the standard model, perhaps informed by a piece of information from an unexpected source.

References

Blake, C., 2011, Mon. Not. R. Astron. Soc. **415**, 2892.

Carroll, S. M., W. H. Press, and E. L. Turner, 1992, Annu. Rev. Astron. Astrophys. **30**, 499.

Chandrasekhar, S., 1931, Astrophys. J. **74**, 81.

de Bernardis, P., *et al.*, 2000, Nature (London) **404**, 955.

de Sitter, W., 1917, Mon. Not. R. Astron. Soc. **78**, 3.

Efstathiou, G., W. J. Sutherland, and S. J. Maddox, 1990, Nature (London) **348**, 705.

Einstein, A., 1917, Sitzungsberichte Berl. Akad. **1**, 142.

Eisenstein, D. J., *et al.*, 2005, Astrophys. J. **633**, 560.

Filippenko, A. V., 1992, Astron. J. **104**, 1543.

Filippenko, A. V., 1997, Annu. Rev. Astron. Astrophys. **35**, 309.

Filippenko, A. V., *et al.*, 1992, Astrophys. J. **384**, L15.

Friedmann, A. A., 1922, Z. Phys. D **10**, 377.

Gaither, C. C., and A. E. Cavazos-Gaither, "Gaither's Dictionary of Scientific Quotations," 2008.

Goobar, A., and S. Perlmutter, 1995, Astrophys. J. **450**, 14.

Guth, A., 1981, Phys. Rev. D **23**, 347.

Guy, J., *et al.*, 2010, Astron. Astrophys. **523**, A7.

Hamuy, M., M. M. Phillips, N. B. Suntzeff, R. A. Schommer, J. Maza, and R. Aviles, 1996, Astron. J. **112**, 2391.

Hanany, S., *et al.*, 2000, Astrophys. J. **545**, L5.

Hicken, M., W. M. Wood-Vasey, S. Blondin, P. Challis, S. Jha, P. L. Kelly, A. Rest, and R. P. Kirshne, 2009, Astrophys. J. **700**, 1097.

Hubble, E., 1929, Proc. Natl. Acad. Sci. U.S.A. **15**, 168.

Humason, M. L., N. U. Mayall, and A. R. Sandage, 1956, Astrophys. J. **61**, 97.

Kessler, R., *et al.*, 2009, Astrophys. J. Suppl. Ser. **185**, 32.

Kim, A., A. Goobar, and S. Perlmutter, 1996, Publ. Astron. Soc. Pac. **108**, 190.

Kirshner, R. P., and J. Kwan, 1975, Astrophys. J. **197**, 415.

Komatsu, E., *et al.*, 2011, Astrophys. J. Suppl. Ser. **192**, 18.

Krauss, L. M., and M. S. Turner, 1995, Gen. Relativ. Gravit. **27**, 1137.

Leibundgut, B., 1988, Thesis (University of Basel).

Leibundgut, B., R. P. Kirshner, A. V. Filippenko, J. C. Shields, C. B. Foltz, M. M. Phillips, and G. Sonneborn, 1991, Astrophys. J. **371**, L23.

Leibundgut, B., *et al.*, 1993, Astron. J. **105**, 301.

Lemaître, G., 1927, Annales Société Scientifique de Bruxelles A **47**, 49.

Norton, J., 1984, Historical studies in the physical sciences **14**, 253–315; 1984, reprinted in Howard, D., and Stachel, J. (eds.), Einstein and the History of General Relativity: Einstein Studies (Birkhauser, Boston), Vol. I, pp. 101–159.

Ostriker, J. P., and P. J. Steinhardt, 1995, Nature (London) **377**, 600.

Percival, W., *et al.*, 2010, Mon. Not. R. Astron. Soc. **401**, 2148.

Perlmutter, S., *et al.*, 1997, Astrophys. J. **483**, 565.

Perlmutter, S., *et al.*, 1998, Nature (London) **391**, 51; 1998**392**,311(E).

Perlmutter, S., *et al.*, 1999, Astrophys. J. **517**, 565.

Phillips, M., 1993, Astrophys. J. **413**, L105.

Phillips, M. M., *et al.*, 1992, Astrophys. J. **103**, 1632.

Riess, A. G., W. H. Press, and R. P. Kirshner, 1996, Astrophys. J. **473**, 88.

Robertson, H. P., 1928, Philos. Mag. **5**, 835.

Sandage, A. R., 1961, Astrophys. J. **133**, 355.

Schmidt, B. P., R. P. Kirshner, R. G. Eastman, M. M. Phillips, Suntzeff, B. Nicholas, M. Hamuy, J. Maza, and R. Aviles, 1994, Astrophys. J. **432**, 42.

Schmidt, B. P., *et al.*, 1998, Astrophys. J. **507**, 46.

Slipher, V. M., 1917, Proceedings of the American Philosophical Society **56**, 403.

Spyromilio, J., W. P. S. Meikle, D. A. Allen, and J. R. Graham, 1992, Mon. Not. R. Astron. Soc. **258**, 53.

Sullivan, M., 2011, Astrophys. J. **737**, 102.

Tinsley, B., 1972, Astrophys. J. **178**, 319.

Wood-Vasey, W. M., 2007, Astrophys. J. **666**, 694.

Directors of Mount Stromlo Observatory

W. Geoffrey Duffield: 1 January 1924–1 August 1929

Richard v.d.R. Woolley: 4 December 1939–7 December 1955

Bart Bok: 7 March 1957–19 March 1966

Olin J. Eggen: 1 July 1966–30 September 1977

Donald S. Mathewson: 5 April 1979–4 May 1986

Alexander W. Rodgers: 12 June 1992–12 December 1993

Jeremy Mould: 13 December 1993–30 January 2001

Penny Sackett: 22 July 2002–14 May 2007

Harvey Butcher: 28 September 2007–6 January 2013

Matthew Colless: 7 January 2013–6 January 2018

Timeline of major events

1905	International Union for Cooperation in Solar Research meets in Oxford, discusses needs of solar research	Duffield sees opportunity for solar observatory in Australia
1907	Resolution advocating Australian solar observatory is adopted by International Union for Cooperation in Solar Research, Meudon	Resolution forwarded to Colonial Office, London; to Governor-General in Australia; by Duffield to many observatories across the world; Farnham estate offers 15 cm Grubb telescope
1909	Australasian Association for the Advancement of Science forms Solar Physics Committee to lobby for observatory	Minister for Home Affairs agrees to match private donations; Deakin government subsequently approves in principle; James Oddie offers 22 cm Grubb telescope; others offer coelostat, pyrheliometer
1910	Government committee considers observatory sites in new Federal Capital Territory	Committee recommends Mount Stromlo site
1911	Lease signed for use of Mount Stromlo early in 1911	Oddie telescope carries out site testing observations from September 1911
1912	7.5 cm transit telescope goes into operation	Prime Meridian (reference longitude) for all of Australia initially at Mount Stromlo
1913	Site testing completed	Baracchi (Government Astronomer, Victoria) concludes site is very suitable for observatory
1914	British Association for the Advancement of Science meets in Australia; First World War breaks out	Fisher government agrees to establish observatory, but only when war is over

1922	Duffield working at Reading University in UK	Returns briefly to Australia to lobby Hughes government
1923	Firm commitment by Australian government to establish Commonwealth Solar Observatory	Australian government asks a committee of British astronomers to recommend a candidate Director
1924	Commonwealth Solar Observatory formally instituted	Duffield appointed first Director from 1 January, returns to Australia with family in November 1924, taking up residence in new Hotel Canberra; John Reynolds offers 76 cm telescope
1925	Construction works begin on Mount Stromlo, continue through 1926	Temporary observatory quarters are set up in Hotel Canberra; relocation to Mount Stromlo at end 1926
1928	Installation of scientific equipment and telescopes	First photoelectric measurements made with 15 cm Farnham telescope
1929	Influenza epidemic in Australia; Great Depression	Duffield dies of influenza on 1 August, is buried at Mount Stromlo; Rimmer appointed 'officer-in-charge; Memoir No 1 is prepared with first research results, but no funding available to publish until 1935
1931	Solar tower telescope completed and operational	Observations leading to an atlas of the solar spectrum are begun by Cla Allen, not the only but arguably the most important work of the 1930s at the Observatory
1935	First ionospheric sounding recording equipment in Australia	Observatory begins research program in ionospheric physics; would provides radio communications predictions in WWII; would become Ionospheric Prediction Service in 1947
1937	Government-appointed committee reviews performance of Observatory	Usefulness of research program confirmed, search for Director recommended
1939	Search committee recommends R. Woolley as new Director; Second World War breaks out	Woolleys arrive in December, are welcomed personally by Minister for the Interior
1940	Optical Munitions Panel formed	Woolley commits Observatory to war effort
1941	Observatory becomes in effect an optical munitions factory	Site closed to visitors for security; staff expanded from 10 to 70; workshops extended and optical shop added
1942	Army Inventions Directorate established	Woolley appointed Chief Executive Officer of AID
1944	Commonwealth Time Service moves to Mount Stromlo	Observatory staff provide Time Service for Army from early in war; Commonwealth Time Service moves from Melbourne in 1944 and remains at Mount Stromlo until 1968

1945	Melbourne Observatory is closed	Great Melbourne Telescope is sold to Commonwealth, transferred to Mount Stromlo
1946	CSIR Radiophysics discovers radio emission from sunspots	Observatory staff begin regular measurements of solar radio noise from Mount Stromlo; survey of galactic radio noise also undertaken
1947	Observatory Advisory Committee discusses future research program with Astronomer Royal	A change of direction to astrophysical research is agreed; the recommendation for a large telescope is immediately accepted by government; the 1.9 m telescope is contracted to Grubb-Parsons; name is to be changed to Commonwealth Observatory
1947	Mary Lee Woods joins staff at Mount Stromlo	Woods brings mathematical and computer programming expertise; in later life she is mother of Tim Berners-Lee, inventor of World Wide Web
1949	First formal connection with the new Australian National University	Woolley appointed (honorary) Professor of Astronomy with right to appoint research fellow and supervise students; other staff receive appointments in 1951
1951	Distance to Large Magellanic Cloud is estimated	Ben Gascoigne uses Cepheid variable stars to estimate the distance to the Large Cloud to be 56 kiloparsecs, very close to the Hubble Space Telescope value published in 2001 of 50 kiloparsecs; this distance is a fundamental step in determining the expansion rate and age of the universe
1951	US astronomers Kron and Eggen visit	Beginning of photoelectric photometry as a major research tool at Mount Stromlo
1952	Bushfires started by lightning	Workshops and records destroyed, a major calamity for the Observatory
1952	Observatory's Visiting Committee meets with CSIRO Executive	It is decided that optical astronomy will be the province of the Commonwealth Observatory at Mount Stromlo, radio astronomy of CSIRO Radiophysics Division
1954	First PhD degree awarded by ANU	Observatory astronomer Antoni Przybylski is awarded the first ANU PhD for his dissertation, *The Maximum Effect of Convection in Stellar Atmospheres on the Observed Properties of Stellar Spectra*
1955	New telescopes at Mount Stromlo	1.9 m telescope (formally dedicated by Governor-General on 8 November 1955); refurbished Great Melbourne Telescope (henceforth the '1.27 m telescope') in operation; Uppsala Schmidt installed; Yale–Columbia 66 cm astrometric refractor (operational from 1957)
1955	Systematic survey of southern galaxies is completed	Research Fellow de Vaucouleurs makes a major photographic contribution to southern hemisphere astronomy with the Reynolds 76 cm telescope
1955	Woolley appointed Astronomer Royal	Woolley's departure triggers Act of Parliament to transfer Observatory to ANU; Hogg appointed Acting Director
1957	Observatory formally transferred to ANU	Bart Bok appointed Director, Head of Dept of Astronomy; name changed to Mount Stromlo Observatory

1957	Sputnik launched	Bok addresses joint gathering of both Houses of Australian Parliament to explain the significance of the event; photograph with Uppsala Schmidt aids accurate orbit determination; Bok holds monthly open house for MPs, senior public servants, diplomatic corps etc.
1957	Field station at dark sky site is considered	Both US and Mount Stromlo astronomers start looking for a better observing site free from city light pollution
1958	Number of PhD students starts to increase	Bok campaign to attract top students includes summer scholarship program, personal visits to state universities
1959	Road to Observatory sealed	Visitors increase to tens of thousands per year
1960	Coudé spectrograph added to 1.9 m	High-resolution spectroscopy becomes a new tool for research
1960	Spectral energy distributions a major advance for observational astrophysics	Photographic spectra replaced by photoelectric hyperspectral scanners: first with a visitor instrument in 1960, then on 1.27 m in 1964 and with 33-channel instrument on 1.9 m in 1972
1960	First digital computer	Mount Stromlo acquires an IBM 610 computer, the first digital computer at the ANU; upgraded to an IBM 1620 in 1964
1962	Observations of the polarisation of starlight begin	Visvanathan takes first polarisation measurements with 51 cm site survey telescope at Mt Bingar; 1.27 m outfitted for polarisation in 1963; 60 cm rotatable telescope at Siding Spring commissioned in 1966
1962	Siding Spring mountain selected for new field station	New suburbs under construction at Woden make a new observing station urgently required
1963	Holmium star discovered	Mount Stromlo astronomer Antoni Przybylski uses 1.9 m Coudé to discover one of the strangest stars ever, HD101065, whose strongest spectral lines are of the rare Earth element, Holmium; it becomes known as Przybylski's star
1964	Duffield building completed	Expanded office space for expanded astronomical staff becomes available
1965	First program of astrophysical theory is started	Don Faulkner joins staff and begins work on Mount Stromlo stellar evolution code, with the evolution of stars to become a major research theme
1965	Siding Spring Observatory starts operations	101 cm and 40 cm telescopes are installed, the 60 cm telescope follows a year later
1965	Anglo-Australian Telescope considered	Australian Academy of Science and British Royal Society agree to approach their respective governments for finance for a new large telescope in Australia
1966	Bok finishes his term as Director	Bok returns to US; Olin Eggen is appointed as his successor; name changed to Mount Stromlo and Siding Spring Observatories

1967	First image intensifiers installed	Visitor Kent Ford brings Carnegie image tubes to 1.9 m; major new research programs on quasars, galaxies, stars and interstellar medium made possible
1968	Time Service leaves Observatory	Time Service incorporated into National Mapping Service
1969	Infrared astronomy comes to Mount Stromlo	Harry Hyland joins staff, collaborates with University of Melbourne to start a program of infrared astronomy on Mount Stromlo telescopes
1970	Unseen matter in galaxies considered	Ken Freeman is first to suggest that rotation curves of spiral galaxies imply the presence of invisible material
1970	Galactic magnetic field revealed	Mathewson and Ford measure the polarisation of light from stars in the Milky Way and Large Magellanic Cloud, interpret the results in terms of the magnetic field structure in the interstellar medium
1970	Anglo-Australian Telescope (AAT) at Siding Spring	Anglo-Australian Telescope Agreement Act is passed in Australian Parliament, the last step in the approval process; Mount Stromlo astronomer Ben Gascoigne is appointed astronomical advisor and in 1974 Commissioning Astronomer, Mount Stromlo engineer Herman Wehner as telescope engineer and from 1973 as Project Manager
1972	Echelle optics added to 1.9 m Coudé laboratory	Very high spectral resolution henceforth efficiently achievable with image intensifiers
1973	Magellanic Stream discovered	Mathewson and Cleary use 18 m and 64 m radio telescopes at Parkes to reveal tidal tails from interaction of Milky Way with Magellanic Clouds
1974	AAT construction completed	Prince of Wales performs the formal opening ceremony
1974	Commonwealth and higher education	Federal government assumes partial responsibility for all Australian higher education, loses interest in maintaining a specifically national mandate for the ANU; years of expansion come to an end
1977	Eggen's term of office finishes	Eggen accepts a research position at Cerro Tololo Inter-American Observatory, Chile; Don Mathewson made Acting Director
1978	New-generation, powerful mini-computer	The first DEC VAX 11/780 in Australia is installed at Mount Stromlo, a second at Siding Spring
1979	Search for a new Director	After an 18-month effort, Mathewson is confirmed as fifth Director of Observatory
1980	Budget tax	ANU Vice-Chancellor imposes a 1% tax on all internal School budgets to allow some innovation, making possible finance for the 2.3 m Advanced Technology Telescope; telescope is to be designed and built in-house at Mount Stromlo
1981	Uppsala Schmidt telescope to Siding Spring	The move from Mount Stromlo is motivated by increasing light pollution from Canberra city lights

1983	Space project and Auspace spin-off	Mathewson organises a formal Letter of Intent by Australian government to participate in international STARLAB project with US and Canada; Auspace Pty Ltd spun off from Observatory's engineering team as Australian prime contractor, later to become Australia's premier space company
1983	Universal expansion measured	Visvanathan concludes a careful measurement of the Hubble Constant, obtaining 73±11 km per sec per Mpc, to be compared with 72±2(statistical)±7(systematic) km per sec per Mpc to be reported by the Hubble Space Telescope Key Project in 2001
1984	2.3 m telescope is completed	Prime Minister Bob Hawke travels to Siding Spring on 16 May 1984 to formally open the 2.3 m telescope
1984	'Oldest' star discovered	Bessell and Norris find that CD-38°245 has 30 000 times less heavy chemical elements than the Sun, a record that will stand for 20 years
1985	2.3 m telescope receives Engineering Excellence Award	Three cost-saving innovations are celebrated: alt-azimuth mount and computer control; thin primary mirror; minimal, co-rotating building
1986	Observatories separate from ANU Research School of Physical Sciences	Observatories become an independent centre within the ANU, but without the same membership rights on academic and funding committees as Research Schools
1986	Government establishes Australian Space Program	Mathewson's space advocacy is influential in setting up the Australian Space Board and Space Office; policy emphasis, however, is on telecommunications and remote sensing from space, not science
1986	Mathewson's term of office finishes	Alex Rodgers appointed Acting Director; becomes sixth Director in 1987
1987	Suite of instruments for 2.3 m telescope	A visitor infrared instrument is available initially; the first new instrument commissioned is a double beam spectrograph in 1987; the second is an imager and low-resolution spectrograph in 1991; the third an infrared imager, CASPIR, in 1993
1990	MACHO MoU signed with University of California and Lawrence Livermore National Lab	A major project is initiated to search for dark matter particles in the Milky Way Galaxy; the 1.27 m telescope refurbishment is undertaken; observations begin in 1992; first detection merits the cover of *Nature* in 1993
1992	Australia invited to join international Gemini Observatory	Invitation is refused; site testing is undertaken in 1993 for a potential large telescope to be located in Australia, but conclusion is that no site can compete with Chile or Hawaii
1992	Rodgers' term as Director finishes	Don Faulkner is appointed Acting Director; ANU Electoral Committee decides to advertise widely; Rodgers applies; Jeremy Mould is appointed seventh Director with a five-year term from 1993
1994	Comet Shoemaker-Levy 9 collides with Jupiter	McGregor's CASPIR on 2.3 m telescope captures infrared images that make the news worldwide
1994	Joint Australian Centre for Astrophysical Research in Antarctica	Mount Stromlo joins with University of New South Wales to establish JACARA; a site testing campaign including Mount Stromlo staff at South Pole is supplied with instrumentation; other commitments force Observatory withdrawal in 2001

1995	Major review of Australian astronomy	*Australian Astronomy Beyond 2000* gives top priority given to joining the European Southern Observatory; two government committees subsequently also recommend an ESO investment; Prime Minister Paul Keating overturns all advice and rejects the bid for joining ESO
1995	Endeavour Space Telescope flies with Astro-2 mission on NASA Space Shuttle	A collaboration between the Observatory and Auspace Pty Ltd, the mission is a successful technology demonstrator for Stromlo photon-counting detectors in space
1995	Brian Schmidt arrives at Mount Stromlo	Schmidt immediately starts setting up the High-Z Supernova Search with colleagues in Chile, Europe and US; the aim is to use supernovae to measure how the expansion of the universe has evolved over cosmic time
1997	2dF galaxy survey is started	Mount Stromlo astronomer Matthew Colless is co-leader of an international team aiming to measure the clumpiness of matter in the universe using the AAT; the data will give a clue to the total amount of matter and ratio of dark to visible matter in the universe; observations run until 2002
1997	Gravimetry station begins measurements	The Research School of Earth Sciences installs a precision gravimeter in the unused solar spectrometer lab of the Commonwealth Solar Observatory building; the instrument measures periodic gravity variations to a part in 10^{12} and thereby probes the inner structures of the Earth
1998	Acceleration of universal expansion announced	Schmidt's team publishes the discovery that not only is the universe expanding but that the expansion is accelerating, an observation not understandable with known physics; *Science* calls it 'the Breakthrough of the Year'; it will lead to a Nobel Prize in Physics
1998	Gemini International Observatory	Through the Australian Research Council the funds are made available for a 4.76% share of the observing time on two 8.1 m telescopes, one in Hawaii and the other in Chile; Australian Gemini Project Office set up at Mount Stromlo; Mount Stromlo astronomer Gary Da Costa is named Project Scientist
1998	Observatory granted ANU Research School status	Following three years of scrutiny, ANU Council finally agrees to upgrade the centre's status; Observatory renamed the ANU Research School of Astronomy and Astrophysics (RSAA)
1998	Satellite Laser Ranging observatory	EOS Space Systems Pty Ltd establishes a precision SLR station near the Oddie telescope
1998	ANU extends Mould's term as Director	Appointment is for a second five-year term, but Mould departs in 2001 to take on directorship of US National Optical Astronomy Observatory with its seven major facilities
1999	Gemini Observatory contracts a new instrument	Mount Stromlo astronomer Peter McGregor wins a competition to develop a $4.5 million Near-infrared Integral Field Spectrometer, with Jan van Harmelen as Project Manager; NIFS is to work with adaptive optics and provide the same image sharpness in the infrared as the Hubble Space Telescopes provides in visible light

2000	Mount Stromlo Visitor Centre opened	Centre includes an exhibition on astronomy, a restaurant and a solar imager for viewing the Sun using Duffield's original Sun Telescope lens
2000	Miss Duffield provides endowment	First Director Duffield's daughter, Miss Joan Duffield, endows the Duffield Chair with Ken Freeman as inaugural holder; she also institutes a student scholarship and donates to the Mount Stromlo Visitor Centre
2001	Science Citation Laureates	Academy of Science honours 33 Australians designated by the Institute of Scientific Information as Science Citation Laureates; nine are astronomers and six are from Mount Stromlo: Mike Bessell, Matthew Colless, Michael Dopita, Ken Freeman, Jeremy Mould and Bruce Peterson
2001	ANU joins the Australian Research Council grants program	Finance for ARC participation is obtained by reducing Research School budgets; RSAA does exceedingly well in initial grant rounds, but grant rules do not allow recovery of lost finance required for operational activities; beginning of a period of annual reductions of operational resources
2001	Directorship vacant	As Mould departs, John Norris is made Acting Director; ANU Electoral Committee recommends Penny Sackett as new Director, she takes up the post in mid 2002
2002	Galactic archaeology	Freeman and Joss Bland-Hawthorne (AAO) publish a major review of work on the formation of the Milky Way Galaxy; their approach is called 'near-field cosmology' but becomes popularly known as 'galactic archaeology'
2002	Gemini Observatory contracts another new instrument	Peter McGregor and team win a $6.3 million contract to build the Gemini South Adaptive Optics Imager; interrupted by the 2003 bushfire devastation, the instrument was delivered in 2005
2002	Planetary Science Institute	RSAA and the Research School of Earth Sciences create a new joint institute for Planetary Science; RSAA is to focus on discovering and observing exoplanetary systems, RSES to investigate the early Solar System using isotopic analyses of meteorites and lunar samples
2003	Bushfires devastate the Mount Stromlo Observatory	On 18 January, all telescopes (except the 15 cm Farnham) are destroyed, as are the workshops (including the NIFS instrument being readied for shipment overseas), the original Commonwealth Solar Observatory building (including the library with its historical manuscripts) and the heritage Director's Residence; modern engineering drawings and astronomical data are in digital form and are not lost
2003	NIFS contract rescued	Auspace Pty Ltd contracted to reconstruct the NIFS instrument, allowing it to be delivered to the Gemini Observatory in 2005
2005	New instrument for 2.3 m telescope	Dopita receives initial funding for Wide-field Integral Field Spectrograph; the design promises a speed gain of 10–100 times over previous spectrographs; WiFeS delivered in 2009
2005	SkyMapper telescope	To replace the 1.27 m telescope at Mount Stromlo, Brian Schmidt receives finance for a new telescope at Siding Spring to survey the southern sky over a five-year period; SkyMapper began operation in 2011

2006	ANU joins Giant Magellan Telescope (GMT) project	ANU provides an initial $1 million to join an international consortium of universities planning to design, build and operate this 25 m class, next generation telescope
2006	Workshops replaced	A new engineering lab and precision mechanical workshop, called the Advanced Instrumentation and Technology Centre, is opened by the leader of the government in the Senate, Minister for Finance and Administration, Nick Minchin
2006	Sackett concludes term as Director	Gary Da Costa appointed Acting Director; in 2007 the search committee recommends Harvey Butcher as ninth Director; Butcher arrives in September 2007, consults with staff and sets out strategic goals: new young staff, completion of WiFeS and SkyMapper projects, and securing full finance for participation in the GMT project
2008	Sackett becomes Chief Scientist of Australia	Penny Sackett is first female Chief Scientist
2008	First Stromlo Fellow appointed	Stromlo Fellowships are granted for five years, are research-only positions and are open to any research field of astronomy; first Fellow is Chiaki Kobayashi from Japan
2009	Thomson-Reuters citation ranking	RSAA and Mount Stromlo are rated number 10 in the world in space science for 1998-2008 on the basis of worldwide citations to published research
2009	Global financial crisis	Australian government sets up economic stimulus program; ANU agrees to support RSAA bid for infrastructure funding for participation in GMT project and extension of Mount Stromlo engineering laboratories; full funding is approved and a four-year contract with government is signed early in 2010
2009	MoU with EOS Space Systems Pty Ltd	A joint investment, shared risk R&D project is agreed to develop adaptive optics technologies, both for use on GMT and for improved space debris management; ANU approval required, takes 22 months to realise; work begins in 2011
2009	New tenants at Siding Spring	Discussions begin to attract new non-ANU observing facilities to Siding Spring; interest from astronomers in US, Korea and Poland lead to concrete plans for new telescopes; goal is to ensure RSAA students have access to some telescopes should ANU have to abandon Siding Spring for financial reasons
2010	Refurbishment of 2.3 m telescope	Original team of engineers responsible for designing the 2.3 m Advanced Technology Telescope at Siding Spring is recalled (mostly from retirement) to help refurbish the telescope after 30 years of operation
2010	GMT Integral Field Spectrometer contract	McGregor and team win a major contract to build one of the two or three first-light instruments for GMT; named GMTIFS, the project aims to design and build an advanced NIFS-like instrument working at 10 times the angular resolution of the Hubble Space Telescope
2010	GMT Adaptive Optics contract	Rod Conan and team win a first contract to do adaptive optics design studies for GMT; subsequent contracts follow
2010	Australian Space Research Program	Harvey Butcher embraces engineering support for non-astronomical space science with the argument that existing funding streams for astronomy are unlikely to be able to finance an internationally competitive engineering program

2011	National Space and Astronomy Museum	ANU Vice-Chancellor Ian Chubb signs MoU with Smithsonian National Air and Space Museum in Washington DC for collaboration to develop an Australian National Space and Astronomy Museum at Mount Stromlo; an international workshop is held at Mount Stromlo to start planning; a proposal for finance as part of the 2013 Canberra Centenary Celebrations is well received by government but no funds are forthcoming
2011	Extension of AITC dedicated	Minister for Innovation, Kim Carr, formally dedicates the second phase of construction of the new engineering laboratory; it will be specially fitted out for both GMT and space research engineering
2011	Faulkner Court opened	New student living accommodation is completed on site; the facility is named after Mount Stromlo astronomer Don Faulkner and is formally dedicated by his widow, June Faulkner
2011	ANU wins insurance case	Long-running litigation in connection with the 2003 bushfires is settled; Stromlo Endowment is consequently augmented to a total of about $10 million; five Stromlo Fellowships can proceed on a rotating basis of one appointment per year
2011	Nobel Prize for Physics	Brian Schmidt is a co-recipient of the 2011 Nobel Prize for Physics for the discovery of the accelerating universe; the award is Australia's second, the first Australian Nobel Prize in Physics was won in 1915 by William and Lawrence Bragg
2012	Prime Minister's Prize for Science	Ken Freeman receives Prize from Prime Minister Julia Gillard for his entire oeuvre
2012	New Master Plan	Work on a Master Plan for managing the future development of the Mount Stromlo campus is started; heritage buildings, additional accommodation, hosting of external organisations are discussed with ANU planners
2012	Butcher's term as Director finishes	Selection Committee recommends Matthew Colless as 10th Director; Butcher's appointment is extended until when Colless arrives in 2013

Index

www.ingramcontent.com/pod-product-compliance
Lightning Source LLC
LaVergne TN
LVHW060614110826
845154LV00003B/87

9781486300754